Surface and Interface Science

Online at: https://doi.org/10.1088/978-0-7503-4916-1

Surface and Interface Science

David L Allara
Pennsylvania State University, University Park, PA, USA

Robert L Opila
University of Delaware, Newark, NJ, USA

IOP Publishing, Bristol, UK

ISBN 978-0-7503-4916-1 (ebook)
ISBN 978-0-7503-4914-7 (print)
ISBN 978-0-7503-4917-8 (myPrint)
ISBN 978-0-7503-4915-4 (mobi)

DOI 10.1088/978-0-7503-4916-1

Version: 20260201

IOP ebooks

British Library Cataloguing-in-Publication Data: A catalogue record for this book is available from the British Library.

Published by IOP Publishing, wholly owned by The Institute of Physics, London

IOP Publishing, No.2 The Distillery, Glassfields, Avon Street, Bristol, BS2 0GR, UK

US Office: IOP Publishing, Inc., 190 North Independence Mall West, Suite 601, Philadelphia, PA 19106, USA

Contents

Authors biographies

David L Allara

David L Allara holds the title of Distinguished Emeritus Professor of Chemistry and Materials Science and Engineering at Pennsylvania State University. He joined the University in 1987 after a career at Bell Laboratories in Murray Hill, NJ, where he had achieved the title of Distinguished Member of Technical Staff and subsequently served as the head of the Department of Physical Chemistry in the newly established offshoot, Bell Communications Research, formed in 1984. Over the years his interests have included polymer materials, modeling of gas phase kinetics reaction systems, including atmospheric reactions, pyrolysis and combustion, and surface spectroscopy, but his primary research for most of his career has been in molecular surface science. While at Bell Labs he developed the area of molecular self-assembly at surfaces. This work has gone on to spawn many applications including chemical and biological sensors and molecular electronics. In addition, in work on recombination rates in silicon he developed wet etch methods for hydrogen passivation of the silicon surface. His recent research ranges across molecular electronics, including single molecule devices, semiconductor surface chemistry, interface spectroscopy, and chem/bio sensing. From the American Chemical Society, he received the Arthur Adamson Award in Surface Chemistry and the Spectrochemical Analysis Award in Analytical Chemistry. He was awarded an Honorary Doctor of Science degree from Linkoping University in Sweden, hold the title of a Fellow of the American Association from the Advancement of Science (AAAS), the AVS Society for Science and Technology and the Royal Society of Chemistry (UK). In 2022 he received the Kavli Prize in Nano Science from the Norwegian Academy of Science. In the commercial sector he is a co-founder of three startup technology companies, has served on scientific boards of two other companies and has been an active consultant in technology areas ranging across biosensors, energy conversion and nanotechnology.

Robert L Opila

Robert L Opila received a PhD in Chemistry from the University of Chicago. He then joined Bell Laboratories, where he investigated the role of surfaces and interfaces in electronic and photonic materials and devices. He was named Distinguished Member of Technical Staff and promoted to Technical Supervisor. Since 2002, he has been with the Materials Science Department at the University of Delaware, where his research now includes applications of surface science in photovoltaics and processing of electronic and photonic devices. Opila also has had appointments in the Departments of Electrical Engineering and Chemistry. Opila is a Fellow of the American Vacuum Society and a former editor of Applied Surface Science. He has completed terms as Fulbright Fellow at Bilkent University in Ankara, Turkey, and as Visiting Professorial Fellow at the University of New

South Wales in Sydney, Australia. In 2015–16 he served as president of the faculty senate at the University of Delaware. Opila was a rotating program officer with the National Science Foundation in the Division of Materials Research, Electron and Photonics Materials Program 2018–20. Opila is also founder and CEO of start-ups in the renewable energy sector.

IOP Publishing

Surface and Interface Science

David L Allara and Robert L Opila

Chapter 1

Introduction to interfaces

In this chapter surface chemistry is distinguished from bulk solid-state behavior. A direct analogy is made to chemistry that occurs with nanometer scale particles. Surfaces are interfaces between condensed matter and gas-phase matter, thus in many situations the ideal example of a vacuum in contact with a crystalline solid is used as a starting point. Interfaces occur when two condensed phases come into contact, and understanding interfacial phenomena is simply an extrapolation of vacuum/crystalline surface chemistry.

1.1 Definitions of a surface and interface, and associated basic concepts

Surface: The collection of points that defines the termination of a material object in space.

Interface: The collection of points that defines the extent of intimate contact of an object with a contiguous medium, which also could be an object.

The definitions are summarized pictorially in figure 1.1. Objects can be comprised of pure substances or homogeneous mixtures of pure substances (e.g. solutions or alloys). The objects must consist of condensed phases which are held together by cohesive forces; both liquids and solids fit this requirement but obviously gases do not. If the object is a mixture of substances and the mixture is not homogeneous, then the constituent substances themselves are objects and the overall object formed by the mixture consists of multiple objects and interfaces.

It is not always simple to tell if an object is a pure substance, e.g. a clear liquid which contains well dispersed, nanometer scale particles, often termed a colloid dispersion. In this case the liquid may appear clear and uniform but actually contains two independent substances, the nanoparticles and the solvent, with interfaces between. If the nanoparticles were to dissolve the interfaces would vanish

doi:10.1088/978-0-7503-4916-1ch1

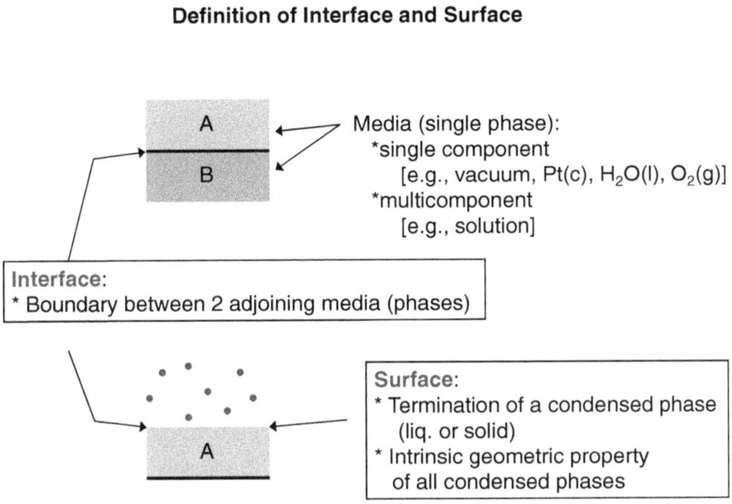

Figure 1.1. Definition of interface and surface.

leaving only a mixture of molecules and soluble atomic scale species, which could include both ions and neutral species.

An object always has both a surface and an interface. The interface defines the region of contact between an object and the surrounding medium. The contacting medium can be another object or a non-cohesive substance, which includes gases, which go to a vacuum in the limit of vanishing gas concentration. If the medium is a vacuum, then the surface and interface are identical.

Although one can consider 'interfaces' between two gases, they are not well defined and at equilibrium would disappear as the gases mixed. Such transient 'interfacial' regions are of great importance in the Earth's atmosphere and in the cases of interface regions of humid, warm air in contact with dry, cool air can be responsible for severe weather phenomena.

The fundamental behavior of a substance is governed by the laws of physics and chemistry. For example, a solid object formed from the substance would be expected to have the reproducible characteristics of melting point, vapor pressure, density, elastic modulus, interatomic distances between constituent atoms and/or molecules, vibrational and electronic spectra, refractive index, dielectric constant, and heat of reaction with some other substance. Each of these properties can be understood in fundamental terms by some combination of quantum mechanics, chemical thermodynamics, mechanics, electrodynamics and other similar approaches.

The surface of an object consists of the same substance as the interior and thus its chemical and physical characteristics must be very similar, perhaps almost identical in some aspects to the interior portions. This provides a simple approach to describing the properties and behavior of surfaces in terms of the *differences* from the bulk. Thus, we can approach the study of surfaces from the point of view of finding the perturbations of the bulk properties that create the surface. So this is good news. We do not have to learn much that is new in terms of fundamental physical laws and chemical principles, and we can use the same format and conventions that are used to describe bulk behavior. In the process of learning to describe the corrections needed for surfaces we will develop some new terminology, but the fundamental groundwork has already been laid in what we know about bulk substances.

1.2 Thicknesses of surfaces and interfaces

How thick is a surface or interface? The surface conceptually consists of a collection of points where the object terminates in space, but then where does the surface stop within the object? Of course, an atom or molecule cannot be divided so the depth must be at least one constituent in length to be free of the surface. This makes the surface a region, a shell of finite thickness. In fact, one might guess that to really be free of the corrections or perturbations describing the surface two to three constituents deep might even be needed. This is a pretty good guess and is close to the actual reality for surfaces of most objects. We will see soon why your intuition works so well. Of course, getting accurate numbers becomes a non-trivial problem and depends very much on the exact type of substance. The above thinking extends very straightforwardly to the thickness of an interface, the key being what is the size of the atomic and/or molecular constituent moieties? Obviously, this gets a bit complicated in detail, in particular for big molecules such as polymers which have huge numbers of possible configurations filling space with different shapes. It is convenient to define a parameter δ which describes the approximate average thickness of a surface or interface (figure 1.2).

Note that the interface between many materials or media often will change if the components can react, e.g. the interface between air and a bare metal surface which can corrode, forming an oxide overlayer. In this case the new material region

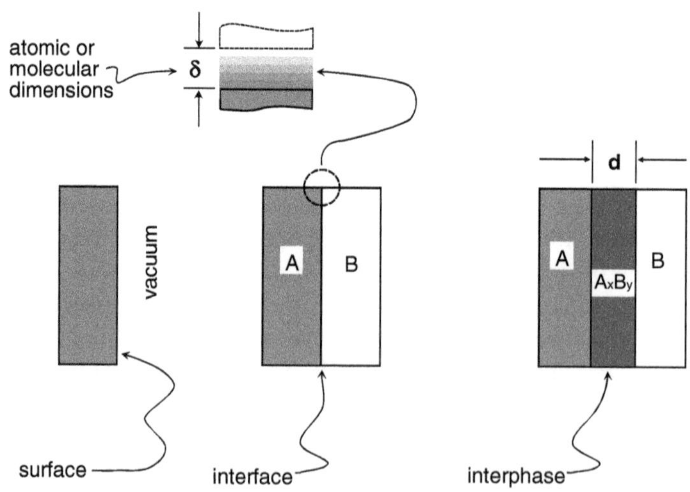

Figure 1.2. The interface thickness parameter δ for surfaces and interfaces.

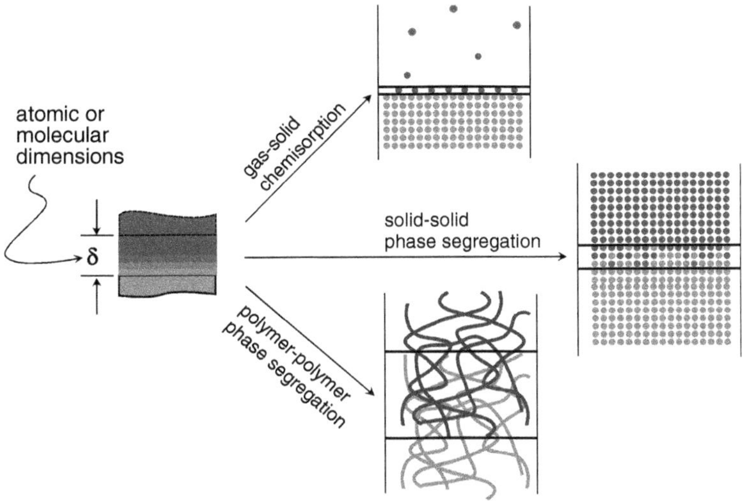

Figure 1.3. A schematic illustration of the range of interfaces thicknesses varying according to the type of medium constituents.

between the two media can be termed an *interphase*. The interphase can grow to be quite thick (e.g. some oxide layers can grow continually until they finally flake off, such as the case for copper or iron metals). In other cases the layer forms between two solids, e.g. upon raising the temperature of the sharp interface between silicon and a gold overlayer film a new compound, gold silicide, will form as an interphase which will grow in thickness with continued heat exposure as Si and Au atoms diffuse from the pure phases into the interphase region and react.

As illustrated in figure 1.3, atomic interfaces, both gas/solid and solid/solid (e.g. metal alloy) interfaces are sharp with δ at the angstrom (Å) scale whereas for polymer interfaces, with average sizes of a statistically random arrangement of the chain repeat units, δ ranges over tens of nanometers (nm).

Figures 1.4 and 1.5 show, respectively, a transmission electron microscope (TEM) atomic resolution image of several inorganic crystal interfaces formed by atomic layer-by-layer deposition and a computer simulated picture of an interface between two incompatible polymers. Note the large difference between the interface thickness parameter δ for the two systems, a few Å versus tens of nm. Note also the statistical variation of chain interpenetration at the polymer interface. The incompatibility parameter δ in the caption of figure 1.5 is not to be confused with the interface thickness parameter.

1.3 Common types of interfaces

Some common types of interfaces of interest for both technology and research are listed in figure 1.6. The listing includes the types of substances (A and B), the common descriptive terms used when discussing these interfaces and a rough scale of

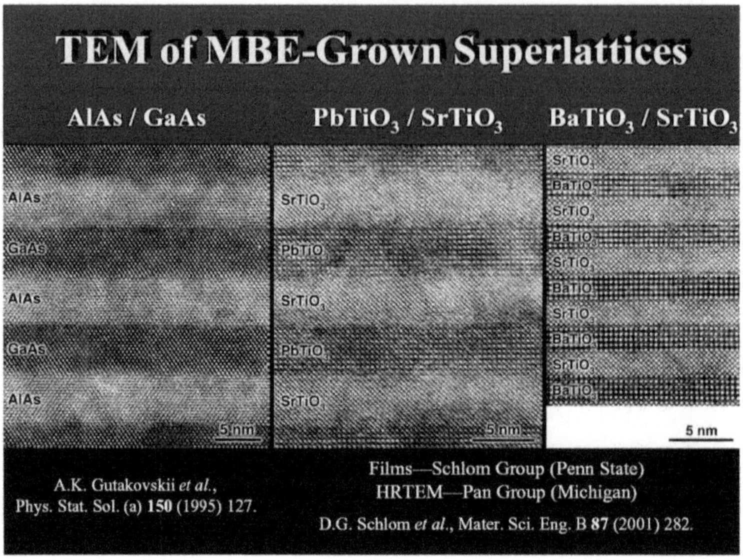

Figure 1.4. TEM images of atomically sharp interfaces of thin film stacks grown by atomic layer-by-layer deposition [1, 2]. [1] John Wiley & Sons. Copyright © 1995 WILEY-VCH Verlag GmbH & Co. KGaA

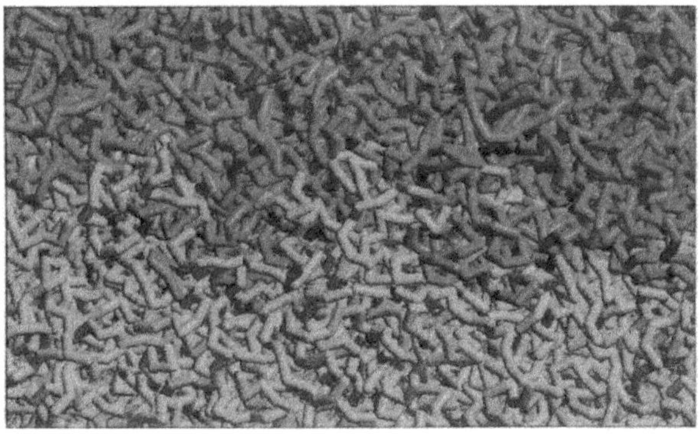

Figure 1.5. A computer simulation of the interfacial region of two incompatible polymers. Reproduced from [3], with permission from Springer Nature.

the magnitude of the interface thickness parameter δ. Notice that for the case of inert molecules or atoms adsorbed (bound to) a surface (often called a substrate) the non-chemical interaction is termed physisorption ('physical adsorption') whereas the case of chemically bonded species is termed chemisorption. In both cases the adsorbed species are termed adsorbates. There are many other types of interfaces, including complicated ones involving biological systems and incompatible liquid mixtures (e.g. a hydrocarbon and water). The interfaces listed though are common systems studied in surface chemistry.

Some Common Cases of Interfaces and Related Definitions

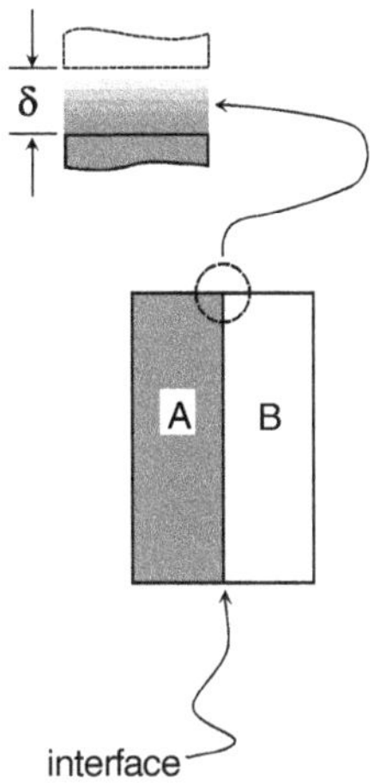

A	B	Common Description	δ
Inert solid	Inert gas	physisorption	< 1 layer
Reactive solid	Reactive gas	chemisorption	~2 atomic or molecular layers
Reactive solid	Reactive liquid	Corrosion (interphase)	Grows with time
Phase-segregated polymer	Phase-segregated polymer	Phase boundary	Typical: 1 – 15 nm

Figure 1.6. A list of some common types of interfaces with their associated δ parameters and the standard descriptions of the phenomena underlying the formation of the interface.

1.4 The surface to volume ratio of nano-objects and its effect on properties

1.4.1 What happens when an object becomes so small that most of the constituent moieties are in the surface region?

Again we can use our intuition to advantage. If the surface region thickness is one to three constituent units then we could guess that there would be significant, and quite observable, changes in the behavior of the object as the volume approaches that of the surface region shell. Since one does not usually know *a priori* the exact surface region thickness for objects, the surface area to volume ratio (S/V) becomes a convenient geometric parameter for tracking the changes in properties. In the limit as S/V increases, a single atom or molecule, or to be general, single particle, is reached. Further, one might expect the properties of the object to jump around a bit since the shape will vary as different small integral numbers of particles are clumped together, e.g. two atoms might form a dumbbell, four a plane and six a cube. A sphere never really occurs until a large enough number are combined evenly around a center to make the intrinsic surface roughness small compared to the object diameter. So overall it is clear that the S/V is a very useful guide for characterizing the point at which some changes might be expected in the properties and behavior of a nano-object as it shrinks towards the discrete atom or molecule limit.

1.4.2 Approximate dimensionality classification of high S/V objects

Another aspect of interest is the overall shape of the shrinking object and which dimension is shrinking. Here are some typical geometrical shapes of high S/V objects and how we can think about them. Illustrations are given in figure 1.7.

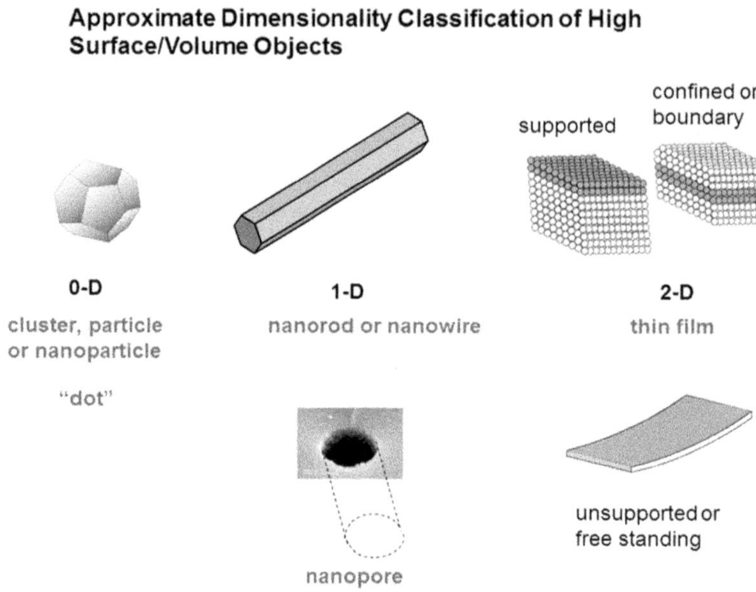

Figure 1.7. Common types of high S/V objects and their classification in terns of dimensionality.

- *Thin film*: A free standing, supported or confined layer. As the thickness (z) shrinks, the two parallel, opposing surfaces approach each other and the surface properties increasingly contribute to the properties of the object associated with the thickness dimension, e.g. electrical conductivity between the opposing surfaces of the film or sheet. On the other hand, properties along the planes of the films may be constant, e.g. the electrical conductivity in the x–y directions. In the limit, the film approaches a two-dimensional (2D) object with the changing property confined to the short z distance across the film thickness but with the approximate bulk properties back and forth in the film plane.
- *Nanorod or nanowire*: The wire radius shrinks relative to a constant length so this object approaches a 1D behavior with the changing property confined back and forth across the diameter (or equivalent x–y plane) but with the approximate bulk properties along the rod axis (z).
- *Spherical cluster or nanoparticle*: Sometimes referred to as a *dot*. As the radius shrinks the surface effects increasingly influence the properties in all directions, with all bulk characteristics changing, so this is a zero-dimensional (0D) object in the limit.

Note that these classifications of the property changes along different directions are only approximate. This makes sense because one might expect that at large S/V ratios the object can change its structure compared to the bulk, e.g. different crystal packing, and properties change even in the extended, non- shrinking directions. Also note that one property, e.g. electrical conductance, may change significantly along a given direction while another, e.g. interparticle spacings, may only change slightly.

Notice the nanopore, an interesting case of a confined geometry object which is actually just a simple hole which bounds an interior surface. The interesting properties of the hole include the ability to constrain flow of molecules diffusing down the pore. A typical form of the nanopore is in a hollow silica optical fiber. Various forms of these fibers provide very useful ways to guide light beams along the fibers with high efficiency (low scattering losses). In addition they can provide nanoscale diameter channels in which nanorods of materials such as silicon can be grown or chemical reactions in confined spaces studied. The waveguide aspect also allows some characterization probes such as Raman vibrational spectroscopy. Even the interior spaces of carbon nanotubes are essentially a nanopore and in this sense the interior space of hollow particles such as C_{60} (Bucky balls) can be considered as closed nanopores. Nanopores are often formed in the processing of small particles and these pores can be both open and accessible or closed. Extremely high surface areas in materials such as carbon nanoparticles can be achieved when the particles have large fractions of pores accessible to vapor or gases and thus can act as efficient absorbers per unit weight of material.

1.4.3 Calculating numbers of constituent particles (atoms or molecules) at the surface of a object

Given the importance of surface regions in affecting the properties of low dimensionality objects, it is very important to be able to calculate the number of particles in a unit area of surface (surface density). If the average bulk density is known a good estimate of the average surface density can be made. In the case of liquids this is the best we can do without specific experimental measurements (and/or theoretical calculations). For a solid if details of the arrangement of particles in the bulk structure are precisely known (e.g. from a crystal structure) it is usually a straightforward matter of making a good estimate of the surface density. Since the surface layer may not always have the particles arranged exactly like they are in the bulk the calculations may not always be exact but can give a very useful approximation. Illustrations of making estimates are shown in figure 1.8.

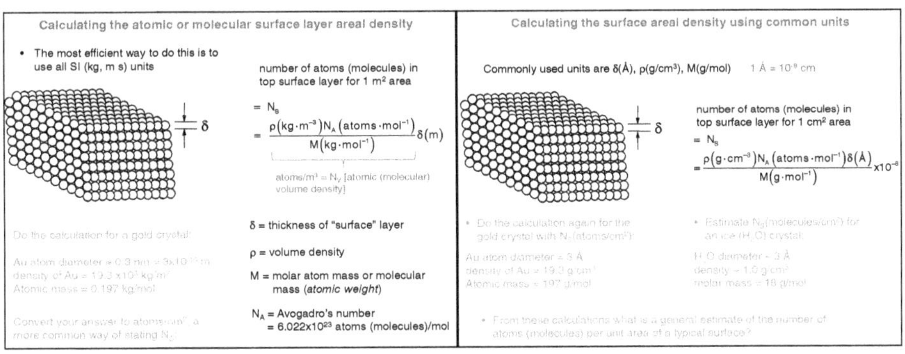

Figure 1.8. Methods for making estimates of the number of atoms per unit area in the top surface layer of a solid. The method also is directly extended to liquids. Left: calculation using SI units of atoms (or molecules) m^{-2}. Right: calculation using common units of atoms cm^{-2} and particle sizes in Å.

Problem: The typical number of atoms in the surface of an atomic solid is $\sim 10^{14}$–10^{15} cm^{-2} of surface. Make an estimate of the surface atom density ρ_s in atoms cm^{-2} and in atoms nm^{-2} for elemental silicon. Bulk density $\rho_b = 2.33$ g cm^{-3}, atomic mass $M_A = 28.1$ g mol^{-1}, atomic radius Si in silicon crystal $r = 110$ pm (0.110 nm).

1.4.4 Effects of the S/V ratio on the physical and chemical properties of small objects

Now we look at some varied examples of how the properties of different substances change as the S/V ratio increases towards the infinite limit of a single constituent atom or molecule. The simple trend of S/V as a function of decreasing size is illustrated for a hypothetical cube of an atomic solid in figure 1.9.

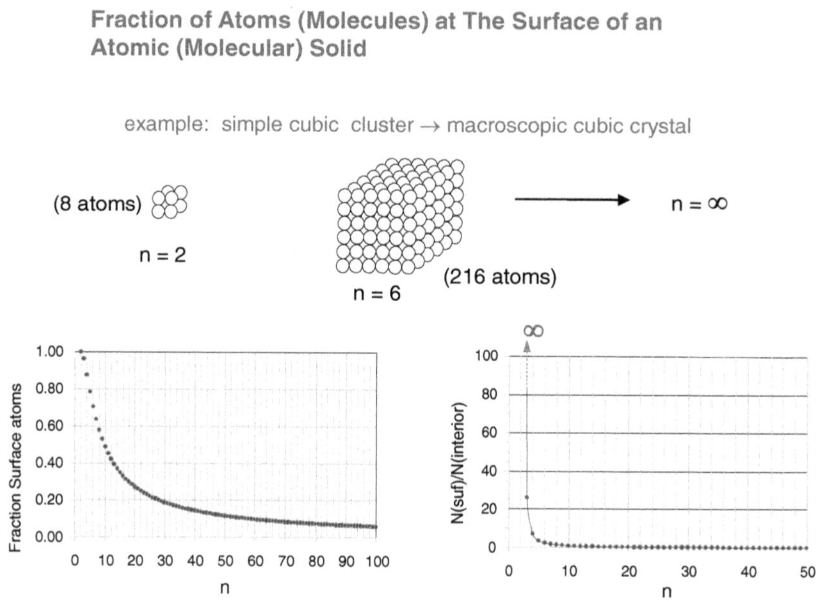

Calculation details for atom fraction at cube surfaces:

Total number of atoms on all sides (6) $= N_{sides} = 2n^2 + 2n(n-2) + 2(n-2)^2$

Total number of atoms in volume $= N_{vol} = n^3$

Total number of atoms in interior $= N_{vol} - N_{sides}$

Fraction of atoms at surface $= f_S = \dfrac{N_{sides}}{N_{vol}}$

Ratio of surface atoms to interior atoms $= R(s/i) = \dfrac{N_{sides}}{(N_{vol} - N_{sides})}$

Figure 1.9. Upper panel: plots of the approximate fraction of atoms in the surface of a cubic packed atomic solid as the size of the solid decreases. The left plot shows the fraction of surface atoms and the right plot the ratio of surface to bulk atoms. Lower panel: The method for doing the calculations.

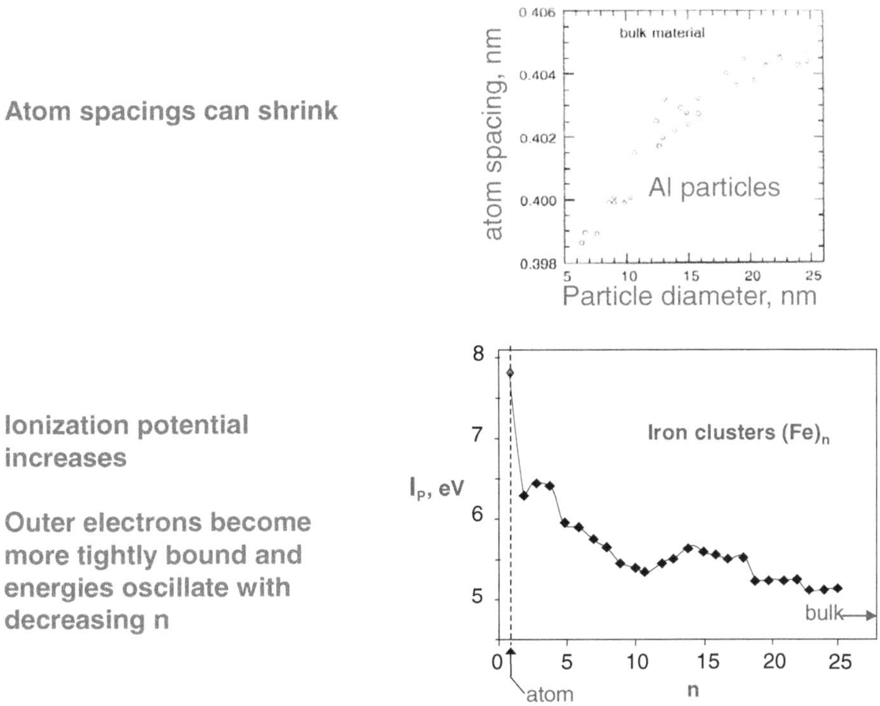

Figure 1.10. Illustration of S/V effects for the average interatomic spacings in Al nanoparticles (upper) and the ionization potential of Fe atom clusters [4, 5]. Reprinted from [4], Copyright (1981), with permission from Elsevier.

In the ultimate S/V limit properties involving collective behavior between the constituent particles vanish or do not make sense, for example, tensile strength, melting point, and heat of vaporization, but others involving electronic effects such as the energy of ionization of an electron (ionization potential) can still remain. We will see that example in the case of the ionization potential of iron clusters, shown in figure 1.10, along with changes in the atom spacings in aluminum metal nanoparticles as the particle sizes decrease. Further examples of changes in intrinsic material properties with shrinking S/V ratios for nanoparticles are given in figures 1.11–1.13, which detail shifts of electronic spectra of CdS nanoparticles, melting points and heat capacities of tin and silicon nitride nano-thickness films, and catalytic activity of oxide supported Au nanoclusters, respectively.

1.5 Fundamental but simple approaches to understanding and predicting properties of surfaces (and interfaces) relative to the bulk—the missing atom (or molecule) model for surface structure, thermodynamics and dynamics

The basic chemical, electronic and structural features of a surface cannot be very much different than those of the bulk. For example, a piece of iron metal reduced to the size of a nanoparticle still consists of only Fe atoms held together by Fe–Fe

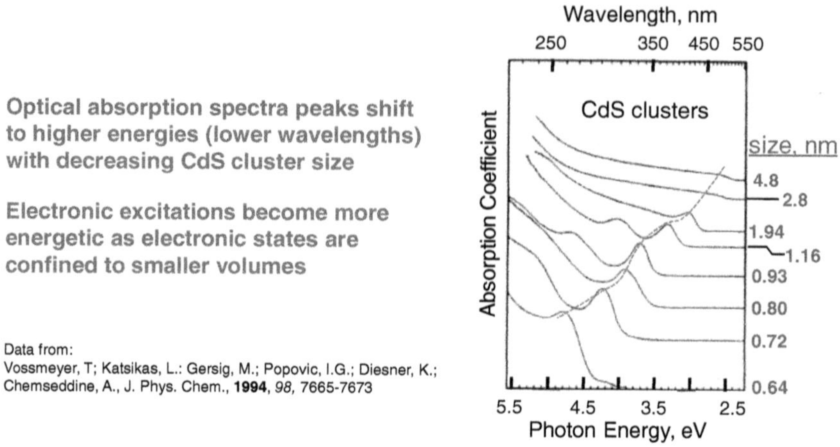

Figure 1.11. Illustration of S/V effects for the optical absorption spectra of CdS clusters. Reprinted with permission from [6]. Copyright (1994) American Chemical Society.

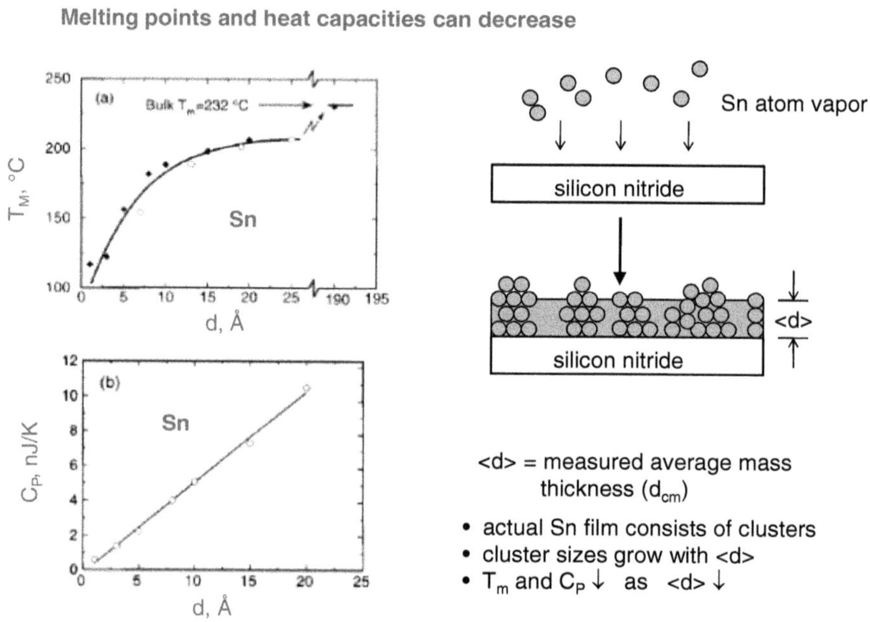

Figure 1.12. Illustration of S/V effects for the melting points (upper) and heat capacities of Sn thin films. Reprinted from [7], with the permission of AIP Publishing.

metallic bonds. We can measure or look up bulk properties in a handbook or similar reference source. If we then are able to pinpoint the key differences between the surface and bulk atoms/molecules we should be able to make predictions, in some cases even quantitatively, of the corrections needed to account for the presence of a surface (or interface). We can divide these key differences into two major classes:

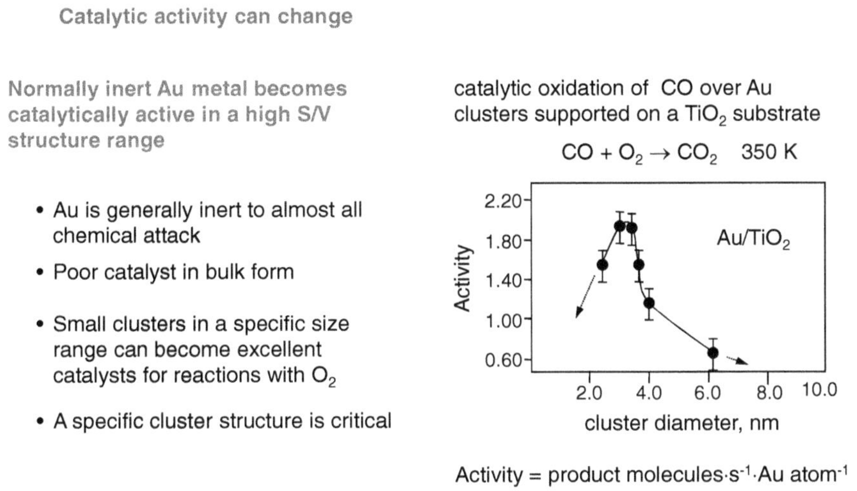

Figure 1.13. Illustration of S/V effects for the catalytic activity of Au clusters in the oxidation of CO to CO_2. From [8]. Reprinted with permission from AAAS.

structure and bonding, and electrical. For each of these classes we can construct a simple picture or model which embodies the underlying fundamental concepts that govern the corrections. From these models we then can make some simple predictions.

Here we will consider a simple *structure and bonding* model, which we term the *missing atom* (or *molecule*) *model*, that will allow us to make useful predictions such as changes in structure, energetics and chemical reactivity in moving from the bulk to the surface. A simple idealized picture of the atoms (or molecules) at a bare surface is shown in figure 1.14. The terminal atoms at the surface suffer a lack of neighbor interactions compared to atoms in the bulk. The missing atoms at the surface are shown as gray circles with dashed lines. The potential energy interactions or forces that ordinarily would exist in the bulk with these two neighbors are gone at the surface thereby reducing the stability of the surface atoms relative to the bulk atoms. The inherent driving force of this instability is to push the atoms back into the bulk but such a reduction of the fraction of surface atoms requires changing the surface area, which is impossible if the shape of the object already is at a minimum of the S/V ratio, which occurs for liquids in a spherical shape (neglecting gravity distortions). In fact, distorting the curvature of a liquid, e.g. pushing on a drop of mercury liquid to flatten it, makes the drop unstable since as soon as the pushing force is removed the drop snaps back to a sphere. It is also possible using this simple model to predict directions of changes of properties at the surface compared to the bulk such as shifting of atom spacings (termed *surface reconstruction*); thermodynamic changes including free energy, entropy and heat capacity; dynamic surface behavior such as surface vibrations and chemical reactivity. This gives us a powerful, yet simple method to gain a qualitative understanding of surface behavior from

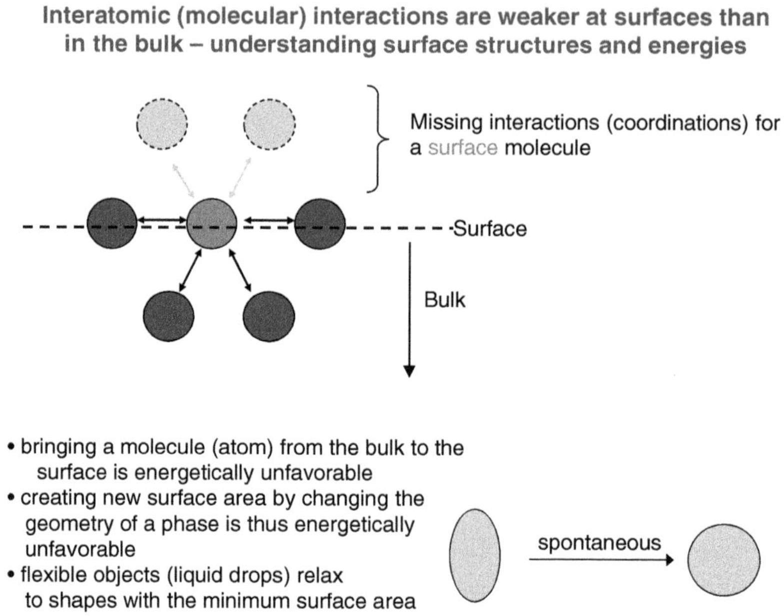

Figure 1.14. Schematic of a simple model of a bare surface for understanding and predicting a variety of differences in structure, bonding and dynamics at surfaces.

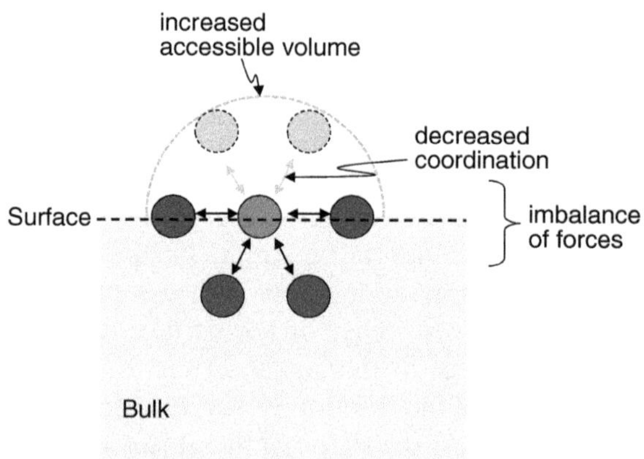

Figure 1.15. The three important aspects that affect surface atoms and their associated properties are the increased accessible volume, the decrease in coordination, and the imbalance of forces above and below.

knowledge of the corresponding bulk characteristics of the object. A quantitative model can subsequently be built upon the estimated behavior by using standard theories of material behavior and by designing appropriate experiments. The three main factors in the model that affect surface atoms and their associated properties are shown in figure 1.15.

1.6 How to approach predicting and understanding trends in electrical and electronic properties of surfaces

In section 1.4 we learned about two types of electronic-based effects that change with S/V ratio for an object: (i) the minimum work required to strip an electron from iron clusters (figure 1.10) is highest for an isolated atom and decreases thereafter with larger sizes, although with a few small ups and down along the way and (ii) the smallest photon energy required for excitation of an electron in a CdS cluster decreases as the cluster size increases (decreasing S/V, see figure 1.11). Is there a simple way to understand and qualitatively predict these trends? Yes, but not as simple as the missing atom model since we need a quick visit to the quantum mechanical world to set our thinking straight. For now we will just amplify a bit on the basic principles and then defer how to make interpretations of the data in figures 1.10 and 1.11 to the later chapter on surface electrical properties.

Let's develop the background a bit more just to get thinking about how to separate physical structure from electronic structure. First, go a bit deeper into the simple missing atom model. It is essentially based on bond energies or strengths. Pulling an atom away from bonding with another atom decreases the bonding force and in the case of a surface atom removing its top neighbor introduces an asymmetry in the force fields. In actual fact, of course, electrons are involved since the bonding is the result of the electron clouds acting like glue (usually with distinct geometric patterns, namely orbitals) between each positive nucleus pulling them together until the distance is so small that repulsive energies push back (even these are quantum mechanical, recall Pauli exclusion). So in fact the character of the object is really more dominated by the electrons in a sense than by the nuclei, which are kind of passive and go where the electrons put them. Nuclei are tiny ($\sim 1 \times 10^{-14}$ m or 10^{-6} Å) compared to the scale of the electron cloud sizes so we essentially neglect them (except for their charge, of course) and focus on the electrons, which can be viewed as responsible for creating the ordered patterns of atoms in crystals, etc.

Here is one simple surface related outcome of this picture. *The topmost layer of an object consists of an electron cloud which spills out into the vacuum, creating a surface dipole, negative charge pointing outward.* Electrons always spill out away from the tiny nuclei so form the outer boundaries of an atom, and thus a surface of atoms. There are two important concepts to consider based on this picture. (i) Unless other factors intervene this means that positive charges or ions are always favorably attracted to a surface. This is certainly true of clean metal surfaces in a vacuum, but other surfaces such as functionalized oxides often have additional dipole contributions from the functional groups, some of which may have their positive charge pointing away from the surface, so each surface must be examined for the specifics. We will visit this rule in detail in the chapter on electrical properties where it will serve well in understanding the bonding of adsorbates on a surface and electrochemical phenomena. (ii) The missing ion in a surface region lead to perturbations of the electronic states of the outermost surface regions, especially the top surface. For now neither of these help us much with the S/V examples in figures 1.10 and 1.11.

For S/V effects of electronic characteristics the main principle, which we will encounter later in the surface electrical properties chapter, is that as an object of a pure substance grows in size more and more atoms are connected to and interact with one another via their electrons. This coupling causes the original sets of identical energy levels (E-levels) of the isolated atoms to split into new sets, each with a stepped series of higher and lower levels, with increased splitting and diverging of the highest and lowest levels as the number of combined atoms increases. Without giving the details now this results in the general type of trends with increasing object size seen in figure 1.10 (easier ionization of the most energetic electron with increasing number of atoms) and figure 1.11 (easier excitation of the most energetic electron into the nearest excited state).

Perhaps you know enough already to make a good guess at the basis of the correlations in figures 1.10 and 1.11? Can you make a simple picture of the change in energy levels with increasing numbers of atoms and use that to make explanations?

1.7 The general areas of surface (interface) chemistry (science)—wet versus dry

Surface chemistry, and to be more general, interface science, extends across broad areas of disciplines and fields. This is to be expected since surfaces are an intrinsic part of objects and the behavior of objects as such is of interest across chemistry, biology, physics, materials science, electrical engineering, geology, to name a few. In general, according to the types of experiments (and theory) that are needed to tackle various interface systems of interest, the field of surface science has traditionally been divided into *wet* and *dry* areas, with wet referring to studies of interfaces involving liquids, e.g. solutions of dispersed nanoparticles, proteins adsorbed on surfaces *in vitro*, lubrication, wetting, electrochemistry and corrosion, and with dry referring to systems typically studied in vacuum or gas environments, e.g. metal surface electronic states, photovoltaic device interfaces, gas-phase catalysis, and fundamental studies of ultraclean semiconductors for electronic device applications. Figure 1.16 gives a brief summary of the interest and characteristics of studies in these two areas.

1.8 Extensions of the traditional concepts of surfaces as the termination of a homogeneous, uniform composition phase object to macromolecular and biomolecular objects

The traditional surfaces of interest in surface science involve objects of substances consisting of identical constituent units, e.g. atoms or molecules, arranged in a crystalline or amorphous form and whose surface properties can be understood in terms of the differences between individual constituent units at the surface and in the bulk., e.g. by the missing atom model. This paradigm has provided a simple approach over the years for predicting how properties of crystalline and amorphous forms of typical materials correlate with surface area and S/V ratio.

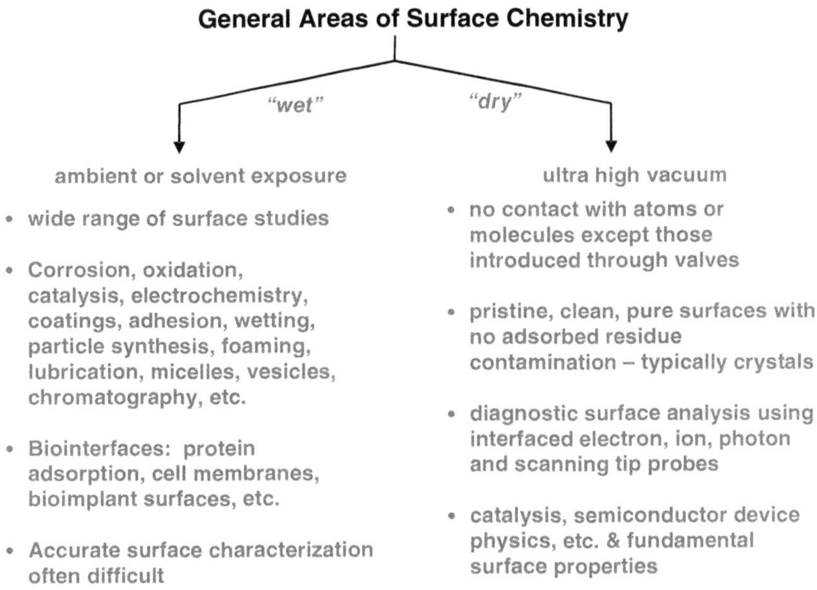

Figure 1.16. A chart summarizing the general areas of surface chemistry and some important topics of research in each.

But what about an object that is just a single huge molecule, say, many tens of nanometers in size? It makes no sense to talk about S/V because changing that would chop the molecule into pieces that become something else completely different. How about objects made up from constituents consisting of large polymeric molecules, each comprised of chemically connected, repeating combinations of small molecular units and which may not repeat in any regular way and whose surface could consist randomly of different subunits? In such cases the chemical connections (bonds) between the adjacent units on the polymer chains are much stronger (perhaps an order of magnitude) than interactions between different chains. Or suppose we have small molecules self-organized into a large object of a specific size and shape?

Examples of these possibilities include synthetic polymer nanoparticles and biomolecular entities such as proteins and viruses. The surface characteristics can be very important for biological function (e.g. surface charge or chemical coordination) and other functions (e.g. ability to disperse polymer nanoparticle agglomerates in liquids). But the surface characteristics do not track S/V ratios and may not be easily predictable from looking at any one region of the bulk since objects such as proteins and viruses have a specific size for their function and cannot be scaled. How should we think about the surface in terms of making predictions and understanding behavior? Or does it even make sense to try to do this? Yet, we cannot avoid the fact that these surfaces are important and should be part of the study of surfaces. The examples in figures 1.17–1.19 will get you thinking.

The main point in this discussion is not a presentation or explanation of details but rather to get you thinking about the broad way that scientists approach surfaces (and

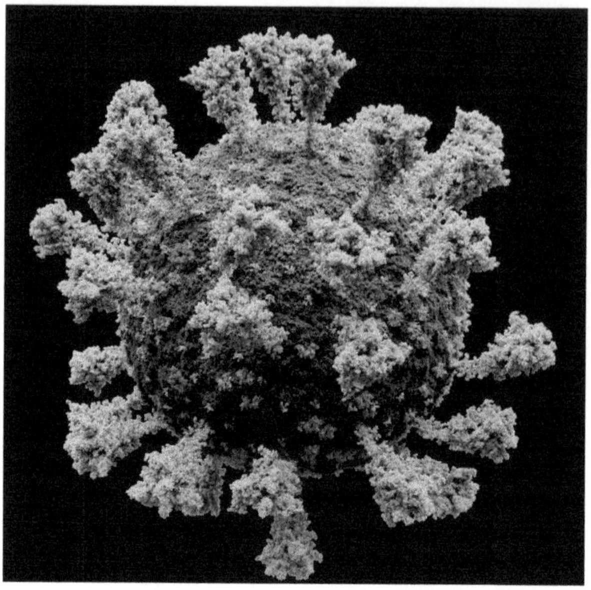

Figure 1.17. Scientifically accurate atomic model of the external structure of the Severe Acute Respiratory Syndrome Corona Virus [9]. (This Coronavirus. SARS-CoV-2 image has been obtained by the author(s) from the Wikimedia website where it was made available by Jul059 under a CC BY-SA 4.0 licence. It is included within this book on that basis. It is attributed to Alexey Solodovnikov (Idea, Producer, CG, Editor), Valeria Arkhipova (Scientific Consultant).)

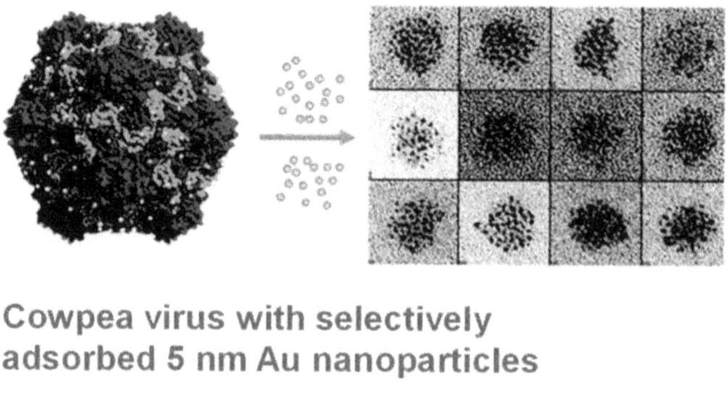

Figure 1.18. A cartoon schematic of an animal virus showing the complex structure with a DNA interior, a caspid shell and structural spikes. (Reproduced with permission from [10]. Copyright (2004) American Chemical Society.)

interfaces more accurately). It is important that you develop approaches to the fundamentals that carry over from one field (and paradigm) to the next without having to drop one way of thinking to replace with another, often with contradictory concepts (a common occurrence). A common set of principles benefits all field of interface science! With that let's launch into some general statements about using what you already know in the most efficient way to charge from one field of surface science to the next.

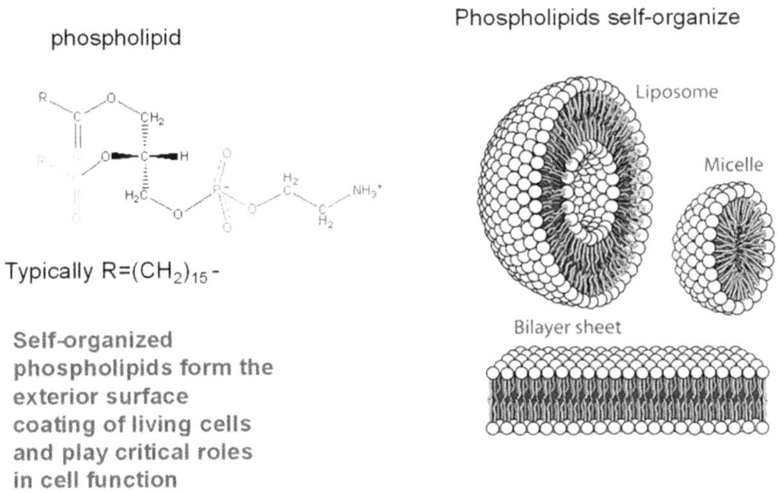

Figure 1.19. Left: the chemical structure of a typical phospholipid molecule. These molecules can spontaneously self-assemble into complex structures. Right: Examples of these structures are given by liposomes, micelles and bilayer sheets, where the white balls represent the polar ionic end groups and the brown the nonpolar alkyl chains. Self-assembled lipid structures are used in nature to form the surface shells around cells [11]. (This Phospholipids aqueous solution structures has been obtained by the author(s) from the Wikimedia website, where it is stated to have been released into the public domain. It is included within this book on that basis.)

1.9 Understanding surface and interface behavior as simple extensions of established fundamental chemistry and physics principles—summary of fundamental relationships

Following the logic established earlier, embodied in the surface model in figure 1.14 as well in quantum mechanics of electronic effects in condensed matter, we will develop a simple approach to understanding surfaces in terms of looking for the anticipated shifts from the known behavior of the bulk. To move forward in our development of the principles of surface chemistry (interface science) we will follow five simple principles:

1. The behavior of atoms/molecules follows the same physical laws and chemical principles in the surface as in the interior of bulk objects
2. Extending a specific behavior to the surface (or interface) region typically requires only minor modifications of concepts and equations
3. Excellent qualitative prediction of surface (or interface) behavior often can be done on the basis of freshman chemistry or beginning physics knowledge
4. There are only ~9 fundamental relationships that underlie the important behaviors of a wide range of chemical and physical systems
5. Accurate predictions often can be made on the basis of analysing the surface (or interface) behavior in terms of a relevant set of these relationships

Summary of Fundamental Relationships (Always Correct)

Particles (atomic/molecular scale)

Conservation Laws:
Energy   Momentum   Mass

Newton's Law ($\vec{F} = \mathbf{F} = m\ddot{\mathbf{x}} = m\frac{d\mathbf{v}}{dt}$)

Coulomb's law $\left(\mathbf{F} = \frac{1}{4\pi\varepsilon_o}\frac{q_1 \cdot q_2}{r^2}\frac{\mathbf{r}}{|\mathbf{r}|}\right)$

- -

Particle-wave duality
• deBroglie eqn. $\left(|\mathbf{p}| = h/\lambda \text{ or } \mathbf{p} = h\mathbf{k}\right)$

Einstein Eqn. ($E = h\nu$)

Heisenberg uncertainty principle $\left(\Delta E \cdot \Delta t \sim \hbar \text{ or } \Delta p \cdot \Delta x \sim \hbar\right)$

Pauli exclusion principle

Thermodynamics (macroscopic)

Elementary Chemistry:
• Equilibrium constant
• LeChatlier's principle

Thermodynamic Laws:
• 1st Law
• 2nd Law
• 3rd Law

Connectors to Particle Behavior:

U = sum of all the various forms of energies of all the particles

S (entropy):
• Boltzmann eqn [$S = k \cdot \ln(W)$]

all the ways the energies can be distributed to keep **U** constant

[Boltzmann distribution is the most probable of these *vs* U]

Figure 1.20. A summary of fundamental relationships that always apply in understanding complex behavior and are useful for generating common rules of thumb for surface behavior. In the table moving down on the left: vectors are indicated as bold type (**F**), m = mass, x = distance, t = time, $\mathbf{v}$ = velocity, ε_o = permittivity of vacuum, q = charge, $\mathbf{r}$ = distance between charges, $\mathbf{p}$ = momentum, h = Planck's constant, λ = wavelength, $\mathbf{k}$ = wavevector, E = energy, $\hbar = h/2\pi$ and on the right, k = Boltzmann's constant and W = statistical multiplicity (all the possible configurations of the system).

This approach relies upon the relentless application of the most fundamental principles of physical behavior that do not depend on special conditions or circumstances and thus always apply to help us towards the goal of unraveling complex behavior. A list is given in figure 1.20.

From these relationships three other associated relationships arise:

1. The diffraction limit for waves and particles according to their wavelength.
2. The Boltzmann probility distribution of populations of energy states in a energy level system according to the equation $\frac{p(E_j)}{p(E_k)} = \frac{g_j}{g_k}\exp\left[-\left(\frac{E_j - E_k}{k_B T}\right)\right]$, where the p and E are the probabilities and energies for the jth and kth states, k_B = Boltzmann's constant and T = temperature.
3. The Schrödinger equation relating the kinetic and potential energy of a system to its wave functions and energy states.

Although we will not be doing much in the way of detailed calculations from these fundamental relationships, a qualitative use of them can help us to develop *rules of thumb* that will serve us well in making simple predictions and developing a basic understanding of complex phenomena at surfaces.

Table 1.1. Nobel prizes awarded for work related to surfaces.

Nobel winner(s)	Year	Field	Country	Discovery or invention
Wilhelm Ostwald	1909	Chemistry	Germany	Pioneer work on catalysis, chemical equilibrium, and reaction velocities
Albert Einstein	1921	Physics	Switzerland	Work in theoretical physics, especially the surface photoelectric effect
Irving Langmuir	1932	Chemistry	US	Discoveries and investigations in surface chemistry
Kai Manne Börje Siegbahn	1981	Physics	Sweden	Development of electron spectroscopy for surface chemical analysis
Ernst Ruska, Gerd Binnig, Heinrich Rohrer	1986	Physics	West Germany	Development of electron microscopes and scanning tunneling microscopy for surfaces
Gerhard Ertl	2007	Chemistry	Germany	Studies of chemical processes on solid surfaces
Konstantin Novoselov, Andre Geim	2010	Physics	Netherlands	Experiments regarding the two-dimensional material graphene

See: http://www.nobelprize.org/nobel_prizes/lists/all/

1.10 Some historical facts about surface science

The birth of an interest in the science of surfaces has been tracked back by Ernesto Paparazzo (see the reference section) to ancient times where this subject was treated by the Greek philosophers. Over the centuries the nature of surfaces was of considerable importance in phenomena such as corrosion, surface tarnishing and other common properties.

Coming to more contemporary times, the importance of surface science, ranging across both physics and chemistry, can be seen by the list of Nobel Prizes awarded for work in this field, as shown chronologically in table 1.1. Note the span of the awards is nearly a century, starting with the award of Ostwald in 1909 for his work in a variety of surface areas, but the most well known typically is nucleation and crystallization with spontaneous joining of small particles to form more stable larger ones (so called Ostwald ripening). This was definitely work involving the nanoscale, although at the time there were really no techniques to directly examine details at the nanoscale. No doubt another prize will be added to the list within the next decade as judged by the intense work in the nanoscale field where surfaces and interfaces dominate important phenomena and are crucial in the evolution of new technologies.

1.11 Units for calculations and fundamental constants

Finally we address the use of units. Following the now standard convention in contemporary science we will use SI units, the system based on meters, kilograms and seconds units (mks). In the area of electrostatic forces and energies we use the

Coulomb unit of charge. For convenience the SI units along with conversions to other common units and some common values of fundamental constants are given below.

Units

- MKS (SI units) --- highly preferred
  - length: m
  - time: s
  - mass: kg
  - force: N (Newton = $kg \cdot m^{-1} \cdot s^{-2} = J \cdot m^{-1}$)
  - charge: Coulombs

- common quantities:
  - length: 1 Å = 10^{-8} cm = 0.1 nm = 10^{-10} m
  - area: 1 cm^2 = $(10^{-2}\ m)^2$ = $10^{-4}\ m^2$
  - volume: 1 cm^3 = $(10^{-2}\ m)^3$ = $10^{-6}\ m^3$ = 10^{-3} L
  - force: 1 N = 10^5 dynes
  - energy: 1 J = 10^7 erg = [1/(4.184)] cal
  - pressure: 1 $N \cdot m^{-2} \equiv$ 1 Pascal (Pa) = 10^{-5} bar ~10^{-5} atm
    = [1/(133.3) = 7.502×10^{-3}] torr (mm Hg)
    or 1 kPa = 7.50 torr

Energy (typical applications using specific units)

J (thermal)
eV (electrical)
cm^{-1} (spectroscopy)
associated eqns to interconvert: $E=h\nu$, $c=\lambda\nu$; $E=Vq$
1 eV = 1.602×10^{-19} J
or on a 1 mol basis: 1 $eV \cdot mol^{-1}$ = 96.5 $kJ \cdot mol^{-1}$ ~ 100 $kJ \cdot mol^{-1}$
1 eV ~8066 cm^{-1} → 1234 nm
1 eV = RT for T = 1.160×10^4 K

Useful Constants (rounded off)

N_A (Avogadro constant): 6.022×10^{23} molecules(atoms)$\cdot mol^{-1}$
h (Planck's constant): 6.626×10^{-34} $J \cdot s$
k (Boltzmann's constant): 1.38×10^{-23} $J \cdot K^{-1}$ = 8.617×10^{-5} $eV \cdot K^{-1}$
R (gas constant or Boltzmann's constant per mol): 8.314 $J \cdot K^{-1} \cdot mol^{-1}$
R (in units for gas eqns): 0.08314 $L \cdot bar \cdot K^{-1} \cdot mol^{-1}$ = 8.314 $Pa \cdot m^3 \cdot K^{-1} \cdot mol^{-1}$
RT (at 298 K) = 2.48 $kJ \cdot mol^{-1}$
c (speed of light): 2.998×10^8 $m \cdot s^{-1}$
e (smallest unit of charge, e^-, etc.): 1.602×10^{-19} C
m_e electron mass): 9.110×10^{-31} kg
m_p (proton mass): 1.672×10^{-27} kg
ε_o (electrical permittivity of vacuum): 8.854×10^{-12} farads.m^{-1} (mks units)
$4\pi\varepsilon_o$: 1.113×10^{-10} farads.m^{-1} (mks units)
F (Faraday constant; charge per mol of electrons): 96,485 coulombs$\cdot mol^{-1}$

Problems

Estimating surface atom densities

1. Following the method in figure 1.8 for estimating the surface atom density, estimate the density for a gold crystal. Give your answer in SI units of atoms/m^2 then convert to the common units of atoms/cm^2 and atoms/nm^2.

   Data in SI units: Au atom diameter = 0.3 nm = 3×10^{-10} m, density of Au = 19.3×10^3 kg m^{-3}, atomic mass = 0.197 kg mol^{-1}.

   Data in other common units: Au atom diameter = 3 Å, density of Au = 19.3 g cm^{-3}, atomic mass = 197 g mol^{-1}.

   Instead of using the known diameter of a Au atom, develop a simple method to estimate the value just from the density of gold metal. Compare the value with the correct one. Is the estimated value smaller or larger? Why?
2. Again following the methods in figures 1.8 and 1.9, estimate the number of water molecules in the surface layer of liquid water in both molecules/cm^2 and molecules/nm^2.

   Data: H_2O diameter ~ 2.8 Å, density ~ 1.0 g cm^{-3}, molar mass = 18 g mol^{-1}.
3. Following problems 1 and 2, estimate the atoms per unit area for a silicon crystal surface in atoms/nm^2.

   Data: bulk density ρ_b = 2.33 g cm^{-3}, atomic mass M_A = 28.1 g mol^{-1}, atomic radius Si in silicon crystal r = 110 pm (0.110 nm).
4. Drawing from the answers in problems 1–3, give a general estimate of the number of surface atoms or molecules in a typical substance.

Applying the missing atom model to explaining the effects of the S/V ratio on the characteristics of nano-objects

5. Based on the simple missing atom surface model, see if you can provide qualitative explanations of the changes in material properties with S/V ratio for the data in figures 1.10, 1.12 and 1.13, and for the trend in atom spacings with distance from the center of the Au cluster for the Au cluster in figure 1.16.

Calculations from basic relationships of particles and waves

6. Use the Einstein equation (figure 1.18) to calculate the energy (in joules) of a photon with $\lambda = 488$ nm (blue light). The frequency of the light can be obtained from the standard wave relationship $c = \lambda \cdot \nu$, where c = speed of light = 3.0×10^8 m s^{-1}, λ = wavelength (m) and ν = frequency (Hz = s^{-1}). For reference (see units and constants tables), h (Planck's constant) = 6.63×10^{-34} J · s.

   Now compare this with a typical bond energy of in an organic molecule (use something simple such as a C–C bond). Do you think blue light would be useful in breaking chemical bonds of molecules adsorbed at surfaces?
7. An electron beam to be used for scattering off surfaces is created in a vacuum chamber by applying a voltage of 50 V between two plates. The electrons

reach a speed of $\sim 4 \times 10^6$ m s^{-1}. Using the de Broglie equation (figure 1.18) estimate the wavelength of the electrons. For reference, h (Planck's constant) = 6.63×10^{-34} J · s, and m_e (electron mass) = 9.11×10^{-31} kg (see the units and constants tables).

Now compare this with the typical spacing of atoms in a crystal. Does this suggest a possible analytical use of this electron beam for characterizing some property of the surfaces of crystals?

1.12 What's ahead?

We will cover lots and lots of different fundamental concepts and relationships in a variety of surface and interface areas ranging from surface structure, interfacial thermodynamics of liquids and solids, surface dynamics and motion, nanoscale forces between objects and colloids, surface electrical and electronic behavior, surface chemical bonding, and heterogeneous catalysis. Hopefully, progressing through these will build a common set of principles and rules that will allow the reader to make useful connections across fields of surface and interface chemistry (and other areas including physics and biology!) that otherwise may not have be readily seen.

Historical

1. Paparazzo E 2005 The elder Pliny, Posidonius and surfaces *Brit. J. Phil. Sci.* 1–14
2. Paparazzo E 2004 How old is surface science? *J. Electron Spectros.* **134** 9–24 10.1016/j.elspec.2003.09.009
3. Allara D 2005 A perspective on surfaces and interfaces *Nature* **437** 638–639 10.1038/nature04234

S/V effects in small objects

1. Roduner E 2006 Size matters: why nanomaterials are different *Chem. Soc. Rev.* **35** 583–592 10.1039/b502142c
2. Shirota M, Ishida T and Kinoshita K 2008 Effects of surface-to-volume ratio of proteins on hydrophilic residues: decrease in occurrence and increase in buried fraction *Protein Sci.* **17** 1596–1602 10.1110/ps.035592.108
3. Teller D C 1976 Accessible area, packing volumes and interaction surfaces of globular proteins *Nature* **260** 729–731 10.1038/260729a0

Bio-interfaces at the nanoscale

1. Nel A E, Mädler L, Velegol D, Xia T, Hoek E M V, Somasundaran P, Klaessig F, Castranova V and Thompson M 2009 Understanding biophysicochemical interactions at the nano–bio interface *Nat. Mater.* **8** 543–557 10.1038/nmat2442

General

1. Somorjai G A and Li Y 2010 *Introduction to Surface Chemistry and Catalysis* 2nd edn (John Wiley). A general advanced undergraduate or beginning graduate level text focused primarily on catalysis and related UHV surface science areas. Chapter 1 covers many of the points given in the current chapter.
2. Adamson A W and Gast A P 1997 *Physical Chemistry of Surfaces* 6th edn (John Wiley). A classic text on surface chemistry with special emphasis on wet surface chemistry aspects and applications.

References

[1] Gutakovskii A K, Fedina L I and Aseev A L 1995 High resolution electron microscopy of semiconductor interfaces *Phys. Stat. Sol.* A **150** 127–40

[2] Schlom D G, Haeni J H, Lettieri J, Theis C D, Tian W, Jiang J C and Pan X Q 2001 Oxide nano-engineering using MBE *Mater. Sci. Eng.* B **87** 282–91

[3] Grest G S, Lacasse M D and Murat M 1997 Molecular-dynamics simulations of polymer surfaces and interfaces *MRS Bull.* **22** 27–32

[4] Woltersdorf J, Nepijko A S and Pippel E 1981 Dependence of lattice parameters of small particles on the size of the nuclei *Surf. Sci.* **106** 64–9

[5] Rohlfing E A, Cox D M, Kaldor A and Johnson K H 1984 Photoionization spectra and electronic structure of small iron clusters *J. Chem. Phys.* **81** 3846–51

[6] Vossmeyer T, Katsikas L, Giersig M, Popovic I G, Diesner K, Chemseddine A, EychmiUler A and Weller H 1994 CdS nanoclusters: synthesis, characterization, size dependent oscillator strength, temperature shift of the excitonic transition energy, and reversible absorbance shift *J. Phys. Chem.* **98** 7665–73

[7] Lai S L, Ramanath G, Allen L H and Infante P 1997 Heat capacity measurements of Sn nanostructures using a thin-film differential scanning calorimeter with 0.2 nJ sensitivity *Appl. Phys. Lett.* **70** 43–5

[8] Valden M, Lai X and Goodman D W 1998 Onset of catalytic activity of gold clusters on titania with the appearance of nonmetallic properties *Science* **281** 1647–50

[9] File:Coronavirus. SARS-CoV-2.png *Wikimedia Commons* https://commons.wikimedia.org/wiki/File:Coronavirus._SARS-CoV-2.png

[10] Blum A S *et al* 2004 Cowpea mosaic virus as a scaffold for 3-D patterning of gold nanoparticles *Nano Lett.* **4** 867–70

[11] File:Phospholipids aqueous solution structures.svg *Wikimedia Commons* https://commons.wikimedia.org/wiki/File:Phospholipids_aqueous_solution_structures.svg

Chapter 2

Structure and defects of periodically arranged materials—background

Structures of crystalline solids and surfaces are described. There is discussion on how defects in these configurations affect their organization.

Part I

Crystal structures

2.1 Description of surfaces (or interfaces) of crystalline materials in terms of the intrinsic bulk crystal structure

Many solid materials with surface and/or interface properties of interest exhibit crystalline structures, e.g. palladium metal in fuel cell electrodes, iron particles in the Haber–Bosch commercial ammonia synthesis process, titanium dioxide in the photo-oxidation processes for organic contaminant removal, and the interfaces constructed in GaAs/AlGaAs compound semiconductor lasers and detectors for optical communications. In these, and many similar applications, the exact structure, composition and chemical state(s) of the surface or interface atoms involved in the operational function can have a profound, and often critical, effect on the efficiency and outcome of the desired process or function. Understanding and controlling this latter aspect requires a precise understanding of the periodic arrangements of the surface and/or interface atoms of the functional material(s).

Since the surface atoms in an object are directly dependent on the bulk crystal structure connected to the surface, achieving an understanding of the surface structures starts with learning about the intrinsic arrangements of bulk crystalline materials. In this sense surface structures are often viewed and discussed in terms of

doi:10.1088/978-0-7503-4916-1ch2

perturbations of the bulk structures. This can be viewed in terms of starting at a region of interest deep within the bulk and then moving toward the outside of the object while looking at the changes in packing forces as the termination boundary of the object is approached. Typically, the onset of the perturbations occurs within a few layers of the surface. Accordingly, it makes sense to adopt descriptions and terminology of the surface structures as simple modifications of those used for the intrinsic bulk structures. The study of bulk crystal structures is an enormous field with extensions ranging across materials science (e.g. metallurgy), earth sciences (e.g. mineralogy) and biology (e.g. protein crystals). Our interest here, however, is very focused and we will proceed with the view that we only learn what is necessary as a background to surface structure so mostly we will examine simple definitions and methods for obtaining simple structural features of bulk crystals and their defects.

2.2 Definition of the unit cell of a periodic lattice and important geometrical characteristics

2.2.1 Periodic lattices and the unit cell

Crystals are built from individual chemical units (atoms, atomic groups or molecules) arranged in space such that the local structure (locations of the nearby units) viewed from some arbitrary point exactly repeats over and over upon moving away along any chosen direction to pass through the crystal. We can think of these structures in terms of a periodic lattice of points with chemical units located at each lattice point. The chemical units may have shapes such that in a perfect crystalline structure they would be oriented in some specific way at each point. An example is given by the CO_2 crystalline structure in figure 2.1. A simple example for a macroscopic, periodic

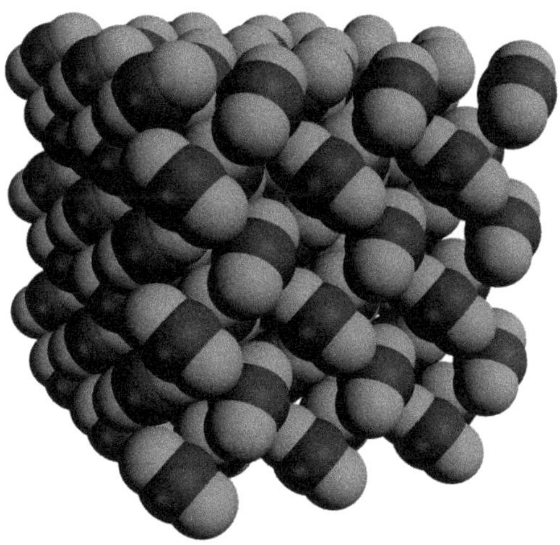

Figure 2.1. A representation of the crystalline structure of the CO_2 molecule. Moving through the crystal in any direction the local structure continues to repeat in an exact periodic manner. Note the extra complexity arising from the orientations of the linear molecules at each lattice point [1]. (This Carbon-dioxide-crystal-3D-vdW has been obtained by the author(s) from the Wikimedia website, where it is stated to have been released into the public domain. It is included within this article on that basis.)

structure is moving along a hallway in a building floor that consists of identical parallel hallways intersecting at regular intervals to other identical perpendicular hallways. As each intersection is encountered, the surroundings look identical. If we divide a crystal (or any periodic structure) into adjacent, identical local regions, the smallest possible repeat region is called the unit cell, e.g. the region around one intersection in the hallways. We can group multiples of these adjacent regions to form a larger region which also will repeat periodically moving from one to the next, but the term unit cell is generally reserved for the one group containing the minimum number of chemical units. Thus we view the crystal as formed from placing unit cells together side by side like bricks in an infinite stack with no gaps. So now the simple question becomes what shapes of bricks are allowed to form a stack with no gaps? Obviously shapes such as spheres or trapezoids do not work since these will not stack in any way without leaving gaps. We can answer this question easily by starting in 2D.

If you pick any point in a crystal and move around in some arbitrarily chosen 2D plane, in order for the 2D pattern to be exactly periodic the points that you encounter must form repeating, adjacent parallelograms. This is illustrated in figure 2.2 where the simplest unit cell for any of the different lattices is a parallelogram (red outline). Sliding (translating, no rotations allowed) the cell in directions parallel to its sides by one side length brings it into perfect superposition with adjacent parallelograms (shown with dashed lines). If continued, an infinite lattice will be generated with no gaps. Special cases of the general 2D lattice include a square cell (equal sides and 90° interior angles in the unit cell) and a hexagonal cell

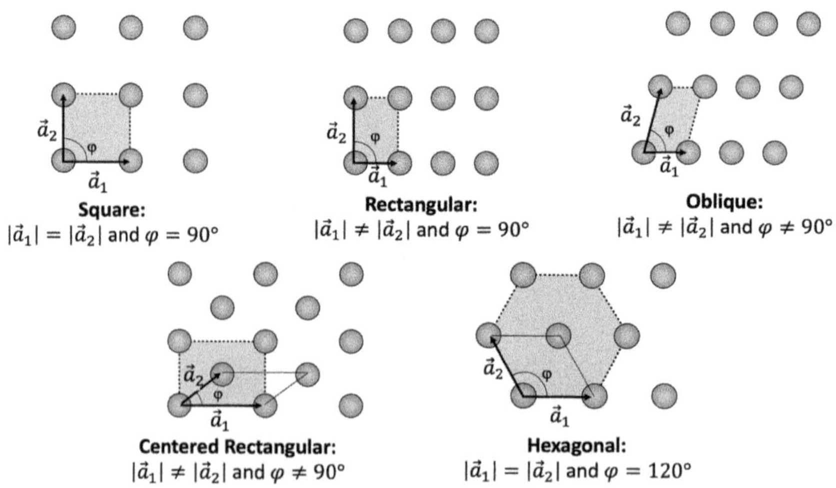

Figure 2.2. Diagrams of periodic lattice points with different aspect ratios and their associated allowed unit cells. The diagram on the upper left is a square as the unit cell. The two other top diagrams show the cases of a square and a parallelogram lattice. In the bottom left lattice, notice the centered rectangle lattice appears, although *a* is actually the unit cell. The lower right lattice is a two-dimensional hexagonal lattice, but again, the unit lattice is oblique, with the lengths of both sides equal. The angle between d_1 and d_2 is 120°. Reproduced from [2].

(equal lengths 60° interior unit cell angles). You can see hexagons in this latter lattice but they are not unit cells since they cannot be moved by the rules above and generate the lattice with no gaps. Try to make a periodic lattice out of other shapes, such as pentagons. It does not work. These shapes cannot be slid along the surface plane to generate adjacent cells that accommodate all the lattice points. There will be gaps in the lattice, i.e. regions where a pentagon will not fit between the ones that are already in place. Only parallelograms work. Large parallelograms formed by grouping adjacent unit cells are allowed and any size unit cell can be placed anywhere on the surface plane relative to the lattice points and still be a unit cell, as shown in in figure 2.2. In all cases translating the cells along their sides by one side length brings them exactly into an adjacent cell position (you should verify this). The diagram also shows a cell (blue) that is not a unit cell since translation by one side length along the axis directions places the cell short of repeating the lattice structure. Clearly, in this case the problem is that the cell does not circumscribe an area sufficient to cover the average area occupied by each point on the lattice; thus translating the cell by a side length brings it short of reaching an exactly matching position on the lattice.

2.2.2 Unit cells in 3D—definition of lattice constants, interaxial angles, basis vectors and location vectors

Now we extend to 3D. From the above discussion it follows that the simplest 3D unit cell must have sides consisting of parallelograms (six in total, one for each surface). An example is given in figure 2.3 where the cell is characterized in terms of a set of three *basis vectors* that extend from any chosen corner point to span the cell, designated as **a**, **b** and **c** (note the convention here that a vector is indicated in bold text; in the figure they are shown with the super arrow convention). Arbitrarily selecting any one corner of the cell as the origin, the vectors are

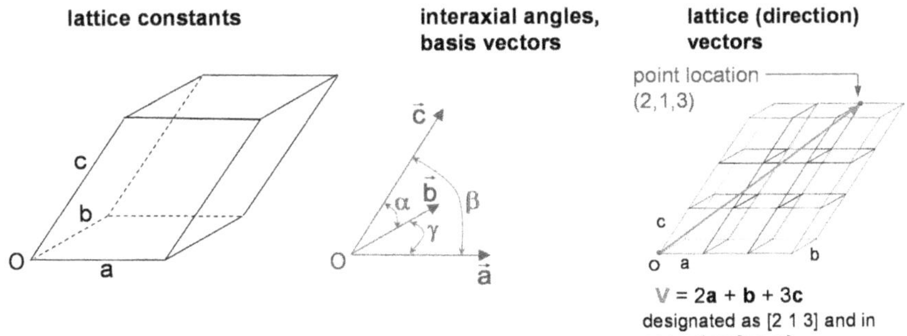

Figure 2.3. Schematic of a simple unit cell of eight lattice points which shows the characteristic unit cell basis vectors $\vec{a}$, $\vec{b}$, $\vec{c}$ (or in bold type convention, **a**, **b**, **c**) extending from the origin O with interaxial angles α, β, γ. The right-hand structure shows a set of adjacent unit cells with a vector **V** extending from the origin to a point located at (2, 1, 3) where the notation (*u*,*v*, *w*) is defined as the distances along the unit cell axes in terms of units of the lattice constants *a*, *b*, *c*, respectively. Thus the lattice vector **V** defines the location of two lattice lengths along *a*, one along *b* and three along *c*.

described in terms of their lengths a, b, c, which are called the lattice constants, and the *interaxial angles* α, β, γ between each pair of basis vectors (**b**, **c**; **a**, **c**; and **a**, **b** respectively). Translating the unit cell along the directions of the basis vectors generates the infinite lattice, as illustrated by the diagram in figure 2.3 (right-hand side). This also gives a method for locating any point in the lattice space from an arbitrarily selected origin, as illustrated by the red *location vector*, designated [2 1 3] which locates the point (2, 1, 3) in the basis vector space from origin O. Note the specific use of the square brackets, parentheses, and commas or no commas. This is a strict protocol designed to avoid confusion between vectors and lattice locations. Also note that the location vector coordinates along the **a**-, **b**- and **c**-axis vectors are typically denoted as [u v w]. It is straightforward to apply these lattice characteristics to 2D lattices since one only needs to consider a unit cell with two basis vectors, one interaxial angle and two coordinates for point location. We will use this characterization for surfaces later.

2.2.3 Direction vectors

The vector which indicates a specific direction from the origin in a unit cell is called a *direction vector*. Direction vectors are very useful in indicating the direction perpendicular to a particular crystal plane, as we will see in the next section. Since the surface of a crystal is a 2D plane you can see how this characterization would be useful in indicating which way the crystal is oriented at the termination boundary relative to a unit cell. Here are the standard conventions for direction vectors.

1. The smallest possible u, v, w coordinate integers are used, e.g. the direction in which the location vector [1 ½ 0] points is designated [2 1 0]. Since we do not care about the length it seems reasonable to just put the direction in terms of a simple unit cell.
2. Negative directions use a bar notation, e.g. $-u = \bar{u}$.
3. All parallel direction vectors use the same [u v w] direction indices.
4. Direction vectors are crystographically equivalent if the vectors are equal in length and point to exactly equivalent atoms in the cell, e.g. [100], [010], [001], [0$\bar{1}$0], [00$\bar{1}$], [$\bar{1}$00] in a cubic unit cell ($a = b = c$, $\alpha = \beta = \gamma = 90°$) represents a family of vectors of the same length that point from the origin at one corner to the three atoms at the three nearest corners along the cube edges. The family is designated by $\langle 100 \rangle$ and is called the cubic edge family.

The definitions and conventions of unit cells and associated vectors are summarized in table 2.1. Note that the directions in which vectors point are specified in terms of physical distances in the coordinate space so are often called real space vectors to distinguish them from vectors based on reciprocal distances which we will encounter soon when discussing crystal planes.

Table 2.1. Characteristics of unit cells, lattice coordinates and vectors.

Characteristic	Definition	Notation
Lattice constants	The lengths of the three sides of the unit cell pointing from the origin.	a, b, c
Basis vectors	Basis vectors span the unit cell and define its volume. Drawn along the cell edges from one arbitrary corner as the origin to each of the three cell corner lattice points.	**a**, **b**, **c**
Interaxial angles	Angles between the three basis vectors.	α, β, γ, for (**b**,**c**), (**a**,**c**) and (**a**,**b**) angles, respectively
Lattice location	Location within the entire lattice with coordinates specified as the distance along each lattice basis direction in units of the lattice constants.	(u, v, w) where the numbers are in units of the basis set lengths a, b, c and can be fractional
Location vector	Vector from the unit cell origin to any location (u, v, w) in the entire lattice, whether there is a lattice point (atom) there or not.	**V**, specified as (u, v, w) or $\mathbf{V} = u\mathbf{a} + v\mathbf{b} + w\mathbf{c}$
Direction vector	A vector used only to specify direction, not location. Since all direction vectors for the same direction are parallel the origin is not specified and the same [u v w] is used for all parallel vectors.	Specified in terms of the smallest unit cell as [u v w] where u, v, w are chosen as the smallest possible integers to indicate the direction. Negative directions are specified by a bar notation (e.g. $-u = \overline{u}$).
Crystallographic family direction vectors	All direction vectors in a unit cell that point to exactly equivalent cell lattice points (e.g. between cube corners along an edge) can be specified by any one of the three lattice vectors.	[u v w] (note: no commas)

2.3 Bravais unit cells

2.3.1 The seven Bravais lattice classes and 14 unit cells

With this background we have the foundation in place to analyse real crystals. To keep things simple we will only consider monatomic crystals. Since the atoms pack together in some way that maximizes the overall attractive energy, either by van der Waals forces (e.g. inert gas atoms), electrostatic charge interactions (metal–metal bonds) or by direct covalent bonding (e.g. silicon), while balancing Pauli exclusion repulsive forces, the exact fashion the atoms will pack in terms of neighbor distances and angles will vary from substance to substance. Regardless of the specific element, however, monatomic crystals pack in only 14 basic unit cells as shown in table 2.2, with one exception, the

Table 2.2. The 14 Bravais lattice cells classified in terms of the seven basic lattice symmetries and the placement of extra atoms within or on the cell. Atom types: red, corner; dark blue, visible from the front view; light blue, hidden from the front view.

Basic lattice shape	Lattice parameters	Unit cell type			
		P	I	C	F
Cubic	$a = b = c$ $\alpha = \beta = \gamma = 90°$	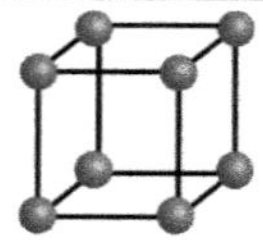	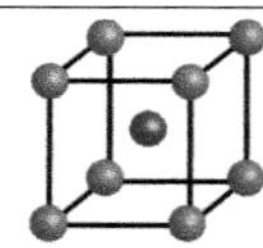		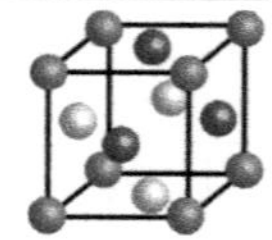
Orthorhombic	$a \neq b \neq c$ $\alpha = \beta = \gamma = 90°$	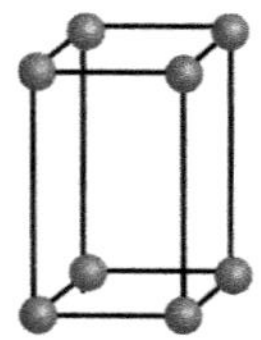	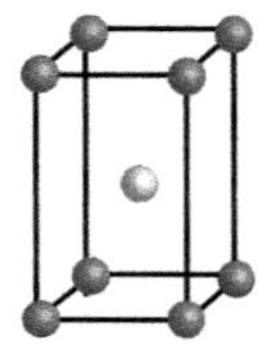	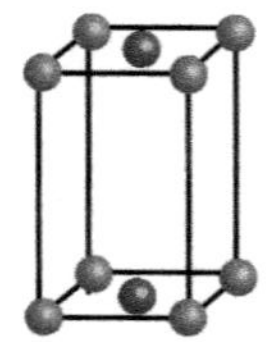	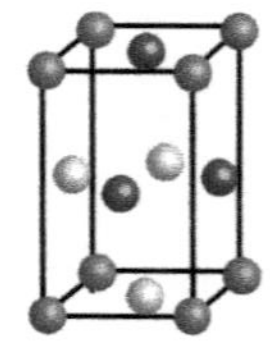
Tetragonal	$a = b \neq c$ $\alpha = \beta = \gamma = 90°$	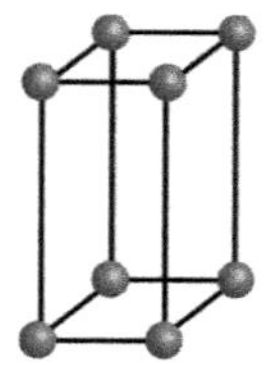	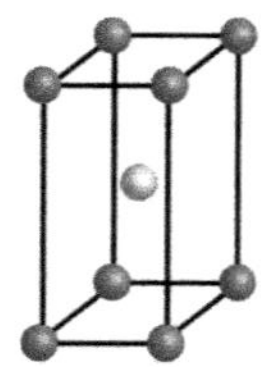		
Monoclinic	$a \neq b \neq c$ $\alpha = \beta = 90°$ $\gamma \neq 90°$	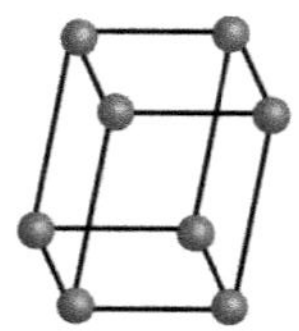		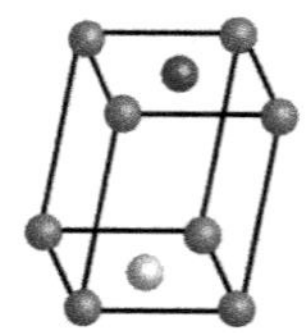	

(*Continued*)

Table 2.2. (*Continued*)

Basic lattice shape	Lattice parameters	Unit cell type: P	I	C	F
Simple hexagonal	$a = b \neq c$ $\alpha = \beta = 90°$ $\gamma = 60° (120°)$	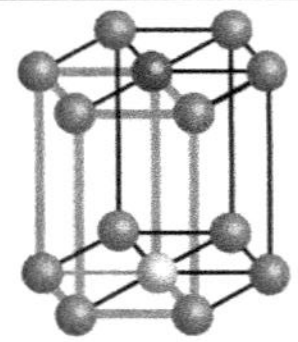			
Trigonal or rhombohedral	$a = b = c$ $\alpha = \beta = \gamma \neq 90°$	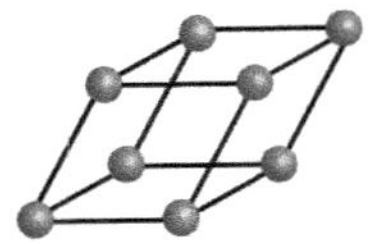			
Triclinic	$a \neq b \neq c$ $\alpha \neq \beta \neq \gamma°$	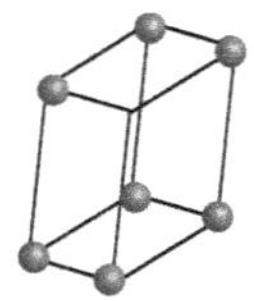			

hexagonal close-packed lattice which we will encounter soon. These cells are known as *Bravais lattices* and can be further classified in terms of seven basis lattice shapes determined by the setting of the cell corner atoms and several sub-types of cells with additional atoms besides those at the corners. For those basic cells which have space to accommodate additional atoms the possible cell types are classified as:

P: Primitive centering, lattice points on the cell corners only.
I: Body centered, one additional lattice point at the center.
F: Face centered, one additional lattice point at the center of each of the faces.
C: Base centered, one additional lattice point at the center of the top and bottom faces.

Note that the full hexagonal cylinder structure shown for the hexagonal unit cell is not the allowed unit cell; rather the subunit indicated by the green lines is the actual unit cell. The larger hexagon shape is shown to emphasize the hexagonal symmetry intrinsic to the the crystal lattice. Only the smaller green cell can be translated in any of the basis vector directions by a lattice constant length to generate the infinite lattice. Looking over the possible combinations of basic lattice types and places to accommodate additional face or interior atoms, it would seem that there are more types of allowed lattices. By careful examination, however, you would find that these each turn out to be just another way of looking at one already in the set of the 14 Bravais lattices in table 2.2. The exception is the hexagonal close-packed cell which actually is two interpenetrating Bravais hexagonal lattices.

2.3.2 The special coordinate system for hexagonal cells

Because of the hexagonal symmetry set on the three axes of a hexagon, the larger hexagonal unit, rather than the simple Bravis unit cell, is typically used to describe the basis vector set and associated lattice locations and directions. The convention can be seen in figure 2.4 which depicts the standard $\boldsymbol{a}$, $\boldsymbol{b}$, $\boldsymbol{c}$ basis vector of the Bravais unit cell (left, shaded box) and the commonly used $\boldsymbol{a}_1$, $\boldsymbol{a}_2$, $\boldsymbol{a}_3$, $\boldsymbol{c}$ basis vectors and associated lattice constants a_1, a_2, a_3, c for the hexagon cell (middle), which is not a unit cell. The a coordinates are useful for locating a point in the hexagon plane with the c coordinate locating the position a point above or below the plane. The use of four coordinates provides a redundant description of locations since only three are actually needed. The six equivalent directions around the cell going counterclockwise around the origin (center of the bottom hexagon) and starting at the $\boldsymbol{a}_1$ axis are given by $\boldsymbol{a}_1$, $-\boldsymbol{a}_3$, $\boldsymbol{a}_2$, $-\boldsymbol{a}_1$, $\boldsymbol{a}_3$, $-\boldsymbol{a}_2$. A direction vector is thus defined by the simplest integer combination of the four coordinates denoted as $[u\ v\ t\ w]$. Illustrations of four different direction vectors are shown in figure 2.4 (right with green vectors) with their $[u\ v\ t\ w]$ notations. Again, notice the bar notation, e.g. $[\bar{1}2\bar{1}0]$ is formed from $u = -1, v = 2, t = -1$ and $w = 0$. Also note that the three in-plane coordinates u, v, t overdetermine the in-plane direction; by inspection you can see that $u + v = -t$ for the vectors. The direction vectors are shown pointing in both directions rather than generating another vector in opposite direction.

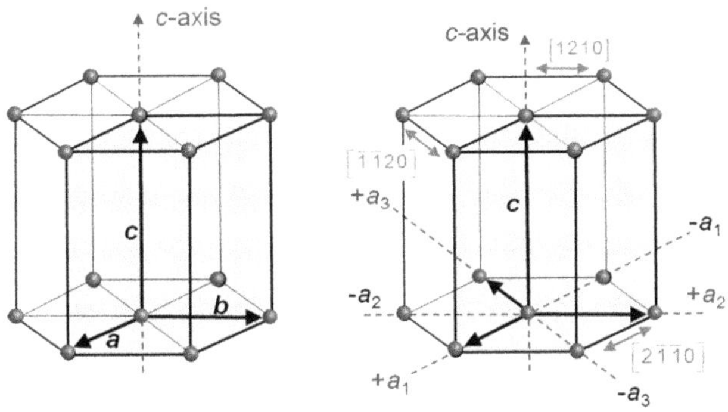

Figure 2.4. Illustrations of a hexagonal crystal structure. The unit cell (not a Bravais cell) is shown as the gray shaded parallelogram in the two figures. Left: the unit cell characterized by an ***a***, ***b***, ***c*** basis vector set with ***a***, ***b*** in the hexagonal symmetry plane and ***c*** along the perpendicular axis. Right: the unit cell and surrounding hexagonal structure characterized by the redundant $\boldsymbol{a}_1$, $\boldsymbol{a}_2$, $\boldsymbol{a}_3$, ***c*** basis set used to show directions around the hexagonal plane. Also shown are examples of direction vectors with the direction vector is defined as $[u\,v\,t\,w]$ where u, v, t, w = smallest integer lengths in lattice constant units along the $\boldsymbol{a}_1$, $\boldsymbol{a}_2$, $\boldsymbol{a}_3$, ***c*** axes, respectively. Note the standard bar convention for negative directions. Reproduced from [2].

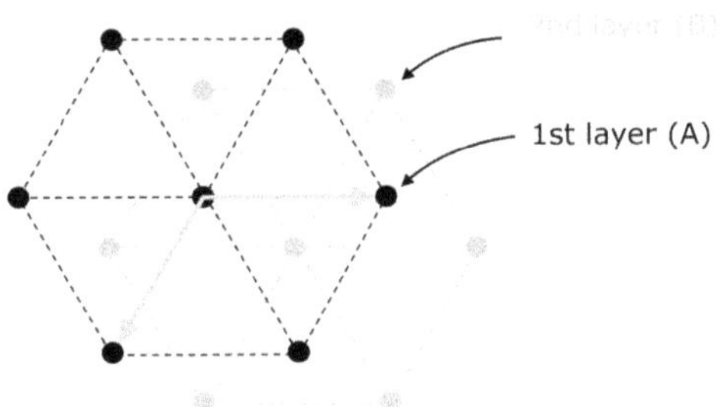

Figure 2.5. Illustration of an HCP structure in terms of the standard hexagonal cylinder cell structure (black atoms) with additional atoms (orange) inserted under the hexagonal structure. The inserted atoms are part of an offset hexagonal plane. Reproduced from [2].

2.3.3 The special case of hexagonal close-packed crystals

The closest packing between atoms that can be obtained with a lattice of underlying hexagonal symmetry is called the hexagonal close-packed (HCP) lattice. An illustration of the basic elements of the structure are shown in figure 2.5, where the orange colored atoms have been inserted symmetrically under the standard hexagonal cell. This forces the atoms into a uniform densest possible packing. This is not a Bravais lattice since the local arrangement around a given atom relative to the c-axis changes in going from one hexagonal plane to the next. By convention the unit cell is taken as the hexagonal unit cell (shaded cell in figure 2.4). The HCP structure is treated as two interpenetrating hexagonal Bravais unit cells with the lattice constant specified as the NN or closest approach (contact) spacing. We will have more to say about HCP packing shortly when we compare it with FCC packing.

2.3.4 Important structural characteristics of simple Bravais unit cells and lattices

In analysing the characteristics of crystal structures there are several aspects that deal with the packing geometry that are valuable for understanding the physico-chemical properties of the crystal and it surfaces. We will define these characteristics in terms of a Bravais cell for a monatomic crystal. Extension to more complicated crystals with mixed compositions of different atoms and chemical groups is usually fairly straightforward. The essential idea here is to see how efficiently the atoms of a given element will pack together in terms of the number of nearest neighbors in contact (at the Pauli repulsion limit) around a given atom (coordination number) and the fraction of geometric volume of the unit cell that is filled by the atoms (or conversely, the fraction of free space remaining). The main features of interest are the atoms per unit cell, the fraction of volume filled by the atoms in the unit cell (atomic packing fraction, APF), the coordination number (CN), and the ratio of the atom diameter to the unit cell lattice constant in a given direction (*R*/*a*, for cubic and hexagonal cells). These quantities are calculated from simple trigonometry and geometry relationships as shown below, with the relationships summarized in table 2.3. Calculations of atom distances for the most common cases of a cubic

Table 2.3. Summary of important characteristic quantities for crystal packing in terms of the unit cell geometry and atom sizes.

Characteristic	Description
Fractional atom volume	FV = 1/(total number of adjacent unit cells that share one selected atom) = fraction of the total atom volume contained in one unit cell.
Atoms per unit cell (volume basis)	Sum of the fractional volume contributions of all atoms in the unit cell of interest or $\sum_{n=1}^{N} \mathrm{FV}_n$, where FV_n = fractional atom volume of the *n*th atom in the unit cell with *N* total atoms associated with the unit cell.
Atom volume	$V = 4/3\pi r^3$, where r = atomic radius of the atom of interest (more generally, in the relevant oxidation state for polyatomic crystals)
Atom diameter to unit cell lattice spacing ratio	$2r/a$, $2r/b$ or $2r/c$ for the selected basis vector direction.
Atom coordination number (CN)	For a given atom, the total number of nearest neighbor (NN) atoms at the closest approach distance.
Atomic packing factor (APF)	Ratio of the total volume contribution of all atoms of a unit cell to the total unit cell geometrical volume; or $\mathrm{APF} = \frac{\sum_{n=1}^{N} \mathrm{FV}n}{V_{\mathrm{cell}}}$, where V_{cell} = geometric unit cell volume.

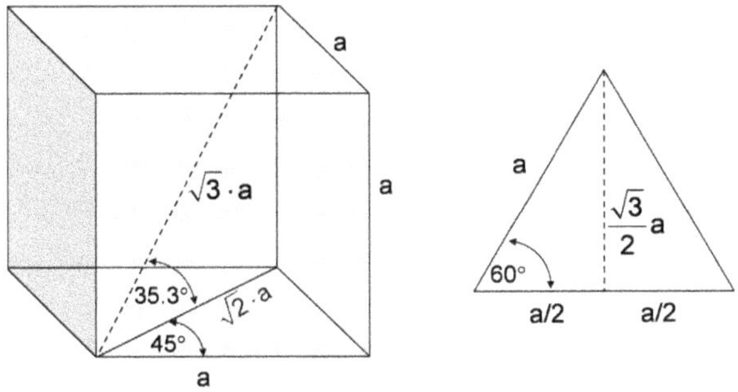

Figure 2.6. Schematic of the simple trigonometric relationships for common distances and angles in unit cells based on cubic and hexagonal symmetries.

symmetry and a hexagonal symmetry type of unit cell involves the simple trigonometric relationships shown in figure 2.6, which will serve as a convenient reference.

Atoms per unit cell

An equation is given in table 2.3, but here are stepwise instructions:

1. Pick a unit cell and surround it by its complete set of adjacent unit cells.
2. Now pick an atom in the central unit cell and note how many total cells, adjacent + central cell, share this atom. Call that number N. If no adjacent cells share this atom then $N = 0$, if the atom is shared by 1 adjacent cell then $N = 2$, etc.
3. For each atom in the unit cell calculate the fraction of the volume of that atom that sits solely in the central unit cell as FV $= 1/N$.
4. Add up the FVs of all the atoms in the central unit cell to get the total volume contribution of those atoms to the unit cell, called the atoms per unit cell.

For example, for a simple cubic cell there are eight corner atoms and each is shared by eight adjacent unit cells. Thus there is $8(1/8) = 1$ atom in a cubic unit cell. You can see the volume occupied by each of the atoms in the cube is 1/8 by looking at figure 2.7. Similarly, for a BCC cell with another atom inserted in the center of a cubic cell and not shared by adjacent cells, there are $1 + 8(1/8) = 2$ atoms per cell. Extending this method to an FCC cell with one atom inserted in each of the six faces of a cubic cell where it is shared with the face of an adjacent cell, the total atom volume contained within the unit cell is $8(1/8) + 6(1/2) = 4$.

Atomic packing fraction (APF)

This is a simple calculation of the ratio of the atom volume contribution to the unit cell divided by the geometric volume of the unit cell. Just to be complete, the information on how to calculate the geometric volume of all Bravais unit cells is given in table 2.4, based on the vector triple product $V = \boldsymbol{a} \cdot (\boldsymbol{b} \times \boldsymbol{c}) = abc\sqrt{1 - \cos\alpha - \cos\beta - \cos\gamma + \cos\alpha \cdot \cos\beta \cdot \cos\gamma}$. Note that no formula is

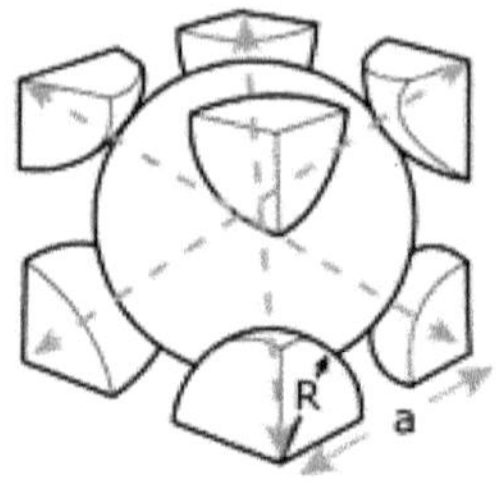

Figure 2.7. Illustration of cubic BCC unit cells showing the corner atoms as truncated spheres with the volumes of the atoms contained solely within the volume of the cube. The remainder of the corner atom volumes is shared by the seven adjacent cells around that atom and for the BCC cell one atom is completely contained in the geometric cell volume. Reproduced from [2].

Table 2.4. Formulas for calculating the volume of Bravais unit cells from the lattice constants and interaxial angles.

Basic unit cell shape	Volume
Cubic	a^3
Orthorhombic	abc
Tetragonal	a^2c
Monoclinic	$abc \cdot \sin\alpha$
Hexagonal	—
Trigonal or rhombohedral	$a^3\sqrt{1 - 3\cos^2\alpha + 2\cos^3\alpha}$
Triclinic	$\mathrm{abc}\sqrt{1 - \cos^2\alpha - \cos^2\beta - \cos^2\gamma + 2\cos\alpha \cdot \cos\beta \cdot \cos\gamma}$

given in the table for the hexagonal cell since there are variations of what is commonly specified as the hexagonal unit cell, including the HCP variation. For this case one would pick the specific unit cell of interest and calculate the volume and distances within the cell by simple trigonometry formulas. But regardless of the unit cell selected, the location and distance of neighboring atoms around a given atom is identical. For this reason, it is not important which unit cell we choose to understand the packing of the crystal.

Coordination number (CN)

This is a very important characteristic which gives the number of bonds of a given atom to its nearest neighbors. In monatomic crystals the NN atoms are placed symmetrically around a central atom so all we need to do to obtain CN is to pick an atom in a unit cell and count the total nearest neighbors including those in adjacent cells if the chosen atom is shared by other cells. It is important in determining CN to select the NN atoms only for counting. There are different ways to count depending on the view of the unit cell. The simple cases are cubic and BCC. For cubic pick a corner atom surrounded by unit cells top and bottom and you will see that there are

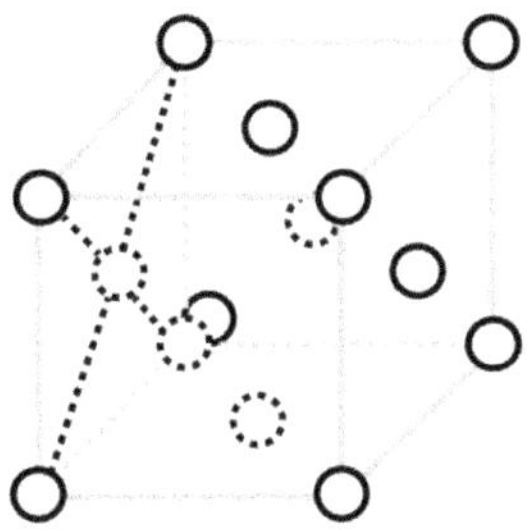

Figure 2.8. Schematics of FCC unit cell showing the NN coordination around a selected face atom in the unit cell. The cells stacked with the NN coordination of the common central atom shown as four NN above, four NN in plane and four NN below to give CN = 12. Reproduced from [2].

four NN atoms in one plane with one NN above and one below to give CN = 6 (verify this) in an octahedral coordination shell. For the BCC cell just look at the central atom and you can see it touches the eight corners as NN atoms to give CN = 8 (verify this). In the simple hexagonal Bravais unit cell there are six NN atoms all in the same plane, as seen by looking at the hexagonal cylinder in table 2.1 (verify this). Note that the hexagonal cylinder consists of three adjacent unit cells to form top and bottom hexagonal planes.

The FCC unit cell takes a bit more work. From the FCC cell in figure 2.8 one can see that the number of CN is 12. In the cell the dashed lines from the center atoms to all NN atoms show four in that face plane. There are four NN in the horizontal plane and four in the other vertical plane perpendicular to the one shown in figure 2.8 to give CN = 12. We also could have counted from any corner atom to its NN atoms in the seven adjacent unit cells to obtain the same answer.

The HCP CN can be easily determined by looking at the hexagonal cylinder cell (figure 2.5). Choosing the red atom in the center face on the top of the hexagonal cell, you can see there are six NN atoms in the hexagon plane (red atoms) and three below (green). There also will be three above that would be located in the dashed line hexagon above to give a total CN = 6 + 3 + 3 = 12 (verify this) and a 3–6–3 coordination shell, similar to the FCC coordination. The only difference between the FCC and HCP is the rotation of the top three atom triangle by 60° relative to the bottom one in the HCP case. We will see this in more detail in the next section.

2.3.5 Packing characteristics of the three principal metal crystal structures: BCC, FCC and HCP

Using the quantities in table 2.3 we can now make direct comparisons of the packing characteristics of BCC, FCC and HCP monatomic crystals, the most three most common types of structures observed in real materials such as metals. The quantities, all specified in terms of lattice constants for the standard BCC and FCC Bravais unit cells, and the HCP trigonal unit cell lattice constant equal to the NN spacing, are summarized in table 2.5. The table shows both the common hexagonal cylinder cell and the simple Bravais trigonal unit cell. The calculations of

Table 2.5. Summary of the packing characteristic for the common BCC, FCC and HCP unit cells.

Quantity	BCC 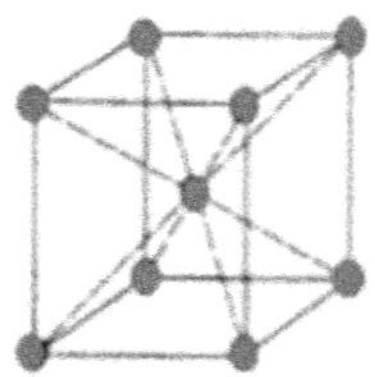	FCC 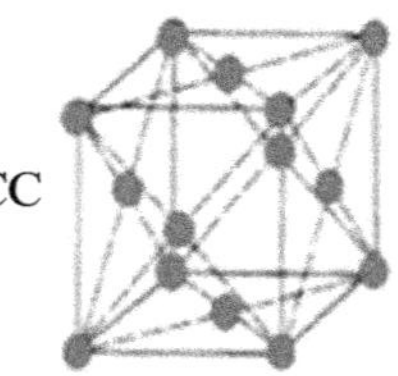	HCP 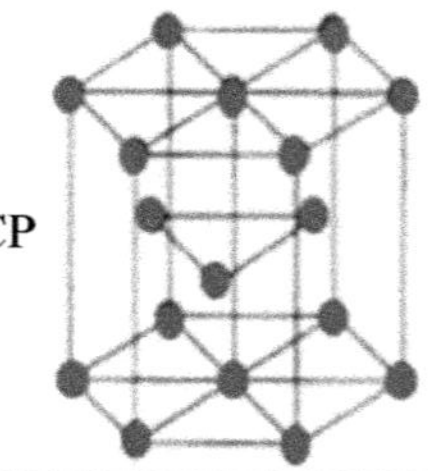 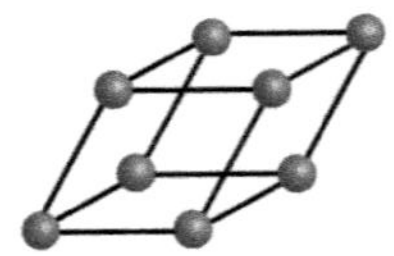
Atoms/unit cell	2	4	1 (for trigonal cell)
CN	8	12	12
PF	0.68	0.74	0.74
R/*a*	$\sqrt{3}/4$	$\sqrt{2}/4$	1

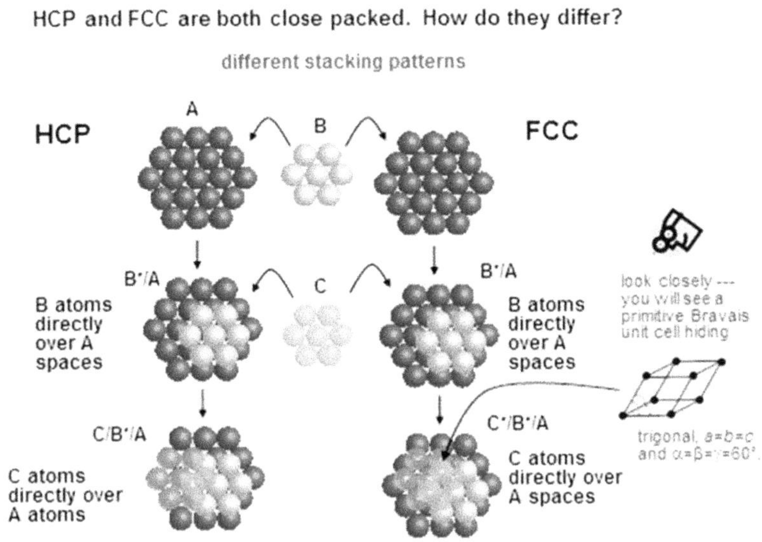

Figure 2.9. Schematic illustrating the different stacking alignments of close-packed hexagonal atom planes for HCP and FCC crystal packing. The HCP arrangement stacks every other layer superimposed (C directly over A) whereas FCC rotates every other stack to place the atoms over the holes of the alternate stack below (shifted with C directly over the holes in A). Notice a Bravais primitive trigonal unit cell hiding in the FCC stack. This cell can be used to generate the FCC lattice as well as the standard cell. Reproduced from [2].

volume and atoms per unit cell are based on the latter. Of course, CN, APF and R/a are identical regardless of the cell chosen since these relate directly to the local atom environment, not the collection of atoms chosen for a cell.

Note that both the FCC and HCP structures have the atoms in the closest packed density with PF = 0.74. But the specific arrangements are not identical. We can see the difference in figure 2.9 which shows the HCP layers stacked with every other layer placed their atoms directly over each other and the FCC with every other layer staggered relative to each other. Thus while each arrangement has CN = 12 with identical APFs the alternate layers are just shifted sideways in alignment to one another giving rise to two different Bravais unit cell symmetries while still placing the atoms as close as is allowed to fill the volume.

2.3.6 Interstitial spaces

Note in table 2.5 that lattices, even the maximum density FCC and HCP cases, have a significant fraction of space distributed periodically between the non-nearest neighbor atoms. These local volumes, called the interstitial or void space, often can accommodate foreign atoms, such as impurities, if the atoms are sufficiently smaller than the native lattice atom spacings. This is often the case for small atoms such as H with $r = 0.5$ Å which have a chemical affinity for metal atoms with much larger radii and associated lattice constants, typically 5 Å or larger in the case of transition metals. Reactive small atoms often can diffuse into lattice voids and be held in place by favorable chemical interactions, often compensating for any unfavorable stresses in the guest atom locale. A good example is H atom insertion into Pd metal lattices with the formation of Pd–H bonding.

Where are the largest spaces located in the host lattice and what geometry of bonding is possible with the NN host atoms? We answer this question for the case of the dense packed FCC lattice and then summarize the results for the common cases of FCC, BCC and HCP unit cells. The FCC unit cell has two types of periodic spaces called octahedral (Oh) and tetrahedral (Td), as illustrated in figures 2.10 and 2.11, with the names associated with the geometry of the location (or bonding) of the guest atom to its NN host lattice atoms. Clearly, guest atoms cannot insert between NN atoms in the host lattice but 12 other spaces exist along the FCC unit cell edges and the center, resulting in 12 Oh sites, as shown in figure 2.10. Similarly, other interstitial spaces also exist along the diagonals, ¼ of the way from each of the cell

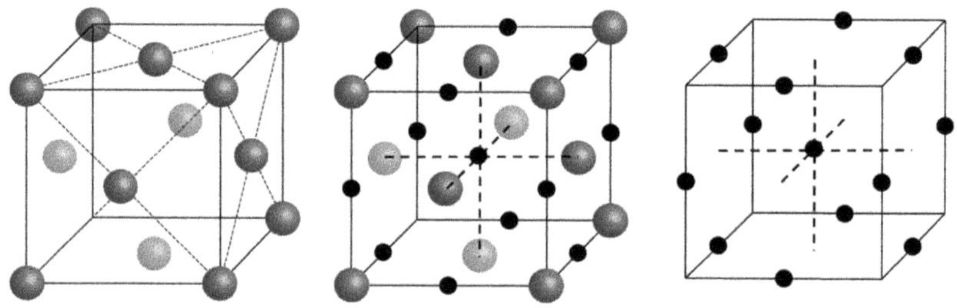

Figure 2.10. Octahedral (Oh) interstitial spaces in the FCC unit cell. Left: The FCC unit cell. Middle: The FCC cell with small black spheres illustrating the location of the Oh interstitial voids. Notice the void locations along the center of each cell edge and one in the middle to give a total of 12 locations. The center sphere shows that each Oh void space has six bonding directions in an octahedral symmetry to the nearest FCC atoms. Right: The FCC cell showing only the Oh void locations.

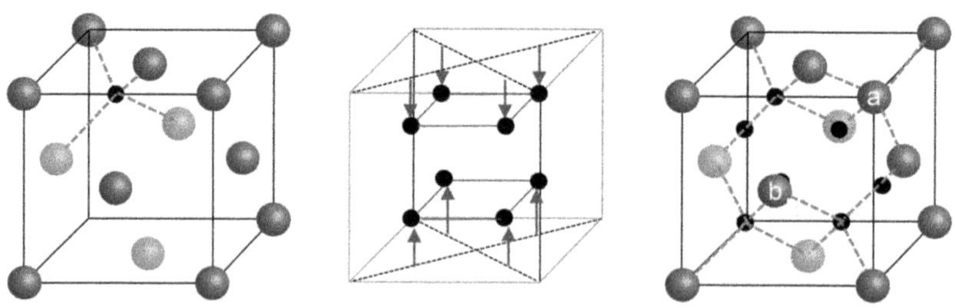

Figure 2.11. Tetrahedral (Td) interstitial spaces in the FCC unit cell. Left: The FCC unit cell with one Td site shown. The NN host atoms form a symmetrical tetrahedral bonding structure with the void space to give CN = 4. Middle: The unit cell outline shown with just the void spaces indicated by black spheres. The Td voids are each located a distance of $a/4$ from the nearest top or bottom and edge faces, where a = FCC lattice constant. The void locations form a small cube entered within the larger FCC cubic cell. Right: The FCC lattice with all the void spaces indicated. Two Td locations are hidden behind FCC atoms a and b. Some of the tetrahedral bonding is shown by dashed lines; for simplicity of viewing some are left out.

Table 2.6. Interstitial site characteristics in three selected unit cell structures.

	Interstitial sites	
Unit cell	Tetrahedral	Octahedral
BCC	12	6
FCC	8	4
HCP	12	6

corners, resulting in 8 Td sites, as shown in figure 2.11. The associated CNs for the Oh and Td sites to their NN lattice atoms are six and four, respectively, as seen in the two figures.

The characteristics for BCC, BCC and HCP lattice interstitial sites are summarized in table 2.6. Notice that the HCP lattice has identical interstitial characteristics to the FCC. This is not surprising given that both lattices are densest packed with identical APF values.

2.4 Crystallographic planes and Miller indices—introduction to reciprocal space

2.4.1 Definition of a crystal plane

Look at a 3D periodic lattice and turn it in different directions to alter the viewing perspective. At certain angles infinitely long channels swing into view where some, or even all, of the lattice points superimpose looking down the viewing direction. For example, for a cubic lattice, the view can be adjusted to be normal to the *a*, *b* face which causes all the corner points to line up on top of each other. For a monatomic crystal think of the alignments along the viewing axis as forming infinitely long corridors bounded by infinite planes of atoms on each side with the inter-planar spacing defined by a characteristic atom–atom distance, or inter-planar spacing, in the crystal lattice. This can be seen in the example of a cubic lattice shown in figure 2.12. The characteristics of crystal planes are very important in determining properties and behavior of the crystal along different directions. For example, many crystalline materials tend to fracture along specific planes and the reactivity of many crystals to chemical species is often more severe when certain planes with wide atomic spacings are exposed at the surface. Also when beams of x-rays are directed at the crystal the diffraction pattern of the x-rays, which arise from interference patterns of the x-rays as they travel between crystal planes, will be very different at different angles. This effect is what makes x-ray diffraction so valuable for characterizing crystal structure. (Just to put a likely question to rest, the x-rays are reflected off the electron density smeared around and between the atoms, not the nuclei, so indeed the crystal planes are actually periodic lumps in the electron density.)

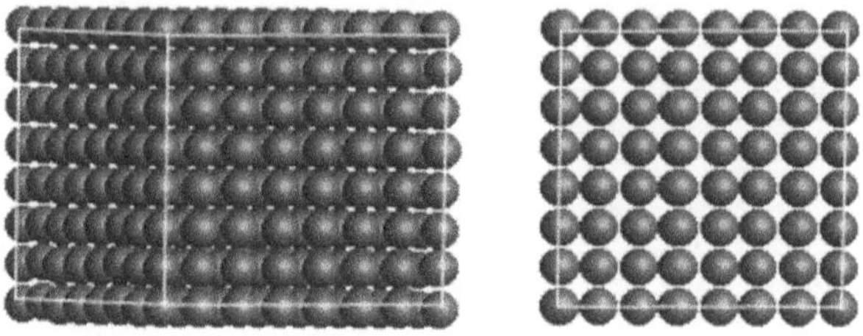

Figure 2.12. Illustration of the appearance of different channels or interplanar spacings as a function of the viewing angle for a cubic crystal. Left: viewed nearly on a corner edge reveals only very small spacings or channels down the view direction. Right: in contrast, viewed directly on a cubic face the atoms create infinite channels down the view direction as all the atoms superimpose. The rows (or columns) of the atoms are part of an infinite plane that runs down the viewing direction (perpendicular to the plane of the page).

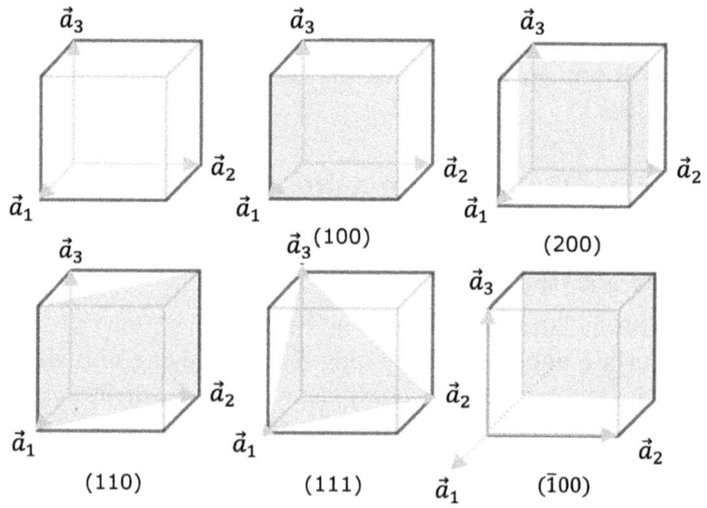

Figure 2.13. Some common crystal planes associated with the cubic unit cell. The planes are named by the inverse of their interception with the ($\overline{a_1}$, $\overline{a_2}$, $\overline{a_3}$). Reproduced from [2].

To develop our discussion of characterizing crystal planes let's start with a simple monatomic Bravais unit cell and look at the alignment of atoms in various planes located at different angles from the origin of the cell. To be complete, of course, you must imagine an infinite number of adjacent cells which will continue the planes through the entire crystal. We start with cubic cells since they have a simple Cartesian coordinate symmetry and we can replace the basis vector directions with x, y, z. For example, figure 2.13 shows selected faces (gray shading) of a cubic unit cell. One way to think of this that will be very useful when we begin studying surfaces is to imagine the crystal is sliced along a direction parallel to the gray faces to expose the atoms in that face. The figure shows that we can slice parallel to a face of the unit cell, diagonally through the cell along the plane or diagonally along the *x–z* or *y–z* planes. In each case a different atom pattern is exposed along the cut plane. Notice that the *y–z* diagonal plane has the atoms located in a hexagonal symmetry with larger spacings than one of the facial planes. The right-hand panels in the figure shows two adjacent cells with parallel cut planes. From this you can extend to the entire crystal to visualize an infinite series of cut planes running through the crystal to form an infinite series of parallel channels. Let's see how these planes are located in terms of coordinates and the common convention for labeling the plane.

The easiest specification to start is just making a list (a_1, a_2, a_3) of the coordinates of the cut or intersection points of the plane of interest along the lattice axis of interest, whether the intersection is in the original unit cell or even 25 unit cells out and fractional intercepts are fine since we are free to select any plane of interest whatever crazy direction it has. After that we can generate a vector from the origin perpendicular to the plane. It is easy to see, as illustrated by the red vector in figure 2.13, that the length of this plane location vector is just equal to the interplanar distance, so this vector seems very useful.

An illustration of naming the planes according to the cut intercepts along the axes is given in figure 2.14 for the case of three different planes in a BCC crystal and illustrates the general approach to cubic unit cells. In the figure three different

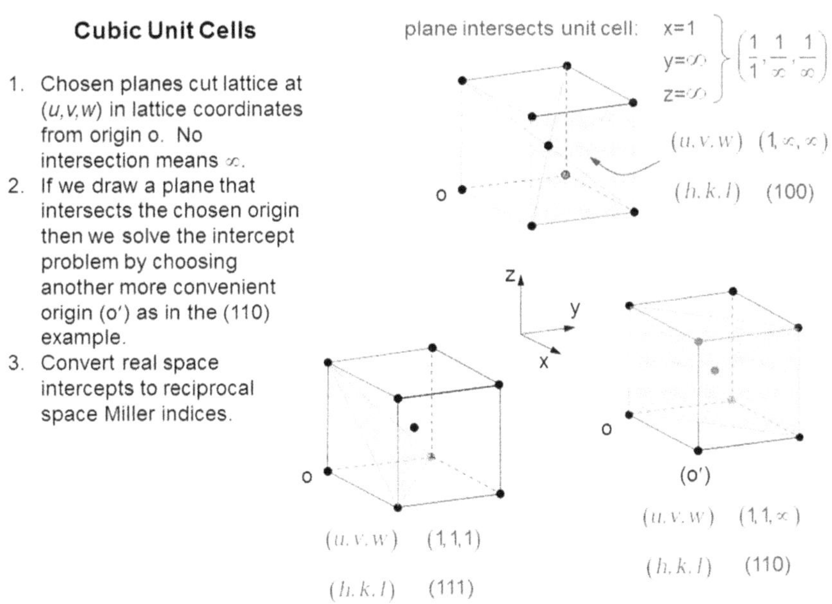

Figure 2.14. Illustrations of three different lattice planes in the cubic cell for the specific case of BCC packing. The selected planes (shaded gray) can be named by their Cartesian intercepts in the unit cell, using ∞ for the case of a plane parallel to a given axis, or by the more common convention of Miller indices which are the inverse of the intercepts. For selected planes which intersect the origin the origin is shifted to allow non zero intercepts to be defined, as for the case of the (110) plane where O is shifted to O′.

direction planes are characterized by their Cartesian coordinates for the intersection of the planes with each axis, designated as the (*u.v.w*) location of the intercept point on the axis. Note that planes which run parallel to an axis never intercept that axis so the intercept point is ∞. We could leave the characterization of planes at that but the development of the field of crystal structures took some turns on this question as the properties of crystal were discovered long ago.

2.4.2 Characterization of the orientation and spacing of a crystal planes by the Miller indices—the reciprocal space notation of planes

A bit of background is useful here. The characterization of crystal planes has its basis historically in the simplest form of the equations that relate crystal spacing to x-ray diffraction patterns, as well as energies associated with electronic states and vibrations of the crystal. In all cases, the simplest structural quantity to deal with is the reciprocal of the spacings. Think of this in terms of waves—higher energy waves have shorter wavelengths (photons are a good example), so if we are interested in energy (or momentum) we look at the inverse or reciprocal wavelength. So it has evolved that we discuss crystal planes in terms of the reciprocals of the coordinates that locate the planes. Thus we view the structure in simple terms as real space, from which we can draw nice structure pictures like we do for molecules, but when it is time to delve into phenomena such as diffraction and electronic properties we turn to reciprocal space, complicating our ability to immediately draw structures, but greatly enhancing our ability to understand physical behavior. With this we launch into trading lattice or direction coordinates ($u\ v\ w$; or $u\ v\ t\ w$) into reciprocal lattice coordinates ($h\ k\ l$; or $h\ k\ i\ l$), which are termed Miller indices for historical reasons. To make life easy we will dispense with reciprocal basis vectors for now (it can indeed get quite complicated but is very useful for diffraction and understanding the energy levels in crystals!). Our goal in the section is to define reciprocal lattice coordinates and show how to use them in determining the spacing between adjacent, parallel crystal planes and their orientation in the unit cell.

2.4.3 Cubic and hexagonal unit cells

Going back to figure 2.14, we rename the cubic planes in terms of their Miller (hkl) indices. The definition is simple—replace the intercepts u, v, w of the plane along the cell axes, with $1/u$, $1/v$, $1/w$ and scale to obtain the smallest set of integers. When the plane runs parallel to an axis then the intercept is infinity. If the plane runs through the origin then another origin is selected to remove this problem. Thus, following the example in figure 2.14, for the selected face plane with intercepts $1,\infty,\infty$ in the BCC unit cell, the reciprocal indices are (1 0 0). As usual, replace any minus signs with super bars. This reciprocal conversion gets rid of the ∞ symbol in specifying a plane, but more importantly, also sets us up to do simple calculations of crystal properties. The first we choose is the interplanar distance. Details are given in the appendix at the end of part I.

Based on simple vector trigonometry relationships the interplanar distance, defined as the shortest distance between adjacent crystal planes, is given by

$$d_{hkl} = \frac{a}{\sqrt{h^2 + k^2 + l^2}}, \tag{2.1}$$

where h, k, l denote the specific plane and a is the lattice constant for the cubic cell. If we draw a vector from the origin of the unit cell to extend perpendicular to the surface of the plane, d_{hkl} is the length of this vector, as you can see in the right-hand cell in figure 2.13.

Now we look at the direction of this vector in terms of the intercepts *uvw* and reciprocal indices. The direction vector perpendicular to the plane turns out to be $[1/u\ 1/v\ 1/w]$ which is just $[h\ k\ l]$, so here we see some simplicity in using reciprocal indices. Just pick a crystal plane in a chosen unit cell, assign the *hkl* indices and you can immediately calculate the interplanar distance and assign a direction perpendicular to the plane in the unit cell.

The simple relationships between interplanar distance and plane direction vectors with reciprocal indices also works for hexagonal planes. One just needs to switch to the a_1, a_2, a_3, c coordinate system and adjust the intercepts accordingly. For example, the various planes in figure 2.15 are defined in terms of the Miller indices for the intercepts of planes on a hexagonal cell. With the origin at the center bottom atom, the simplest plane is the hexagonal one which we denote in Miller indices as (0001) in terms of being located at $c = 1$ and parallel to the a_1, a_2 a_3 axes. The planes

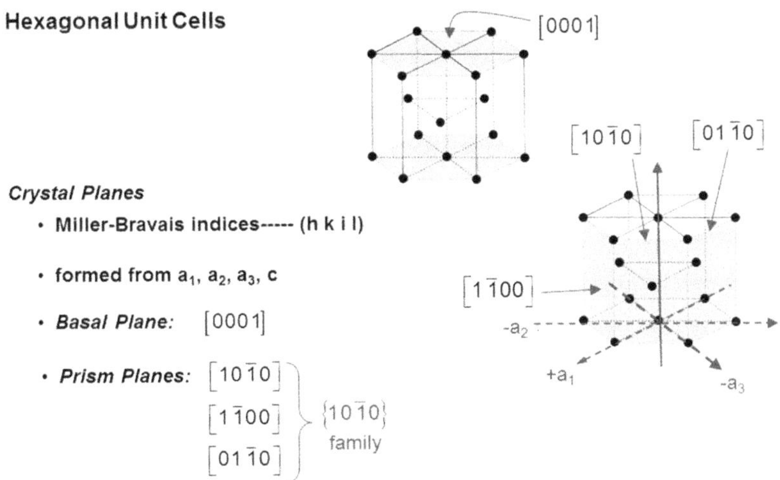

Figure 2.15. Examples of crystal planes for the hexagonal lattice system characterized by the redundant a_1, a_2 a_3, c lattice basis system. The real space intersection points of a plane are converted to reciprocal space Miller indices to denote the plane in the common convention. The Miller indices are denoted as (*hkil*) for the reciprocals of the (*uvtw*) intercept location coordinates. The simplest planes are the basal planes which have intrinsic hexagonal symmetry in their atoms and the prism planes defining the faces of the hexagonal cylinder.

which frame the hexagon on six sides are called the prism planes and are denoted in parallel pairs as $(1\bar{1}10)$, $(10\bar{1}0)$ and $(01\bar{1}0)$ (you should verify this). Also note that the six prism planes are symmetrically distributed around the *c*-axis and can be classified as a family, denoted as any one of the planes in brackets, e.g. pick $\{10\bar{1}0\}$, exactly like we defined families of direction vectors (see earlier). It follows from the previous section that the direction vectors perpendicular to the planes are given by the reciprocal indices as $[h\ k\ l\ i]$.

Appendix A: Calculation of interplanar spacings from the Miller indices for cubic cells and a note on reciprocal space

In figure 2.16 we define an arbitrary *hkl* plane intersecting the *xyz* axes of a cubic system at intercepts u, v, w (in units of lattice constant *a*) from the origin O and locate the plane by extending a vector $\mathbf{d_{hkl}} = d_x\mathbf{a_x} + d_y\mathbf{a_y} + d_z\mathbf{a_z}$ from O at angles α_x, α_y, α_z from the *x*-, *y*-, *z*-axes, respectively, to contact the plane at normal incidence. We analyse the length of the vector using simple trigonometry. Referring to the coordinate schematic in the left side of the figure the relationships between the direction cosines, the plane intercepts and length of the vector are given by

$$\begin{aligned}\cos\alpha_X &= d_{hkl}/u\\ \cos\alpha_Y &= d_{hkl}/v\\ \cos\alpha_Z &= d_{hkl}/w.\end{aligned} \tag{2.2}$$

Using the trigonometry relationship $\cos^2\alpha_X + \cos^2\alpha_Y + \cos^2\alpha_Z = 1$, equating the reciprocals of *u, v, w* to the *h, k, l* indices, and collecting terms we have

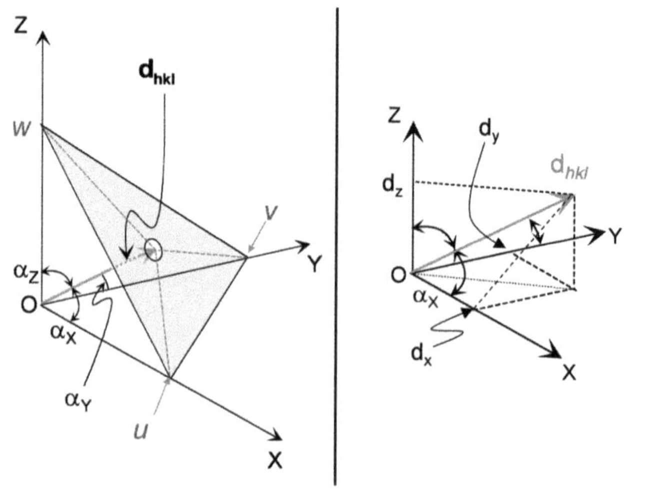

Figure 2.16. A diagram of the relationship between the direction vector and distance of a crystal plane from the origin of a lattice and the Miller indices that describe the plane.

$$d_{hkl} = \frac{1}{\sqrt{h^2 + k^2 + l^2}}, \tag{2.3}$$

where d_{hkl} is in units of the lattice constant. To convert to the actual distance in standard units just multiply by a. This is the shortest distance from the origin of the cell to the surface of the plane but since the planes all intersect the adjacent unit cells in exactly the same way then d_{hkl} must represent the interplanar spacings. Thus characterizing the plane by its reciprocal space indices gives quick access to calculating interplanar spacings which underlie many physical and chemical properties.

What is the direction of d_{hkl}? To figure this our we need to find the x, y, z components of the vector, d_x, d_y, d_z, which come directly from the direction angles. Referring to the right-hand coordinate diagram in figure 2.16 the coordinates are expressed as

$$\begin{aligned} d_x &= d_{hkl} \cos \alpha_x \\ d_y &= d_{hkl} \cos \alpha_y \\ d_z &= d_{hkl} \cos \alpha_z. \end{aligned} \tag{2.4}$$

Combining equations (2.3) and (2.4) gives

$$\left(d_x, d_y, d_z\right) = \left(\frac{{d_{hkl}}^2}{u}, \frac{{d_{hkl}}^2}{v}, \frac{{d_{hkl}}^2}{w}\right) = {d_{hkl}}^2(h, k, l),$$

which is just the coordinates expressed as the vector magnitude squared times each of the h, k, l indices. Scaling by ${d_{hkl}}^2$ gives the simple vector $[h\ k\ l]$ as the direction of $\boldsymbol{d}_{hkl}$. So again, the use of the reciprocal indices and coordinates simplifies the expression of useful quantities. We can add one more extension which brings us to the edge of handling crystal diffraction and solid state physics phenomena.

So to be complete, let's define a reciprocal space vector $\mathbf{g}_{hkl}$:

$$\mathbf{g}_{\mathbf{hkl}} = h \cdot \boldsymbol{b}_x + k \cdot \boldsymbol{b}_y + l \cdot \boldsymbol{b}_z,$$

where we we start with a Cartesian space unit cell with x-, y-, z-axes, invert the cell coordinates to reciprocal coordinates and use these to generate a unit reciprocal cell with basis vectors $\boldsymbol{b}_x$, $\boldsymbol{b}_y$, $\boldsymbol{b}_z$ and dimensions h, k, l to describe periodic sets of planes. This looks like a fast way to make something simple into something complicated but when one needs to treat properties of a crystal such as energy levels, vibrations and x-ray (or other particles) diffraction, all of which depend on inverse distances between the periodic crystal planes, forming the problem in terms of reciprocal space and reciprocal vectors makes the math much simpler.

Part II

Crystal defects

Defects in surfaces often play a critical role in surface phenomena ranging across corrosion, catalysis, solid state electronic devices and mechanical strength of thin films. Thus it is useful to learn a few fundamentals about defects starting with simple, periodic crystal systems. In this section we will take a quick look at the standard defects starting with those that span 3D bulk regions and ending up with those in the final termination layer at the surface of a crystal.

2.5 Bulk defects

In the bulk there are two classes of defects based on the dimensionality of the defect. We will cover so-called zero-dimensional point defects involving single atoms (or constituent chemical groups) and 2D defects involving rows or sheets lines of atoms in the form of crystal planes.

2.5.1 Point defects

There are several different types of point defects, and these depend to some extent on the type of crystal. The simplest defect is a missing atom or constituent chemical group in the lattice. For simplicity let's consider this defect, as depicted in figure 2.17. Note the stress involved in the surrounding lattice atoms. All other factors constant, the stress is outward from the vacancy since the surrounding atoms will tend to pull the red outlined inner shell atoms away from the vacancy due to the loss of the central missing atom force which pulls the atoms inwards.

Given that the missing atom may be buried deep in the crystal, a logical question is how did it get there? Where did the missing atom go? The most common way this happens is by thermal energy near the melting point (T_m). Approaching T_m upon heating, the thermal energy available to force atoms out of an ordered, periodic position becomes sufficient to create holes. The constant battering of atoms against each other from the thermal energy causes a dynamic condition in which a certain

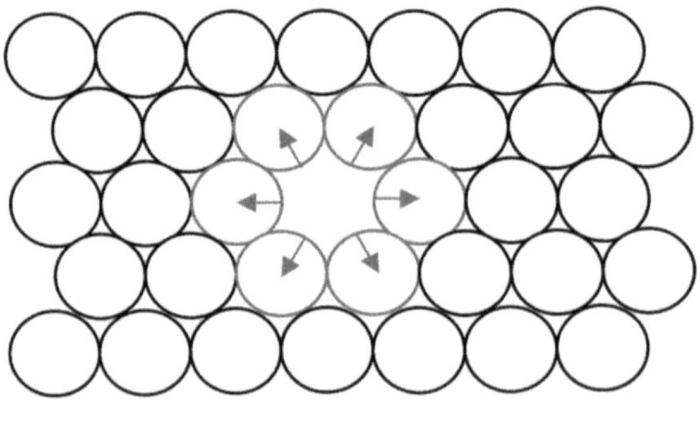

Figure 2.17. Illustration of a simple missing atom vacancy. The stresses on the inner shell of atoms around the vacancy tend to pull the atoms back towards their NN atoms since the opposing attractive force of the missing atom is gone.

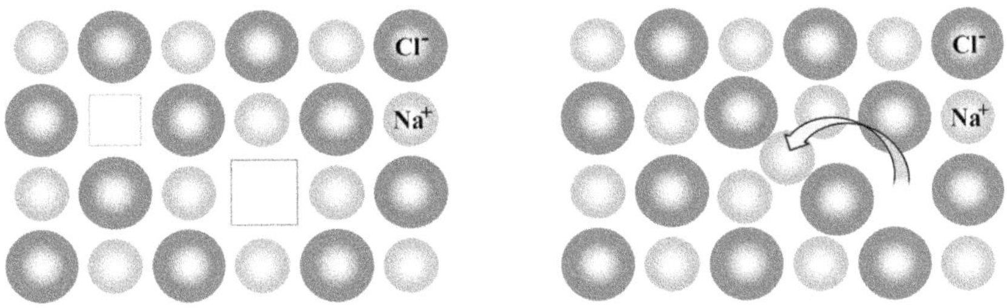

Figure 2.18. Ionic lattice defects illustrated for the case of a simple NaCl lattice. Left: A Schottky defect with the oppositely charged ions each removed to leave a pair of missing ion holes and in an electrically neutral crystal. Right: A Frenkel defect in which a small ion is forced by some energetic process (e.g. high energy radiation) into a nearby interstitial space of the lattice to cause formation of a defect region with local induced stress [3]. (These NaCl - Frenkel defect & Schottky defect images have been obtained by the author(s) from the Wikimedia website where they were made available by VladVD under a CC BY-SA 3.0 licence. They are included within this book on that basis. They are attributed to VladVD.)

fraction of the crystal volume is now occupied by holes popping in and out as fluctuations in the structure. Finally at T_m, the entire lattice structure washes out and a random liquid structure appears. For example, concentrations of holes in common metals such as copper can be of the order of ~0.1%. With this in mind it is easy to imagine that if the molten metal were suddenly quenched some fraction of these holes would be frozen in as static defects.

Hole defects can also appear in multiples in the case of ionic lattices where combinations of oppositely charged ions are present as constituents in the neutral crystal. There are two types of missing atom (ion) defects to consider. Schottky defects consist of missing ions of opposite charge, thereby keeping the crystal neutral, as illustrated in figure 2.18 for the case of a NaCl crystal.

Another type of single atom (or group) defect can be caused by the incorporation of a foreign atom of sufficiently small size into an interstitial void in the lattice, as depicted in figure 2.19. The types of interstitial voids in common lattices were discussed earlier in section 2.3.6.

A similar, but often far more stressful defect, is caused by the substitution of a a foreign atom (or group) directly into the position of a lattice atom (or group), as illustrated in figure 2.20. In this case if the foreign atom is simply inserted by some energetic process into a lattice position or if there is a considerable size mismatch, the lattice can be substantially stressed to accommodate the required volume change. This process can be assisted considerably if the chemical bonding of the foreign atom to the lattice atoms is strong, thereby compensating for the unfavorable stress.

2.5.2 Line defects—dislocations

A line defect arises when the planes of a crystal in some regions are perturbed and are not perfectly aligned. In the case of a *dislocation defect* one of the planes terminates suddenly, as shown in figure 2.20, in which the adjacent planes are formed around the missing atom where the plane terminated. This defect is

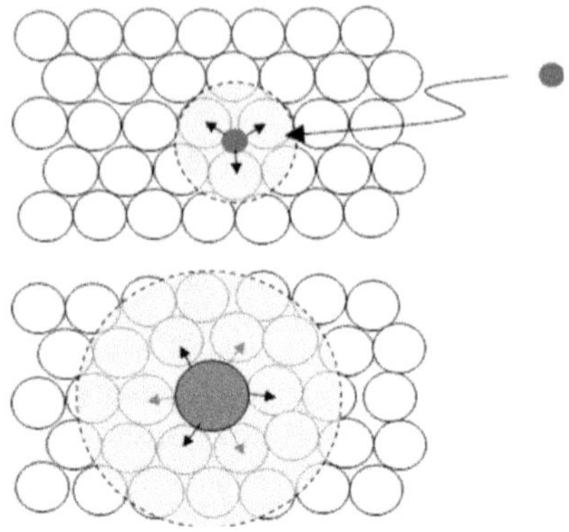

Figure 2.19. Top: Illustration of an interstitial defect in which a small atom (red sphere) inserts into an interstitial void of a larger spacing lattice. Typically, the void volume is not sufficient to accommodate the foreign atom without some local compressive strain on the surrounding shells of lattice atoms. Bottom: Illustration of a substitutional defect in which a foreign atom (red sphere) is substituted into the position of an intrinsic lattice atom. If the sizes do not match well the local region undergoes stress.

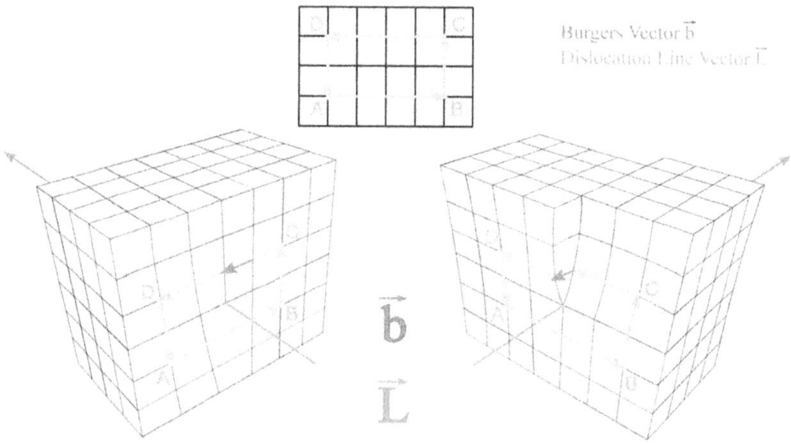

Figure 2.20. Schematics of line dislocation defects. Left: Edge view of a line dislocation. A vertical plane terminates suddenly at the green line. The adjacent planes left and right are dislocated by the presence of the terminated plane. Illustration of a Burger's vector circuit running clockwise around the defect. The circuit cannot complete because of the missing row below the defect a indicated by the extra red vector needed to close the gap. Right: Illustration of a screw defect and an associated Burger's vector test circuit which identifies the screw defect by an extra vector required running perpendicular to the plane of the circuit. The perpendicular green line on the left face marks the location of the start of the screw twist [4]. (This Burgers Vector and dislocations (screw and edge type) image has been obtained by the author(s) from the Wikimedia website where it was made available by Martin Fleck under a CC BY 4.0 licence. It is included within this article on that basis. It is attributed to Martin Fleck.)

commonly marked by a $\perp$ symbol, shown in the figure. In the case of a screw defect, one set of parallel planes are oriented or twisted at a different angle than the adjacent planes, as though someone grabbed the crystal and twisted it, also shown in figure 2.20. The interface where the two sets of planes meet and slide past each other is called the cut or cutting plane. Screw defects are common in the many single crystal materials and often defeat obtaining perfect periodic surface lattices for fundamental surface studies. We will learn more about this in the next chapter.

A common convention in fields such as metallurgy to characterize line defects is the use of a Burger's vector circuit. We are not interested generally in such details, but it is useful to know what the Burger's vector is in case you run into it in some surface related study of a crystalline material. You can see how the circuit works by following the red vectors in figure 2.20 which run clockwise around the ⊥ defect. It is easy to follow the logic. If you walk a circuit around some point on a perfect x, y grid of steppingstones equal numbers of stones going to the right and left and equal number up and down you should end up where you start. If you do not then someone added or removed some of the steppingstones. In the figure the arbitrary test circuit we chose had the red vector starting out on the top left for four atom steps to the right, then another three down, then four steps to the left and finally three steps up. For a perfect lattice the last vector should meet the start of the first. In the line dislocation defect, however, you can note that there is a gap of one step, marked by a blue vector pointing right. This vector indicates the presence of a single ⊥ dislocation.

The Burger's vector circuit for a screw dislocation, as shown in figure 2.20, has the red path going around with equal numbers of steps in each direction but we see that we cannot get back to where we started without a step up. So the Burger's vector points perpendicular to the circuit, the mark of a screw dislocation. The center of the dislocation is indicated by a blue line with a rotation arrow around it in the upper part of the figure.

2.6 Surface defects

In the previous section we looked at some of the most common defects within the bulk of the crystal. For surface studies we need to know what to expect in terms of these bulk defects ending up at the surface, for example, a screw dislocation that leaves a surface with the atoms in adjacent regions at different relative tilts. In general, the most common bulk defects are dislocation defects, and these tend to run through the crystal in loops until they end by chance or intersect a surface. Even annealed metals show high dislocation defect densities. For example, if we imagined a perfect slice of an annealed crystal that exposes an arbitrary face there could be as many as 10^6–10^8 dislocation lines per cm^2 exiting through the face. You can see from this that if we actually cut a crystal (using special tools such as lasers, diamond saws or spark cutters) to expose some surface of interest and removed the roughness by surface annealing, we would still expect to see a significant number of line and screw dislocations across the surface. Thus it takes significant work to make really low defect surfaces for surface chemistry and physics studies.

2.6.1 Massive defects and polycrystalline surfaces

If one views a large area of a typical surface of a crystal that has been cut purposely to expose a specific crystal plane the surface typically contains two types of regions:

terraces with exposed crystal planes of periodically arranged atoms (or groups), which are termed grains, and massive defect regions, called grain boundaries, that separate the grains. The grain boundary regions contain all the major defects that one might expect, including dislocations, trenches and hills, and even amorphous regions with only traces of ordering.

In such surfaces the terrace planes in different grains (and even in the same grain as we will discuss below) are typically not aligned to each other in terms of the intrinsic directions of the atom rows and often such terraces are at different heights relative to each other, like mesas in a desert. This type of surface with different orientation of terraces is called a polycrystalline surface. This is typically what is obtained by various thin film deposition techniques of metals. Even though the terraces may all be (111) in orientation, there is no correlation between them in terms of relative lining up of the rows, so the surface is designated as {111}, meaning a family of (111) regions with no single orientation of the plane axes.

2.6.2 Defects within terrace regions and domain boundaries

Within a given terrace region there are often sudden shifts in atom alignments across a distance of one to a few atoms. This sharp line of demarcation of structure is called a domain boundary. On one side the terrace atoms may constitute a domain with a specific alignment direction of the unit cells while on the other side of the boundary the terrace atoms have some relative shift in alignment so do not line up with those of the other domain. A common example of this is given by so-called *twin boundaries* which are sharp boundary lines in a crystalline surface that separate ordered terraces distributed as two domains, each with a different orientation. This is a common type of defect that plagues attempts to prepare pure single crystals with a highly uniform crystal orientation. For crystals more complex than monatomic ones, e.g. for A–B binary compositions such as NaCl or for molecular materials with oriented molecules, the shifts can involve a mismatch in whether A is next to B or next to another A or whether molecules which are in exactly the right positions have different tilt angles across the domain boundary. You can see these types of defects illustrated in figure 2.21. We will say a bit more about such defects when we discuss self-assembled molecular monolayers and associated structures in later chapters.

2.6.3 Atomic scale defects

Within a single domain of an otherwise perfect terrace a number of atom-level defects can arise. These can be of utmost importance in catalysis mechanisms and semiconductor electronic device performance, and has been of great interest in fundamental chemistry and physics experiments and theory studies of surface behavior. We will encounter these defects many times in the material ahead but for now we just list the defects and their common names so you can add this information to your tool kit. Most of what can be said be found by simply going carefully through the details of figure 2.22, which is based on a hexagonal symmetry lattice. Pay attention to the CN of each type of defect (you should figure these out

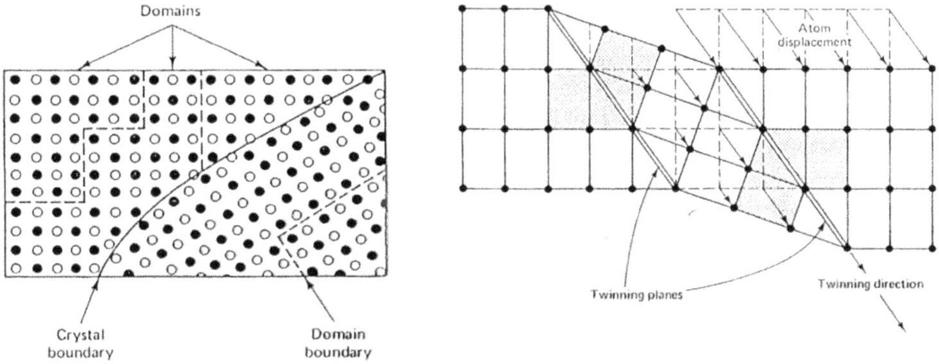

Figure 2.21. Illustrations of two types of domain boundaries in a crystalline surface. Left: A domain boundary in a binary A–B composition lattice. One type of boundary involves switching which atoms are NN, A–A, A–B, or B–B. The other involves a change in the orientation of the surface unit cells. Right: Twin boundaries in which one type of domain has a different relative lattice orientation from the other with the change occurring over a one atom wide distance pivoted on a common atom for both domains.

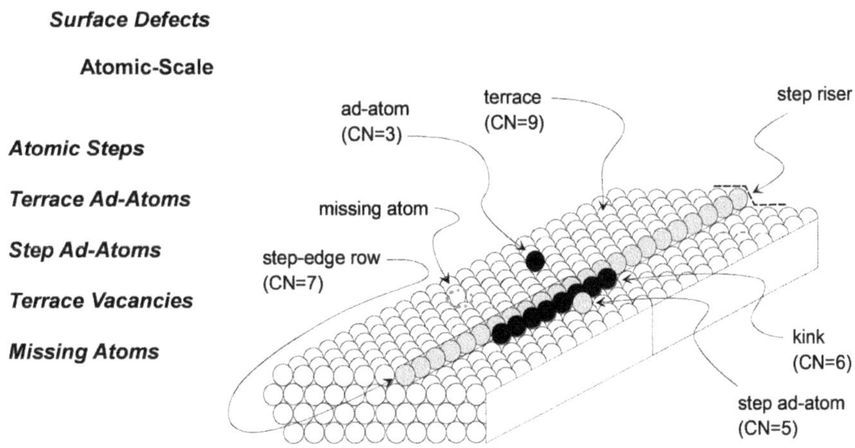

Figure 2.22. Schematic illustration of atomic scale surface defects on a surface crystal terrace of a hexagonal lattice. The illustration is meant to be general but the numerical characteristics of the defects such as CN depend upon the specific type of lattice, e.g. cubic or HCP. The terminology of the types or classes of defects is general.

for yourself by inspection) and to the naming of the defects and associated characteristics, e.g. step riser, step edge and ad-atom.

Chapter 2 Problems

1. Show that the packing of spheres in a body-centered cubic structure fills 68% of available space.
2. Draw the atomic structure of (100) and (110) planes for a face-centered cubic lattice.

3. The angles between the tetrahedral bonds of diamond are the same as the angles between the body diagonals of a cube. Use elementary vector analysis to find the value of the angle.

Further reading

1. Ashcroft N W and Mermin N D 1976 *Solid State Physics* 1st edn (Brooks/Cole). A general introductory graduate level text book on solid state physics. Chapters 4, 5 and 7 cover many of the points in the current chapter.
2. Hofmann P 2015 *Solid State Physics, An Introduction* 2nd edn (Wiley). A general introductory late undergraduate or early graduate level text book on solid state physics. Chapter 1 covers many of the points in the current chapter.
3. Kittel C 1996 *Introduction to Solid State Physics* 7th edn (Wiley). A general introductory late undergraduate or early graduate level text book on solid state physics. Chapters 1 and 2 cover many of the points in the current chapter.

References

[1] File:Carbon-dioxide-crystal-3D-vdW.png *Wikimedia Commons* https://commons.wikimedia.org/wiki/File:Carbon-dioxide-crystal-3D-vdW.png

[2] Leek L M 2023 *Crystalline Solid State Physics: An interactive guide* (IOP Publishing)

[3] File:NaCl - Schottky defect.jpg *Wikimedia Commons* https://commons.wikimedia.org/wiki/File:NaCl_-_Schottky_defect.jpg

[4] File:Burgers Vector and dislocations (screw and edge type).svg *Wikimedia Commons* https://commons.wikimedia.org/wiki/File:Burgers_Vector_and_dislocations_(screw_and_edge_type).svg

IOP Publishing

Surface and Interface Science

David L Allara and Robert L Opila

Chapter 3

Surface structure

In this chapter we will consider the structures of surfaces of condensed phase objects comprised of constituent particles of atoms, inorganic groups and molecules or molecular groups (collectively termed particles for convenience). In learning about surface structure it is convenient to start with ordered surfaces since these are the most amenable to quantitative characterization. With this in hand we can form the basis for understanding the properties of many important crystalline material surfaces such as those of catalytically active transition metals and semiconductors such as silicon used in electronic devices. There are two main cases of interest for physical surface structure: (i) bare surfaces, which are defined as the termination of the substance in a vacuum, and (ii) overlayers of another substance, typically a single layer of atoms or small molecules, either physisorbed or chemisorbed. These classifications can be broken down further in terms of the differences between the ordering in the top surface layer (at the ambient interface) and the bulk.

3.1 Surface layer structures

3.1.1 Classifications of physical structure

In this chapter we will consider the structures of surfaces of condensed phase objects comprised of constituent particles of atoms, inorganic groups and molecules or molecular groups (collectively termed particles for convenience). In general, such condensed substances can be structured at two limits: fully ordered, a perfect crystal, and with no correlations between the positions of neighbors, as found in amorphous solids and liquids. In learning about surface structure it is convenient to start with ordered surfaces since these are the most amenable to quantitative characterization. With this in hand we can form the basis for understanding the properties of many important crystalline materials surfaces such as those of catalytically active transition metals and semiconductors such as silicon used in electronic devices.

doi:10.1088/978-0-7503-4916-1ch3

There are two overall aspects of condensed phase surfaces to consider:

1. The physical structure, i.e. where are the constituent particles (atoms, molecules) located on an *xyz* grid and what is the symmetry.
2. The chemical nature of the atoms or groups at the surface, either stemming from a mixed stoichiometry solid (e.g. GaAs) which may present different atoms at the ambient interface or in molecular films which can have molecules oriented at a surface to present chemically different groups to the ambient.

There are two main cases of interest for physical surface structure: (i) bare surfaces, which are defined as the termination of the substance in a vacuum, and (ii) overlayers of another substance, typically a single layer of atoms or small molecules, either physisorbed or chemisorbed. These classifications can be broken down further in terms of the differences between the ordering in the top surface layer (at the ambient interface) and the bulk, as summarized in figure 3.1.

For atomic solids of a single element, the physical structure is quite straightforward, as we saw in studying the simple crystalline structures previously, but the arrangements can be considerably more complex if multiple elements, inorganic groups or molecules and polymers are involved. For heteroatom crystalline solids such as GaAs the ambient surface may be comprised of Ga, As or both elements in some type of pattern. For crystalline molecular or polymeric solids the constituent units in the bulk are arranged with some specific conformational sequence and/or orientations.

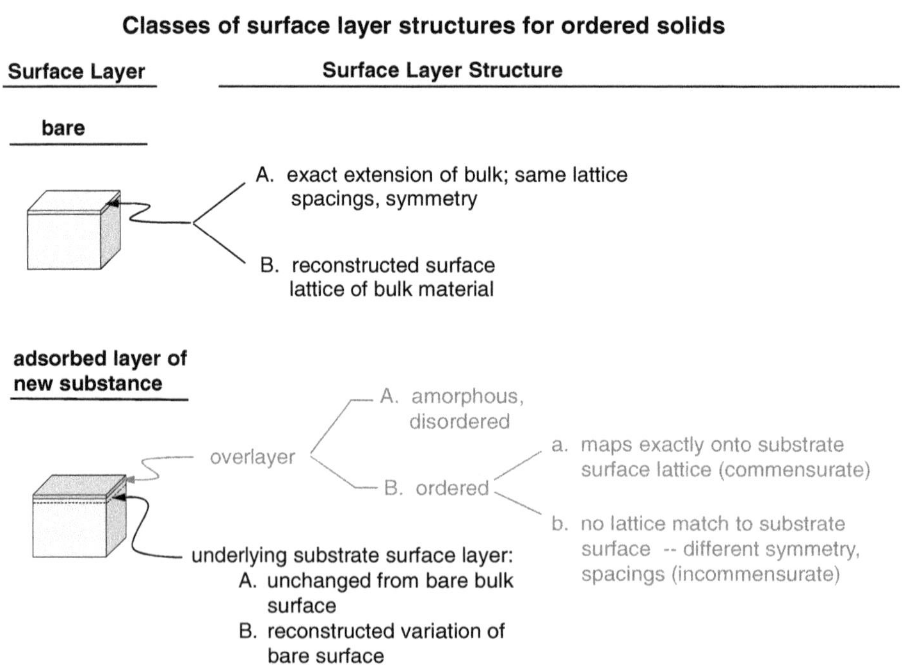

Figure 3.1. Classes of surface structures for ordered solids. The two cases involve a bare surface layer and an adsorbed overlayer of a different substance consisting of atoms or molecules.

For example, the centers of the molecules could all be located on an exact lattice in the bulk but they could be tilted in different ways so while there is translational order (moving the molecules along the *x*-, *y*- or *z*-direction to repeat the structure) there may not be tilt order or conformational uniformity from site to site. In organized molecular substances where the molecules can orient and fit together in different ways, such as liquid crystals (famous for their applications in liquid crystal displays (LCDs)), the question of the order is often a multifaceted one involving a number of different order parameters. Overall we must appreciate that whatever complexity exists in the bulk material will have its consequences manifested in the surface region in some way as the bulk structure is propagated to the surface termination, undoubtedly with perturbations. Further, in the case of surfaces consisting of an adsorbed layer of molecules with several structural degrees of freedom, such as a self-assembled monolayer or a lipid bilayer, several order parameters involving conformations and orientations also can be involved. In this chapter we will focus primarily on the simple aspects of ordering in terms of a translational lattice, such as exists in an atomic solid, and consider a few more complex cases such as self-assembled monolayers in passing.

3.1.2 Basic definitions of reconstruction and superlattices, geometric relationships between surface and bulk ordering and some common superlattices

In cases for which the top layer of an ordered solid is also ordered, we define its structure relative to that of the surface of a parallel plane deep in the bulk solid, as illustrated in figure 3.2. The simplest case is the perfect termination of a bare solid surface with an ideal *hkl* plane, as found deep in the solid and parallel to the surface. Often, however, there are driving forces, as suggested by the simple missing atom model in figure 1.14, that tend to shift the arrangements of the constituent particles in the top surface, and in some cases even several layers below. In these cases the ordered top layer assumes a new unit cell structure, which may or may not be directly related to the reference *hkl* plane. The common term for this rearrangement is *surface reconstruction*. Adsorbate overlayers are treated similarly. If the pattern of the surface plane, namely, unit cell angles and/or lattice spacings, have no simple match to the reference plane the surface layer is termed *incommensurate*. This can occur when the fundamental symmetries are different, namely, square versus

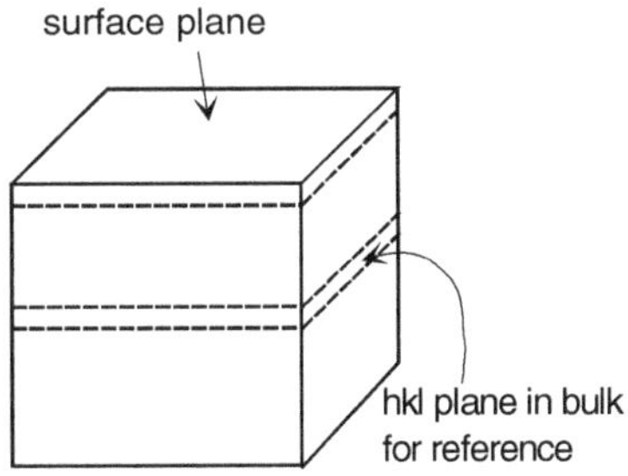

Figure 3.2. Schematic of a solid with a ordered top surface layer which is described with reference to a perfect *hkl* plane deep in the solid.

hexagonal, or, more loosely, for the same symmetries when the surface spacings do not quite allow a simple overlaying of the surface and reference lattices. When there is a good match of the lattices the overlayer is called *commensurate*.

The differences between surface and bulk lattices can be put on a quantitative basis by defining the vector operations needed to transform a unit cell in the reference *hkl* plane into the surface unit cell structure. The basic operation is to scale the *hkl* plane lattice spacings to match the surface plane spacings and rotate the scaled *hkl* unit surface cell as needed to overlay exactly on the surface lattice. If the overlayer unit cell exactly matches the *hkl* reference unit cell the scale factors are unity and no rotation is needed. For example, for a square unit cell this would be expressed as a $(1 \times 1)R_0$, or more simply as a (1×1), superlattice, where (1×1) expresses that the sides of the *hkl* cell do not change and no rotation is needed. An example of a $(\sqrt{2} \times \sqrt{2})R_{45}$ superlattice is shown in figure 3.3. Panel A shows the (001) reference plane of an FCC lattice. The unit cell is shown as gray shaded box with lattice constant a and unit cell vector components a_x, a_y. On this lattice red adsorbate particle are placed on diagonal corners away from the unit cell (perhaps more simply, placed on next nearest neighbors (NNNs)) to create a new square lattice with lattice constant a', as shown in panel B. The juxtaposition of the two unit cells is shown in panel C along with a breakout of the original and new lattice vectors on the right. From simple trigonometry $a/a' = \sqrt{2}$ and the angle required to rotate the original x, y coordinate system into the new x', y' one is 45°. With this transformation we can exactly relate the position of any point in the new lattice coordinates to the equivalent position expressed in the original lattice coordinates, or equivalently, transform any vector

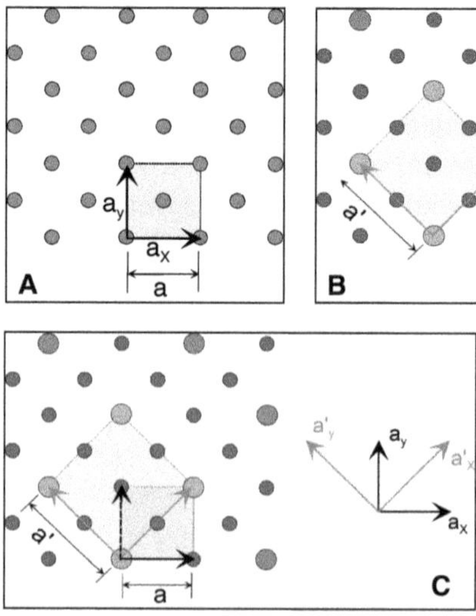

Figure 3.3. Illustration of the operations to connect an FCC unit cell to the corresponding overlayer cell of an overlayer with each particle located on the diagonal corner of each original unit cell. The original cell is rotated by 45° and then stretched by $\sqrt{2}$. Similarly, any lattice vector pointing to some location in the original unit cell is converted to the corresponding lattice vector pointing to the same relative location in the overlayer cell by these operations.

in the original lattice to one in the new lattice so that it points to the equivalent lattice point (such as 3 over and 1 up in each lattice).

The above coordinate transformation can be expressed compactly as the matrix operation $\begin{pmatrix} v_x' \\ v_y' \end{pmatrix} = \sqrt{2}\begin{pmatrix} \cos 45 & \sin 45 \\ -\sin 45 & \cos 45 \end{pmatrix}\begin{pmatrix} v_x \\ v_y \end{pmatrix} = S \times R\begin{pmatrix} v_x \\ v_y \end{pmatrix}$, where $S =$ the scaling operator $\begin{pmatrix} \sqrt{2} & 0 \\ 0 & \sqrt{2} \end{pmatrix}$ and $R =$ the 45° rotation matrix operator. This transformation gives the basis for the $(\sqrt{2} \times \sqrt{2})R_{45}$ superlattice notation. In general, we can extend this notation to any commensurate superlattice referenced to an underlying surface plane unit cell with base lattice vectors $\mathbf{v_a}$, $\mathbf{v_b}$ by writing $\begin{pmatrix} v_a' \\ v_b' \end{pmatrix} = S \times R\begin{pmatrix} v_a \\ v_b \end{pmatrix}$, where $S = \begin{pmatrix} s_a & 0 \\ 0 & s_b \end{pmatrix}$ with s_a and s_b as the scaling factors for the a and b lattice vectors of the unit cell and $R = \begin{pmatrix} \cos\theta & \sin\theta \\ -\sin\theta & \cos\theta \end{pmatrix}$ with θ as the rotation angle required to align the overlayer and reference unit cells. This general notation includes the cases of both rectangular and hexagonal unit cells. Overall, this leads to the notation $(s_a \times s_b)$ R_θ. To distinguish the cases of a simple surface layer of the solid substrate and an adsorbate overlayer of a new material the notation $(s_a \times s_b)R_\theta - A$ can be used, where A is the type of overlayer. Examples of this notation are given in table 3.1 for some common types of surface lattices and adsorbate systems.

Table 3.1. Examples of coordinate scaling matrices.

Overlayer structure	Scaling matrix (S)	Unit cell rotation angle	Example	Notation
Same material, no reconstruction, ideal termination	$\begin{pmatrix} 1 & 0 \\ 0 & 1 \end{pmatrix}$	No rotation	Bare, unreconstructed Pt(111)	Pt(111) (1×1)
Surface adsorbate; unit cell same as substrate reference unit cell	$\begin{pmatrix} 1 & 0 \\ 0 & 1 \end{pmatrix}$	No rotation	H on Si(111)	Si(111) – (1×1) – H
Surface adsorbate unit cell = 2 × substrate unit cell	$\begin{pmatrix} 2 & 0 \\ 0 & 2 \end{pmatrix}$	No rotation	H on W(211)	W(211) – (2×2) – H
Hexagonal substrate unit cell; adsorbates at NNN positions	$\begin{pmatrix} \sqrt{2} & 0 \\ 0 & \sqrt{2} \end{pmatrix}$	$\theta = 30°$	Alkanethiolate (AT) molecules on Au(111)	Au(111) – $(\sqrt{3}\text{x}\sqrt{3})$ $R_{30°}$ – AT
Square substrate unit cell; adsorbates at NNN positions	$\begin{pmatrix} \sqrt{3} & 0 \\ 0 & \sqrt{3} \end{pmatrix}$	$\theta = 45°$	CO on Cu(100)	Cu(100) – $(\sqrt{3}\text{x}\sqrt{3})$ $R_{45°}$ – CO

A $(\sqrt{3}x\sqrt{3})R_{30}$ superlattice is illustrated in figure 3.4. In this case the substrate unit cell is hexagonal and the base vectors are set at an angle of 60°. Adsorbates located at the NNN positions of the substrate lattice create a new unit cell rotated 30° with the base vectors stretched by a factor of $\sqrt{3}$. A specific example involving organic molecule adsorbates is given in figure 3.5 for the case of an alkanethiolate lattice. The alkanethiol molecules (RSH) bond to the Au substrate by loss of H to from RS–Au interactions but are too large to fit on nearest neighbor (NN) gold atoms (NN spacing = 2.88 Å) so they move to NNN positions (spacing = 5.0 Å) and pack together optimally, minimizing void spaces, by extending the long alkyl chains and slightly tilting ~30° away from being exactly perpendicular to the surface. This example illustrates the variety of order parameters that can be involved in molecular adsorbates where the molecules have flexibility and asymmetrical shapes; in this case exhibiting translational (positions on the substrate), orientational (molecules all tilted at some angle), and conformational (chains all extended, not folded in some

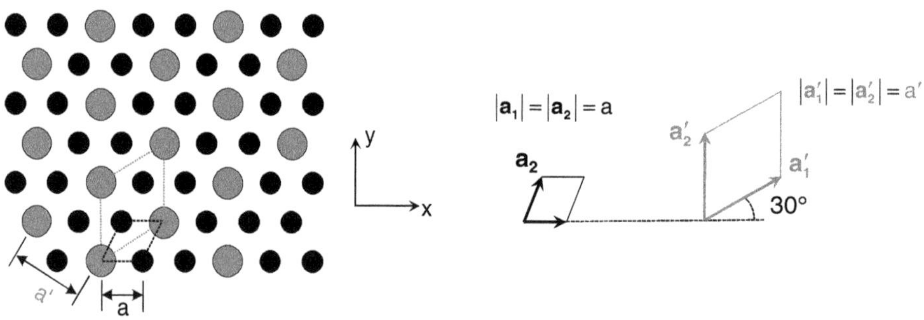

Figure 3.4. Illustration of the unit cell relationships for $(\sqrt{3} \times \sqrt{3})R_{30}$ superlattice on an FCC substrate lattice. The base vectors for the two unit cells are shown on the right with their relative lengths and directions indicated.

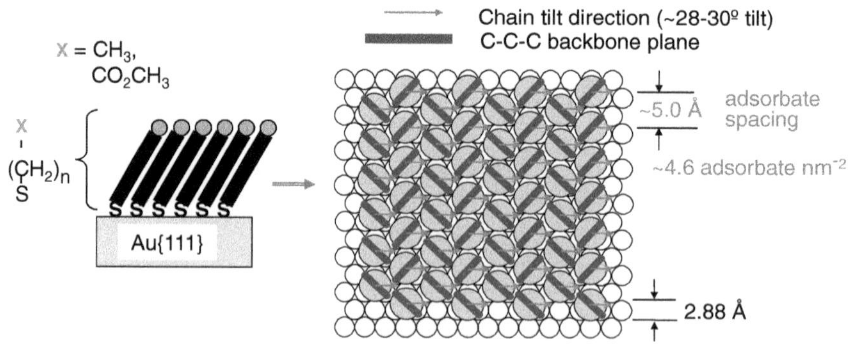

Figure 3.5. Schematic of a self-assembly monolayer formed by chemisorption of alkanethiol molecules on a Au(111) substrate. The molecules are pinned by Au–S bonds at NNN sites in the Au lattice to form a $(\sqrt{3} \times \sqrt{3})R_{30}$ overlayer. When ideally packed the molecules in the chains extend fully away from the surface at a uniform tilt angle of ~30° away from the vertical to the surface and the C–C–C planes of alternating molecules are set at 90° to each other.

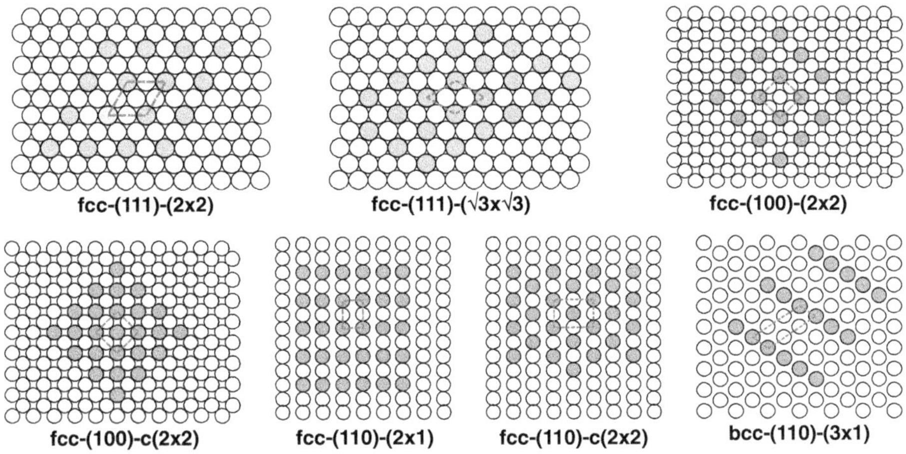

Figure 3.6. Some common examples of adsorbate superlattices on FCC substrates with one example shown for a BCC substrate. The superlattice unit cell is indicated in each case by a dashed red box.

way) parameters. In addition the alkanethiolate molecules can pack together more favorably if every other chain has its C–C–C plane twisted perpendicular to that of its neighbors, as illustrated in figure 3.5, where the top view shows the planes of the molecules, perpendicular to the page, twisting alternately across the lattice. Note that the $(\sqrt{3}\text{x}\sqrt{3})R_{30}$ superlattice structure is often obtained even when the ends of the chains (away from the sulfur atom) are substituted with various functional groups but in some cases whereas the S atoms maintain the superlattice structure the top groups may be disordered (adding another order parameter).

Some examples of typical superlattice structures that arise on common planes of FCC and BCC crystal surfaces are shown in figure 3.6. The superlattice unit cells are shown with red dashed boxes.

3.1.3 High order plane surfaces—periodic surface defects

Whereas the low Miller index surface planes (e.g. 001, 111, 211, etc) we have discussed to this point are always perfectly flat one should consider that a crystal can be sliced at specific angles to expose high index surfaces that exhibit a regular pattern of corrugations consisting of periodic steps or even kinks (cusps or corners). The atoms in these corrugation features have fewer neighbors than in a completely flat terrace and thus the surface can be considered as having an infinitely repeating set of defect sites or features. As one might expect from the missing atom model, these defects can be selectively active as catalytic sites and exhibit different local electronic properties. For this reason high index surfaces often are good models for the behavior of irregularly shaped small catalyst particles whose activity depends on the presence of high index faces and thus the study of periodic defect surfaces has become quite common in surface science.

Examples of different FCC surfaces of high index planes with periodic step and step-kink defects are shown in figure 3.7. The repeat patterns across a (755) and a

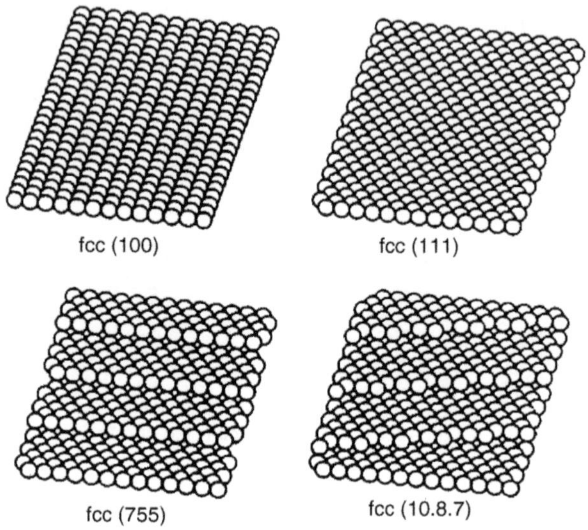

Figure 3.7. Illustrations of different high index planes of an FCC crystal that produce periodic steps. Note the appearance of periodic kinks along the step edges in the higher index planes. Reprinted from [1]. Copyright (2005), with permission from Elsevier.

(10·87) face (the dot is added in the index notation to avoid confusion arising from the two-digit number) of an FCC solid are shown in detail in figure 3.8. The main features to note are: (i) the atoms in the step edges are part of the atoms that exist in the plane of a lower index surface and (ii) the repeat patterns of the steps can be characterized by the number of terrace atoms between each step in moving from step to step. These features can be used to define the shorthand notation $S - n_t(h_tk_tl_t) \times n_s(h_sk_sl_s)$, where S indicates a stepped surface, n_t is the number of atoms across the terrace between each step edge, $(h_tk_tl_t)$ is the terrace index, n_s is the height of the step in atoms and $(h_sk_sl_s)$ is the step index. These notations are indicated on the left for the surfaces in figure 3.8.

3.1.4 Preparation of periodic surface defects by cutting crystals at selected angles

Periodic defect surfaces are usually prepared by orienting a crystal along a chosen direction in a precision holder and then cutting along a chosen angle parallel to the desired plane. Cutting can be done in various ways including precision diamond saws and spark cutters. The freshly cut surface will have large number of massive defects and residue since it is impossible to cut the crystal cleanly and with atomically sharp precision, but if the angle is correct within some small tolerance (typically ± 1°) the average surface will run acceptably parallel to the desired plane to have large regions with the correct structure. For appropriate materials, typically hard metals and non-metals, e.g. platinum and silicon, after washing the surface with pure water and solvents the sliced sample can be placed in a vacuum, bombarded with energetic ions such as those of a rare gas (typically Ar at 1–10 keV energies) to sputter off impurities and protruding defects and then heated in a

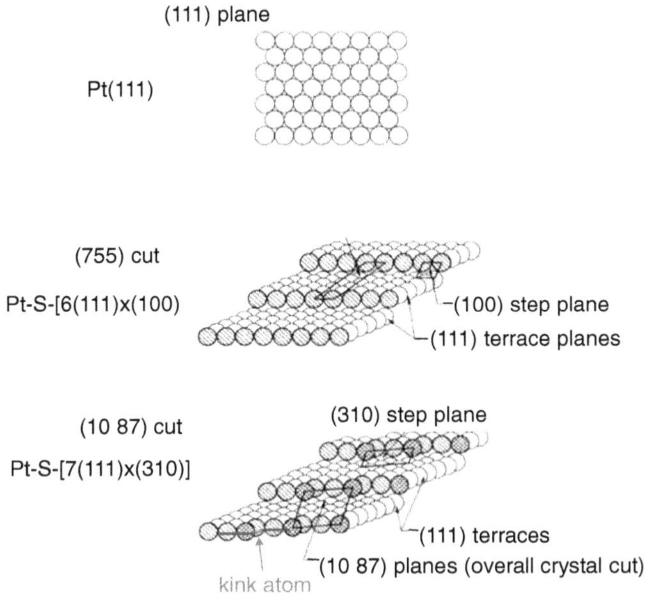

Figure 3.8. A detailed look at the step and kink features of the (755) and (10·87) high index planes of a Pt FCC surface with the (111) terrace shown for comparison. The atoms running across the (755) and (10·87) step edges are members of the (100) and (310) planes of the crystal, respectively. The notations on the left below the (755) and (10·87) labels indicate the specific number of repeat atoms in the terrace between each step and the step planes. For details see the text.

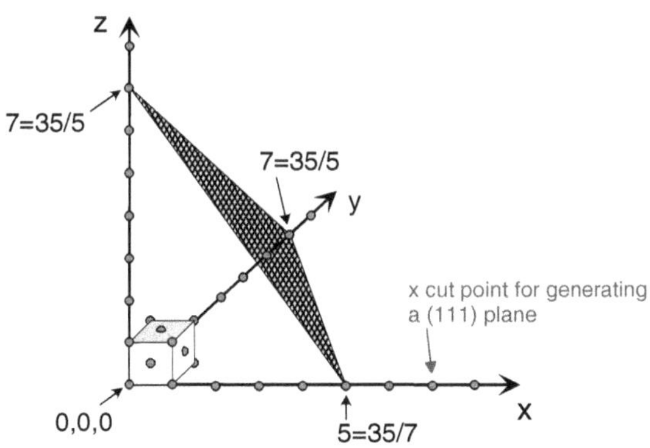

Figure 3.9. The coordinate frame shows an FCC unit cell at the origin aligned along the cell axes and a plane intersecting at the x, y, z positions 5, 7, 7.

vacuum to just below the melting point so that the bulk remains solid while the surface starts annealing, letting the surface atoms relax towards the desired surface structure. A number of these sputter–anneal cycles can result in a high quality stepped surface.

It is instructive to see the details of aligning a crystal for cutting to expose a stepped surface. Using the (755) surface as an example, figure 3.9 shows an FCC unit cell

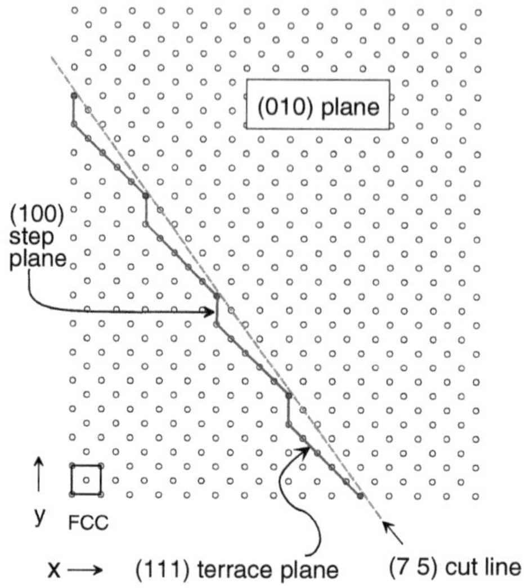

Figure 3.10. A diagram of an FCC lattice with unit cell aligned along the x, y, z coordinate frame as seen in the x–y plane. The red dashed line is the (75) index direction and runs along the tops of atoms which define step edges with 100 step planes and reveals exposed (111) terrace atoms below the steps.

aligned along the x-, y-, z-axes. The plane intersecting the x-, y-, z-axes at 5, 7, 7 defines the cutting plane to produce the (755) Miller index plane (see figure 3.10). In more detail figure 3.10 shows a side view of the edge of the cut plane in the x–y plane (a red dashed line); you can imagine that this is the line across which a saw, for example, would cut with the saw blade perpendicular to the x–y plane. The cut line runs on the upper parts of the step edges and exposes terrace atoms below running at an angle to the cutting surface. In order to produce a kinked surface the Miller indices must all be different, for example, the case of the (10·87) surface in figure 3.8. This corresponds to tilting the saw in a plane away from perpendicular to the x–y plane (figure 3.10).

3.2 Surface reconstruction

3.2.1 Imbalance of surface and bulk forces drive surface reconstruction

The abrupt termination of a surface produces an imbalance in the forces on the atoms in the surface region relative to the bulk, as captured in the simple missing atom model. While this effect is general for all condensed phases of constituent particles we will restrict the discussion here to an atomic solid for simplicity.

The simplest form of reconstruction that the missing atom model would predict is a compression of the top atoms towards the underlying bulk. The top atoms only feel forces between each other and from the layer below so so one expects some compression to pack the top layer more tightly to the second layer. An example of this surface contraction is seen in figure 3.11 which correlates the per cent compression of the top of different crystal faces, as determined from both experiment and theory, as a function of the openness or void fraction of the surface crystal plane for a variety of different atomic solids. The surface contraction shows some

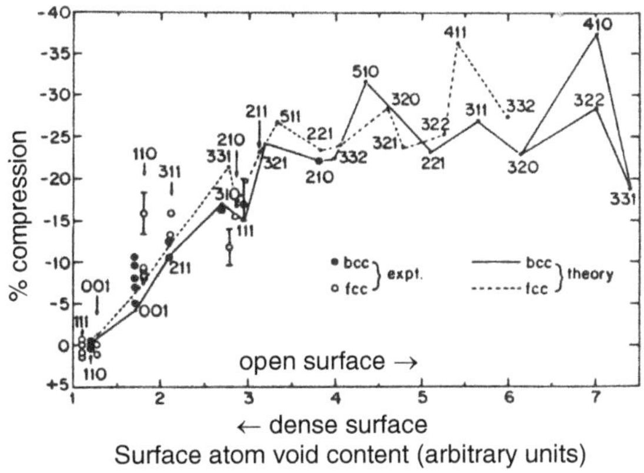

Figure 3.11. Correlations between the per cent vertical compression of surface atoms into the solid versus the void content or openness of the surface plane (arbitrary units) for different faces of a variety of atomic solids. The points are from experiments and the lines represent theory calculations.

oscillation but clearly the surfaces with fewer atoms per unit area are much more highly compressed than the very dense surfaces. This makes good sense since the more open surfaces provide more open space into which the top atoms can be pulled to increase their bonding or interactions with neighbors below.

We might anticipate that compression could even go into the second layers and beyond for a few layers since while the second layer is fully bonded to nearest neighbors it is lacking the longer range forces from missing NNN atoms above the surface that are present below in the bulk. Since these NNN forces are far weaker than the NN bonding, the compression effects typically die off quickly after the top layer but small perturbations can extend a layer or two further for some solids. In the cases of liquids, e.g. Hg, the decay length for spacing perturbations extends ~3 layers into the bulk. Note that this depth corresponds to the surface region thickness parameter δ discussed in chapter 1.

In the cases of atomic solids with extremely strong lattices with directed, covalent bonds, e.g. silicon and carbon (diamond), the imbalance of surface and bulk forces can cause severe reconstruction in the surface plane as well as vertical compression effects. A classic case is the Si(100) surface. Each silicon atom in the bulk is covalently bonded in an sp^3 tetrahedral geometry to neighbors with very strong bonds (vaporization: $T_{vap} = 3538$ K, $\Delta H_{vap} = 359$ kJ mol^{-1}). A unit cell showing the tetrahedral bonding is shown in figure 3.12. Note the two missing bonds at each atom in the (001) family of the crystal faces. The energy gained by distorting the surface lattice to allow the Si atoms to reach over to form a dimer with a neighboring surface atom is sufficient to overcome the strain induced and the surface reconstructs, as shown in figure 3.13.

At this point, based on what we now know about the main factors that drive reconstruction for atomic solids, we can propose two general predictions or rules of thumb:

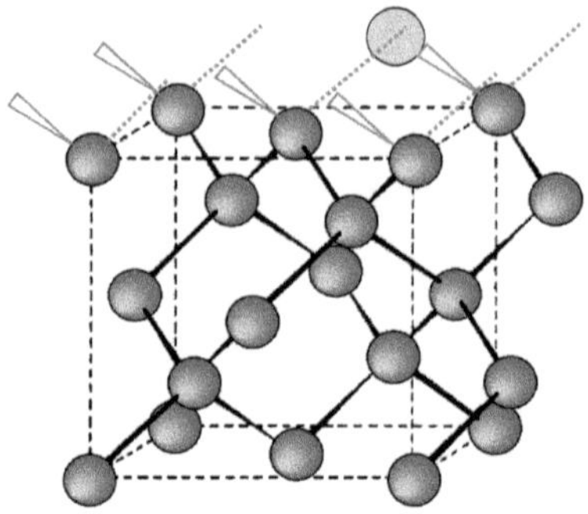

Figure 3.12. Representation of the diamond structure unit cell of a Si crystal with tetrahedral bonding. Each atom in the (001) face has two missing bonds. A missing atom present above the surface is shown (gray, shaded circle). Reconstruction compensates for the missing bonding with this atom.

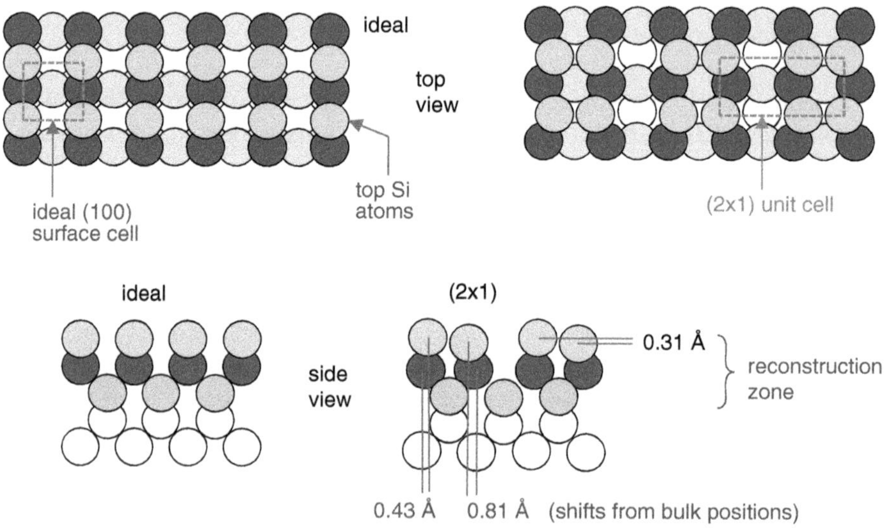

Figure 3.13. The reconstruction of a Si(100) surface to a (2 × 1) dimer structure. A top view of the surface is shown at the top of the figure and a side view at the bottom. Both in-plane and out-of-plane shifting of the atoms occurs with the reconstruction zone extending to three layers deep into the crystal.

Rule 1: Dense faces with small spacings and high coordination numbers suffer less atom shifting that the corresponding open faces with larger average spacings and lower coordination numbers.

Rule 2: Solids with strong lattice bonding are more prone to reconstruction than those with weaker bonding.

Rule 1 follows from the fact that reconstruction forces drive the surface region to become more densely packed with an increase in atom coordination, thereby stabilizing the surface as best as possible relative to the bulk, and more open surfaces have lower CNs for the atoms and more space available to shift positions.

Rule 2 similarly follows in that more stabilization is to be gained when the new coordinations that form have high energies.

These rules (or predictions) work very well in general, though one must be cautious in extending such simple ideas too far. Supporting examples are given by the reconstruction of iridium surfaces. Iridium is a dense (22.42 g cm^{-3}), hard, refractory metal with extremely strong bonding ($T_{vap} = 4403$ K, $\Delta H_{vap} = 604$ kJ mol^{-1}). As expected from the high density, it has an FCC structure which allows the closest packing of atoms. The densest face (111), is not prone to reconstruction. The second densest (100), does undergo reconstruction, as illustrated in figure 3.14 (left), strong enough to induce a collapse of the square packing into a hexagonal type of lattice which serves to pack the atoms more tightly (recall the packing fraction in unit cells). An overlayer unit cell forms that spans a (1 × 5) area on the unreconstructed (100) lattice, as seen in the figure. In order to pack the top atoms optimally with the underlying layer the top layer is puckered; this also distorts the perfect planar hexagonal pattern and compensates for the mismatch in the square and hexagonal symmetries. The (110) face, with a rectangular unit cell is an even more open structure than the square (100) surface. This leads to severe reconstruction which results in a (1 × 2) missing row structure as shown in figure 3.14 (right, shown by arrows). Although the missing row may appear to leave the surface more open and less stable than the unreconstructed (110) surface the reconstruction does result in a net increase in the packing between the atoms.

We can see that trends in these FCC reconstruction versus surface plane examples fit very well with the trends in the CNs of the surface atoms, as shown in figure 3.15. The figure shows the top surface layer in terms of the surrounding NN (black) and NNN (gray) atoms for a chosen surface atom (cross hatch) for four different surface planes. In moving across (111), (100), (110) and (210) the CNs diminish through the

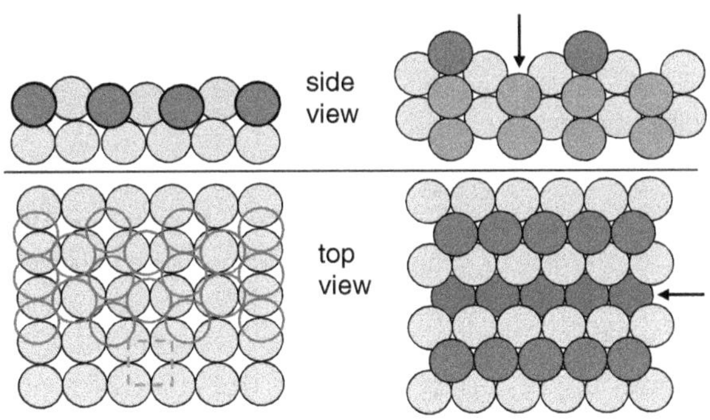

Figure 3.14. Schematic of the reconstructions of the Ir(100) (left) and (110) (right) surfaces. Ir(100): The top view shows a shift from square (unit cell in dashed red line) to hexagonal symmetry (top atoms are blue circles). The side view shows that the original (100) spacings are maintained in the second layer and that the top layer is puckered for better packing of the surface atoms. Ir(110): The top view shows a missing surface row of blue atoms—the dark green atoms are seen two rows below. The side view shows the top blue atoms and the lower rows of green atoms. The arrows point to the missing row in both views.

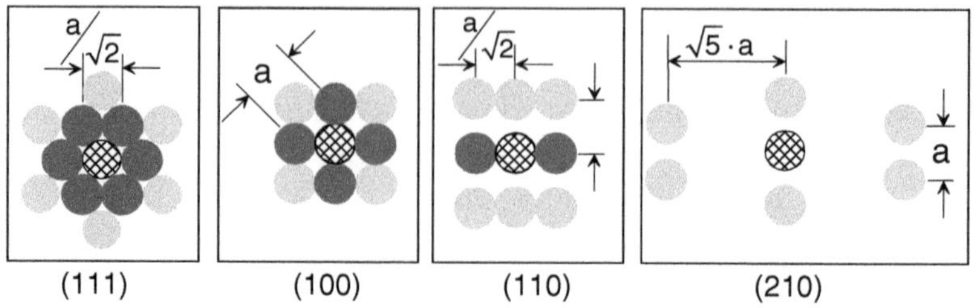

CN	(111)	(100)	(110)	(210)
Surface Layer:	6	4	2	0
Layer below:	3	4	5	5
	9	8	7	5
Missing CN: [Bulk(12) – Surface]	3	4	5	6

Figure 3.15. The diagrams show the NN (black) and NNN (gray) atoms around a central atom (cross hatched) in four ideal FCC surface planes. The atoms below are not shown. As the plane indices increase the NN coordination (CN) decreases. In the case of the (210) surface the CN of the central atom is due to NN atoms in the layer below leaving a highly open surface. The table at the bottom of the figure shows the coordination numbers in each case.

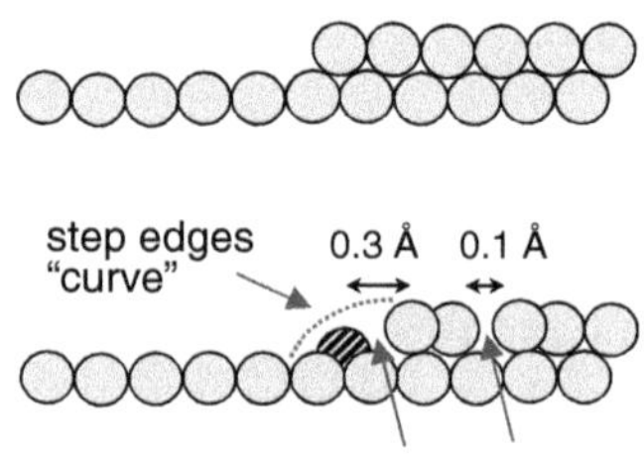

Figure 3.16. Reconstruction of a step edge on a Cu(410) surface. The step edge relaxes to move the shaded atom into the layer below while opening gaps in the adjacent surface layer (blue arrows). The overall effect is to smooth the curvature at the step edge.

sequence 9, 8, 7, 5; the corresponding number of missing atoms compared to the bulk CN of 12 is 3, 4, 5 and 6. Clearly the more open surfaces appear prone to collapse to increase the coordination number of the top layer. Although no data are shown for the (210) surface we can guess that it would be severely reconstructed since it has no NN bonding and only bonds to the layer of atoms below, leaving a very open surface.

3.2.2 Reconstruction of steps and kinks

The atoms incorporated in step and kink defects have lower coordination numbers than those in terraces for a given crystal structure. On this basis one can expect reconstruction to packing with higher coordination if sufficient energy is available to move the atoms into the new positions. An example is shown in figure 3.16 for the reconstruction of a step feature on a Cu(410) surface. The figure shows the side views of the top two layers. The reconstruction of the ideal step (top) results in shifting of one atom downwards to 0.2 Å above the second layer and small spacings (cracks) opening (0.1 and 0.3 Å) in the top layer. This behavior can be understood in general by a another simple 'rule of thumb'.

Rule 3: Surfaces always try to remove sharp edges and replace them with smoothly curving ones.

This rule is just a simple matter of maximizing the contacts (coordination numbers) between constituent particles and moving as many atoms from the surface towards the bulk as is possible. In chapter 4 we will learn how to determine the energy of this process and then put the rule into quantitative form.

Questions for thought

At this point you should have a good intuitive feel for the underlying factors that cause surface reconstruction so here are a couple of simple questions to test this.

1. As the surface atom density ↑ would you expect reconstruction to be stronger or weaker? Why (explain on some fundamental structural and bonding basis)?
2. As the lattice strength of the material ↑ which way would reconstruction effects change, to be stronger or weaker? Why?

3.2.3 Surface reconstruction in soft materials

Up to this point we have considered surface structure and reconstruction for hard materials with strong lattice bonding, e.g. elemental metals and alloys and non-metals such as silicon. In these cases reconstruction generally takes take place by movement of atoms over very short distances of atomic spacing lengths and the strong cohesive lattice forces impose sizeable barriers to atom movement so reconstruction is usually limited to the very top surface. Deep reconstruction into several layers can arise at higher temperatures, approaching the limit of the melting points of the solids. In contrast, soft organic-based materials exhibit much weaker cohesive interactions between constituent units so are far more prone to reshuffling of their surface regions at much lower temperatures. In the case of polymers the chemical repeat units on a single chain are connected by strong covalent bonding (typically ∼300–400 kJ mol^{-1}) but the chain–chain interactions per chain segment are far weaker, typically a few kJ mol^{-1}. Further, the forces along a single chain which resist torsional reorientation (relative local twisting) of adjacent units relative to one another around their connecting bond are easily overcome at temperature thresholds (the so called glass transition temperature or T_g) near ambient temperatures for many polymers. The freedom of each chain to twist local groups and the ability for some slippage between chains allows the chains to snake through the matrix relative to one another, thereby providing a mechanism for the polymer to shuffle different chemical units in the near surface region to and from the outer surface. Such reconstruction allows the surface region to relax to lower energy configurations which can change the chemical nature of the outside surface in cases where there are different types of chemical repeat units in the polymer (co-polymers). A brief summary of homo- and co-polymer structures is shown in figure 3.17 for reference.

Reconstruction frequently occurs for flexible polymer chains with a mix of nonpolar and polar units in block co-polymers and side chain homopolymers. An illustration is given in figure 3.18 for a hypothetical polymer with polar terminated side chains (e.g. $-(CH_2)_n-CO_2CH_3)$). The polymer is originally prepared in a state with a fraction of the polar groups exposed at the top surface. After standing under the polar groups retreat into the polymer matrix where they are more stabilized than at the top surface. Upon further exposure to water some fraction of the buried polar

Types of polymers with functional groups

- homo polymers:
  - with functional segments $[-(CH_2CH_2)\text{-}NH-]_n$
  - with side chains $[-(CH_2CH)-]_n$ (with $CO_2C_2H_5$ side group on CH)
- co-polymers (2 or more different monomers)
  - random (-A-B-B-A-B-B-A-B-A-B-B-B-)
  - repeat (A-B-A-B-A-B-A-B-A-B)
  - block $[-(A)_m-(B)_n-]$

Figure 3.17. Summary of the main types of polymers containing functional groups, represented by homopolymers with functional groups on the main and side chains, and co-polymers, with the functional units distributed in random, regular repeat and block configurations.

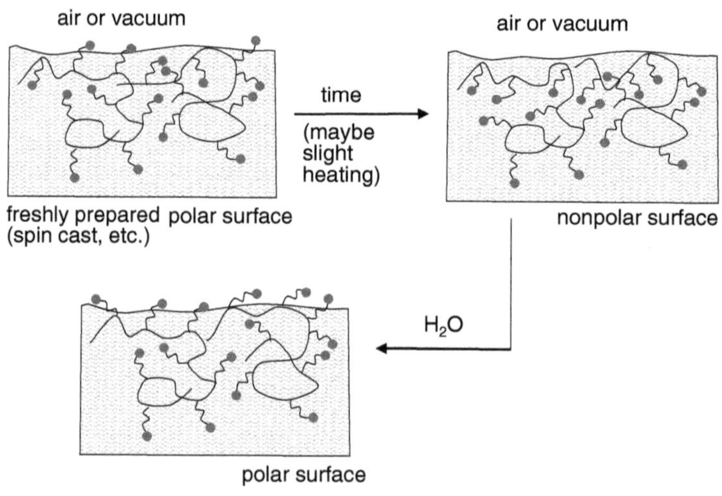

Figure 3.18. Schematic of a solid linear chain polymer with a nonpolar main chain and flexible side chain units with a terminal polar group (red balls).

groups in the near surface region work their way to the surface where the interaction with water provides better stability giving a enhanced polar surface.

An example of polymer surface reconstruction is shown in figure 3.19 for a tri-block co-polymer film with repeat blocks of polystyrene (PS) (A), polybutadiene (PB) (B) and poly(methylmethacrylate) (PMMA) (C), each block of different lengths. The TEM image of a thin section of a film is interpreted in terms of a bulk structure formed as stacked planar layers with repeating, alternating rows of the A and C blocks with interspersed B blocks shaped like ~9 nm diameter balls and an ~47.5 nm repeat distance between each A, C set of rows (see figure). Atomic force microscopy images (see later section on characterization probes) reveal that the surface has reconstructed from the planar layered structure of the bulk to a missing

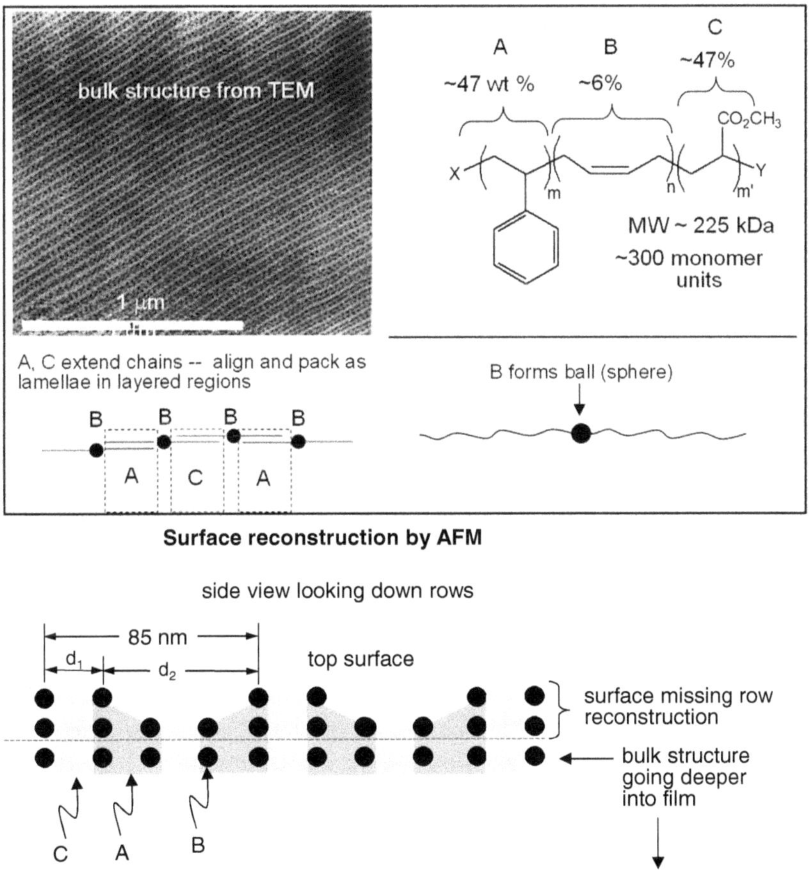

Figure 3.19. Missing row reconstruction in a tri-block co-polymer film. Top: The TEM image on the left shows a top view of the bulk structure which consists of layers of repeating rows (lamellae) of alternating A, C blocks with interspersed C blocks shaped into small spheres (~9 nm), as shown in the lower portion of the panel. The chemical structure of the polymer is shown upper right. Bottom: A schematic of the side view of the film showing the top surface with a missing row structure, as determined from AFM images of the surface. Data taken from [2]. Reprinted with permission from [2]. Copyright (1996) American Chemical Society.

row surface topography, as illustrated in the bottom figure. This provides an interesting example of a missing row reconstruction in soft matter with the reconstruction depth at the ~10 nm scale.

Questions for thought

Go back to the previous rules of thumb for predicting surface reconstruction in atomic solids and think about how you might extend and apply these rules to the cases of polymers and especially the missing row reconstruction in figure 3.19.

1. See if you can keep the usefulness of the missing atom (now molecular group) model. How do you modify the missing atom (molecule) bonding principle in the case of polymers? Generally polymer scientists think in terms of segment–segment interactions and functional group–matrix interactions, sort of like solvation of a group in a surrounding matrix rather than lattice bonding and CN.
2. In the case of the missing row tri-block polymer reconstruction, which of the blocks is the most polar? Which is the least polar?
3. In the missing row reconstruction which block is missing in the outer surface layer? Is a PMMA surface in contact with air less or more stable than a PS or PB block? Why?
4. Looking at the top of the reconstructed surface which blocks show increased area contact with the ambient compared to a fully unreconstructed planar surface? Could that be the basis for a reconstruction driving force?

3.3 Adsorbate overlayer structures and coverage effects

3.3.1 Definition of coverage

When atoms or molecules adsorb onto an ordered surface we need some quantitative way of describing what fraction of the surface is occupied by the adsorbates. Obviously the most general would be to count the number of adsorbate atoms per unit area of the substrate surface. For superlattices, however, where the adsorbates form a commensurate overlayer, we can be more specific by defining a coverage (θ) in terms of the ratio of the number of adsorbate atoms atoms per substrate atoms over a given area of the unreconstructed substrate surface. Thus a fully occupied (1×1) superlattice has $\theta = 1$. In cases in which the fully developed superlattice only occupies a specific fraction of the unreconstructed substrate atoms, e.g. a (2×2) overlayer, we alternatively can define θ as the fraction of possible substrate sites that can be occupied, e.g. a fully formed (2×2) overlayer would have $\theta = 1$ by this definition. While the first definition is the most common, in any case where the term coverage is being used it must first be defined to avoid confusion. Moving forward, we will use the first definition unless otherwise stated.

3.3.2 Forces governing adsorbate overlayer structure and associated coverage effects

The structure of an adsorbate layer depends on the balance of three types of interactions: adsorbate–adsorbate, adsorbate–substrate and substrate–substrate, as illustrated in figure 3.20 which shows molecules attached to the surface at various points on the ordered substrate. A more detailed summary of attractive and

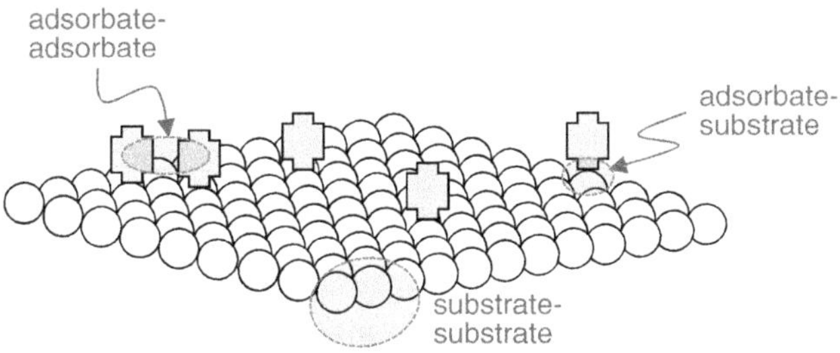

Figure 3.20. Schematic of an ordered substrate surface with adsorbates which illustrates the three main types of possible interactions. The adsorbates are shaped to indicate molecules which could interact at different parts of the structure.

Table 3.2. Summary of the major attractive and repulsive forces for the three types of important interactions in adsorbate overlayers.

	Major interaction forces			
	Attraction		Repulsion	
Interaction	Short range bonds	Long range	Short range	Long range
Adsorbate–adsorbate	H-bonding	vdW	Pauli	μ–μ
Adsorbate–substrate	Covalent, donor–acceptor	μ–μ	Pauli	μ–μ
Substrate–substrate (atomic solid)	Covalent, metal–metal	(Ionic)*	Pauli	(Ionic)*

Notes* For ionic solids, e.g. NaCl.

repulsive forces is given in table 3.2. If the overlayer does not form a superlattice, the molecules need not be pinned at specific substrate lattice locations and could be oriented relative to each other and the substrates in different ways. The substrate–substrate forces which govern reconstruction processes of the substrate were discussed in detail in the previous section. In this section we focus on substrates that do not reconstruct upon forming adsorbate overlayers and leave the subject of adsorption-induced reconstruction for the next section. To be general, also note that if the system were immersed in the liquid phase, liquid–substrate and liquid–adsorbate forces would need to be added. We will discuss these effects later in chapter 4 (Surface and Interface Thermodynamics).

Following the general strategy for understanding surface/interface phenomena in terms of the known behavior of the bulk constituents, we approach the understanding of overlayer structures in terms of the forces that underlie the structures of pure adsorbate phases and then add the new interactions that arise on attachment of the adsorbates to the substrate atoms. All adsorbates in their pure condensed phases interact by weak van der Waals (vdW) attractions, which generally are considered to have no directional behavior, and for molecules further attractions can arise involving electric dipole (μ–μ forces; attraction or repulsion depending on relative molecule orientations) and hydrogen bonding, depending on the specific molecule. Extremely short range Pauli repulsive forces are always operative and in the case of molecules with electric dipoles, Coulombic repulsion can arise when the dipoles are aligned appropriately. Adsorbate–substrate interactions can range from weak, non-directional vdW attractions with no preference for specific substrate sites or locations, to strong site-specific chemical bonds formed at specific substrate atoms. The nature of adsorbate–substrate interactions will be discussed in detail in the chapter on surface chemical bonding, chapter 7. Overall, the balance of all the forces at play determines the final adsorbate overlayer structure and we can consider two limiting cases.

In the featureless substrate limit, a superlattice does not form and incoming adsorbate atoms or molecules are not *pinned* to any particular substrate site but rather have no preference to be at any location, as though on a featureless plane, namely, the substrate lattice does not propagate its structure into the overlayer or, equivalently, does not act as a template. In this case we say that the adsorbate structure is *decoupled* from substrate lattice and the layer of adsorbates can be either frozen in statistically random positions at sufficiently low temperature or, if they have sufficient thermal energy, behave as a 2D gas in which the particles move freely across the surface. Analogous to a 3D gas we can expect that for $\theta \to 0$ the adsorbates would move about as isolated entities, behaving as a gas, but as the coverage increases the layer would form a uniform condensed phase depending on the strength of the interaction between particles and the temperature of the system, resulting in islands or domains of adsorbates, as shown in figure 3.21. Eventually as $\theta \to 1$ a uniform overlayer would form with the final packing density controlled by the repulsive forces and steric interactions between the adsorbate species. This type of behavior is typically seen with the adsorption of inert gases at low temperatures and for many simple molecules such as oxygen or nitrogen on unreactive surfaces such as graphite and silica. The condensed phases which form can be disordered or in some cases ordered, according to the adsorbate structures, intermolecular forces and temperature. For molecules that have sufficiently strong intermolecular interactions relative to adsorbate–substrate interactions, islands can grow in the vertical direction into 3D clusters, eventually forming multilayers. We will discuss these behaviors in detail in the next section and in chapter 4 (Surface and Interface Thermodynamics).

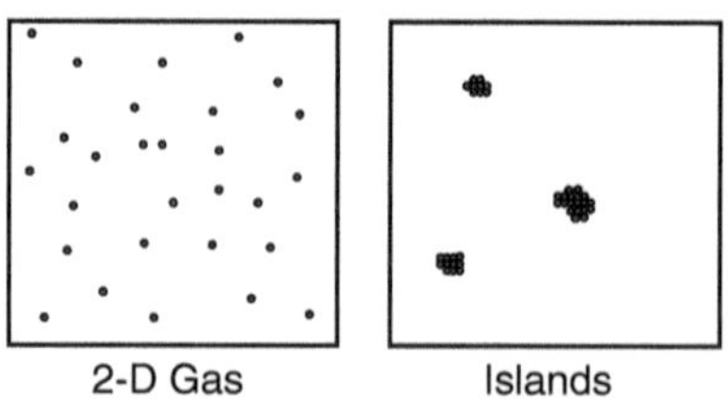

Figure 3.21. Illustration of two limits of adsorbate distribution on a substrate with no templating effect (featureless). Left: 2D gas. Right: island formation.

In the superlattice limit, the incoming adsorbates have sufficiently strong interactions with the atoms (or molecules) of the substrate surface lattice in comparison to the available thermal energy at system temperature, that the adsorbates are pinned at specific substrate lattice sites with the substrate lattice acting as a template to direct the overlayer assembly. Depending on the temperature and the amount of thermal energy available, the adsorbates can be frozen at their attachment sites or can be mobile, hopping laterally from site to site. In the frozen case, one would expect that as coverage increases the empty sites would fill in statistically as incoming adsorbates from random directions impinge on the surface bonding to the nearest site, with the superlattice not reaching the final ordered structure until $\theta \rightarrow 1$. In the mobile case, the overlayer structure would depend on the adsorbate–adsorbate interactions. When attractive there would be a tendency for the densest structure to form in domains until $\theta \rightarrow 1$ when the layer becomes uniform. Much of the behavior of superlattice formation falls between these limits and the correlations between overlayer structure with coverage in each case is a balance of adsorbate–substrate site bonding, adsorbate–adsorbate interactions and the adsorbate surface mobility at the system temperature.

A specific case that shows the balance of forces in superlattice structuring is given by the evolution of structure for O atoms on Ni(100) with increasing coverage, as shown in figure 3.22. In this case O atoms are strongly bonded to Ni atoms and form a superlattice. Defining coverage in terms of the total number of Ni atoms per unit area, initially domains of a (2 × 2) structure form up to $\theta = 1/4$. Thereafter as coverage increases up to $\theta = 1/2$ a limiting c(2 × 2) structure forms. In this case it is likely that the aligned Ni–O dipoles have repulsive interactions and it becomes energetically unfavorable for the O atoms to occupy NN Ni lattice sites.

Another example of the balance of forces is the formation of self-assembled monolayers of alkanethiolates on Au(111) surfaces, mentioned earlier (see figure 3.5). In this system the S thiolate groups have sufficiently strong bonding to specific Au (111) lattice sites that the long molecules are attached by that atom but there is sufficient thermal energy at room temperature to allow some mobility so the molecules can migrate into adjacent $(\sqrt{3} \times \sqrt{3})R_{30}$ superlattice positions as the layer forms, thereby forming domains stabilized by the favorable molecule-molecule vdW attractions of the alkyl chains. The final structure forms with the alkyl

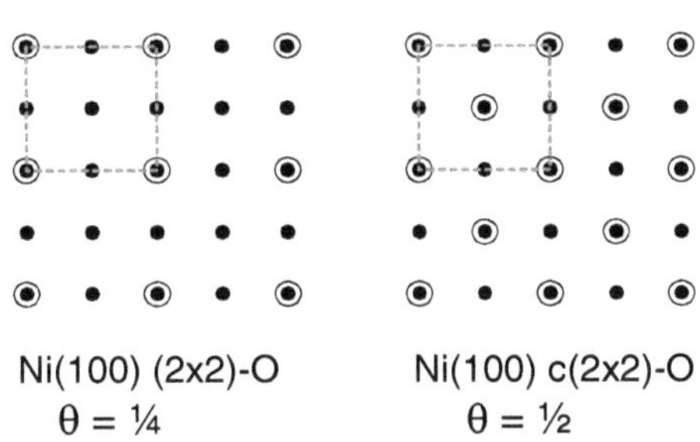

Figure 3.22. Shifting of superlattice structure with coverage for O adsorption on the Ni(100) surface. The superlattice unit cells are shown by red, dashed boxes.

chains in an upright orientation with a tilt of ~30° to maximize packing. This is often typical for molecular adsorbates in which one part of the molecule, e.g. the S in the above case, sets the translational ordering while molecular chains and groups provide a secondary driving force that gives the final overlayer its detailed structure. In this case the overlayer ordering is described in additional terms of orientational and conformational order and the twist of the C–C–C alkyl planes.

3.3.3 Multiple overlayers—supported thin films

When a surface is exposed to adsorbate atoms, molecules or precursors, e.g. O_2 gas to form O atom overlayers or alkanethiols (RSH) to form form alkanethiolate (RS) monolayers, only a single layer forms since the adsorption precursor–overlayer attraction energy is not sufficient to overcome the thermal energy of the system and additional atoms or molecules are prevented from from sticking to the fully covered overlayer. In the case of O_2 gas at room temperature the gas molecules do not condense onto the oxide surface layer while in the case of the RSH molecules, typically dissolved in organic solvent, the stability of the solute molecules is far greater than the adsorption energy on the formed SAM surface with a top layer of inert $-CH_3$ groups so no additional layer forms with sufficient stability to prevent being easily rinsed off by solvent.

In cases where the adsorbate–adsorbate interactions are quite strong compared to the stability of the adsorbate precursor in the gas or solution phases, however, multilayers can easily form. This situation arises in vacuum deposition of metal films, semiconductor films (e.g. Ge) and other atomic composition thin films (SiO_2, etc) onto substrates. These films are called *supported thin films* since the substrate acts as a mechanical support. In many cases, particularly for metal films, the depositions can be done by heating the pure material in an inert boat or basket in a vacuum to produce vapors which then condense on the support surface. In this case the vapors are highly unstable to condensation on any surface below the vaporization temperature so there is no limit to the number of layers that can be formed. In the case of metals, alternatively, depositions can be done from solution in which a metal ion is reduced to form nascent metal atoms which deposit on all the surfaces, including the one of interest. An example of this is given by the formation of silver mirrors on glass substrates from a solution of Ag ions. An important practical consideration in preparing these films, which have a huge number of applications ranging across semiconductor technology to biosensors, is the ability to control the uniformity of the deposition, which can be framed into three general outcomes, listed with their common names: (i) uniform single atom thick, layer by layer growth (Frank–van der Merwe), (ii) an initial uniform single atom thick layer followed by 3D island growth (Stranski–Krastanov), and (iii) all 3D island growth (Volmer–Weber). Schematic representations of the three growth modes are given in figure 3.23.

In some instances it is desirable to grow thin films with highly uniform crystalline order which exhibit specific crystal planes at the surface. It is rare that such growth occurs spontaneously on a typical support. One strategy is to use a single crystal

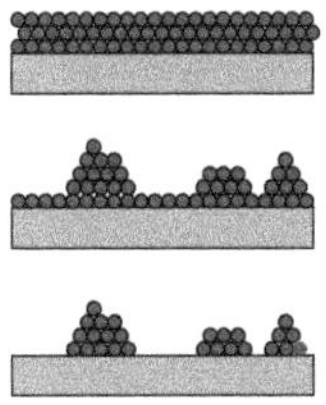

Figure 3.23. Illustration of the three modes of thin film growth and their common names.

support with an ordered surface structure that closely matches the desired thin film surface in order to *seed* the growth. There are two cases to consider: (i) growth on a single crystal substrate of the same material as the films and (ii) growth of a different film material than the support. If the overlayer forms as stacks of ideal (1 × 1) superlattices the growth is called *epitaxial* and if the film and substrate are the same or different materials the growths are called *homoepitaxial* and *heteroepitaxial*. In the latter case if the overlayer and substrate crystal planes match the overlayer structure is called *psuedomorphic*. Epitaxial growth is a difficult process since thermal energies often defeat the continued layer by layer growth by allowing stresses to be relieved, diverting growth to a path that favors some other more stable crystal structure for the thin film. Further, for heteroepitaxy if the substrate lattice has some slight mismatch of spacings to the corresponding spacings of the overlayer lattice planes then the growth often only lasts for a few layers before stresses build up to the point that the preferred structure dissipates and growth follows the Stranski–Krastanov mode. Developing strategies to obtain epitaxial growth is an important area of surface science and many specialized techniques have been developed, particularly molecular beam and atomic layer epitaxy (MBE and ALE). In these techniques fluxes (or beams) of gaseous atoms are directed with selected atomic stoichiometries onto crystalline seed substrates to produce layer by layer growth of the desired film material, e.g. Ga and As vapors to produce highly crystalline GaAs films.

3.3.4 Spontaneous formation of new overlayers in alloys

The different types of atoms in alloys often have different relative stabilities between the surface and the bulk, as one might expect by considering the different sizes and bonding interactions in light of the missing atom model. With this it is not surprising that alloys often show phase segregation between the bulk and the top surface region. An interesting example is the Cu_xAl_y alloy with $x/y = 5.25$ and a structure with no long range order which forms a surface layer with a new 1:1 composition and an ordered (111) structure, as shown in figure 3.24. Though the two atoms nearly have the same diameters (Cu, 2.846 Å; Al, 2.70 Å) and both form FCC lattices, the mismatch is sufficient that the bulk alloy does not form with long range order, although local regions will be FCC. The driving force for the surface segregation apparently is to provide a composition which allows better interactions between the two atoms, leading to an ordered (111) structure, although with some

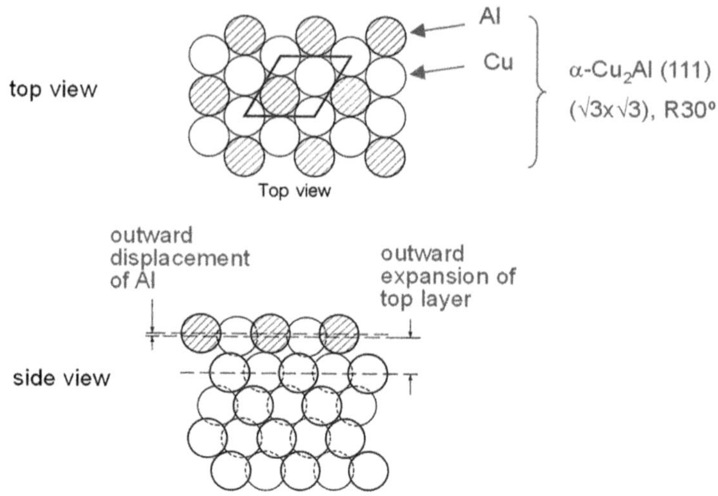

Figure 3.24. Surface enrichment of one component in an Cu(84%) Al(16%) alloy to produce a Cu_2Al (111) ordered overlayer.

slight vertical perturbation of the atom spacing relative to the original alloy underlayer. We will examine the fundamental driving forces for alloy surface segregation in chapter 4.

3.3.5 Chemisorption-induced substrate restructuring

Chemical bonding at substrate surfaces exerts new forces that can be substantial relative to the forces between substrate atoms and cause severe restructuring of the substrate surface region. The effects can include no change in the substrate surface structure for the cases of weak bonding but generally changes do result. Three general cases of effects of chemisorption on the structures of surfaces of atomic solids are listed schematically in figure 3.25. The surfaces of substrates which can form chemical bonds are generally in some state of reconstruction in their bare state, even if very a minor perturbation so we will consider the initial bare substrate surface to be reconstructed in the discussion. For example, Au(111), with a very dense surface plane and generally considered chemically inert except for a few agents such as CN^- ions, exhibits a slight reconstruction of its bare surface compared to the ideal (111) structure. We first discuss some representative examples of cases II and III.

Two example of case II are given by the effects of C and S atom chemisorption on the Ni(100) and Fe(110) surfaces, shown in figure 3.26. In both examples the adsorbate atoms form strong chemical bonds with the metal atoms and distort the surface lattice. C atoms will tend to make four bonds per C if possible, as allowed by the positions of neighboring metal atoms and the ability to strain the substrate lattice as needed to drive the atoms into the surface region. S atoms tend to form two bonds per atom and can satisfy this requirement in the upper parts of the surface layer. These two examples are important since both are involved in catalytic reaction

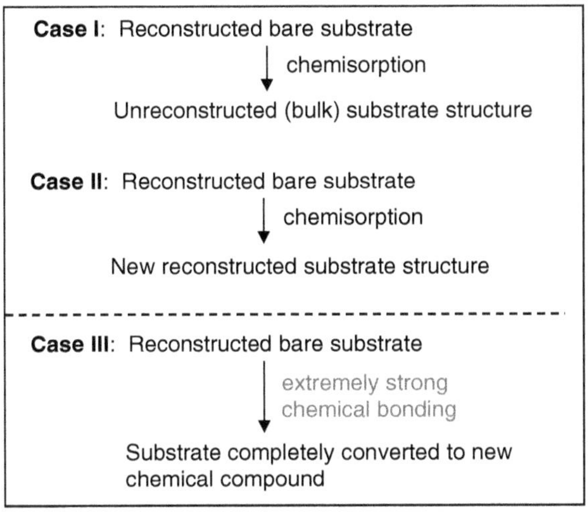

Figure 3.25. Schematic of three general cases of chemisorption-induced changes in substrate surface structure. Case III involves the reaction limit in which the entire bulk substrate is converted to new chemical compound.

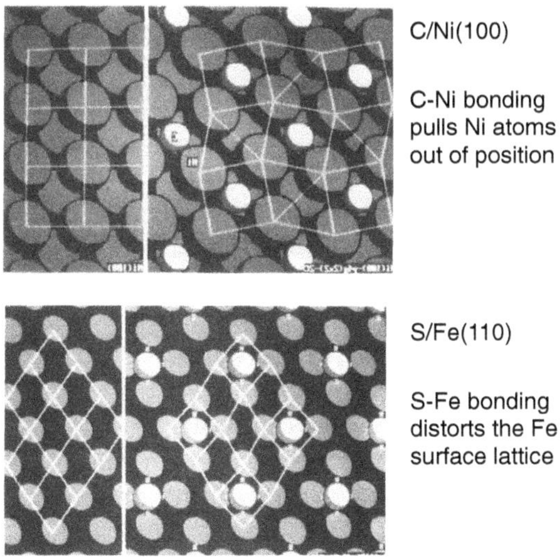

Figure 3.26. Two examples of case II chemisorption-induced surface reconstruction. Upper: C on Ni(100) and S on Fe(110). In both diagrams the initial surface is shown on the left.

systems. Ni is a very active catalyst for many reactions involving molecules containing C, e.g. dehydrogenation of hydrocarbons. In these cases some of the molecules can be exhaustively degraded and C atoms form at the Ni surface. The strong Ni–C bonds leave the C atoms tenaciously bound on a reconstructed surface with the result surface that catalytic activity can diminish. Fe is also used in a number of catalytic reactions, e.g. conversion of N_2 to NH_3. S-containing molecules are often impurities present in catalytic systems (particularly if petroleum based feedstocks are used) and slowly become chemisorbed to produce tenaciously bound S atoms thereby poisoning the catalytic activity.

Case III represents chemical interactions which are so strong that the substrate lattice is literally ripped apart with formation of a new chemical compound. One would guess that as the chemical bonding strength increases starting with weak chemisorption that the adsorbate atoms would be driven deeper and deeper into the surface region, overcoming the substrate material lattice strength and approaching case III chemical conversion. This leads to two more useful rules of thumb.

Rule 4: Chemisorption-induced reconstruction (CIR) ↑ as substrate surface atom coordination number (CN) ↓.

Rule 5: Rate of CIR ↓ as depth into substrate ↑.

Rule 4 follows from the reasoning that substrate surfaces in which the atoms have low CNs both offer more missing bonding sites for adsorbate bonding than high CN surfaces and would be easier to pull apart to allow incursion of adsorbate atoms. Rule 5, which deals with the rates of CIR, similarly follows in that penetration of an adsorbate atom in the top layer which is missing bonds at the top surface is easier than in deeper layers which approach full 3D bonding, versus, the energy barriers encountered in ripping the lattice atoms apart become increasingly higher deeper into the bulk, and accordingly we would expect the rates of penetration to diminish exponentially.

Several examples illustrating these rules for case II chemisorption are given in figures 3.27 and 3.28. Following our earlier rule, for FCC atomic solids, particularly mid-row transition metals such as Ir, Ni, Pt, etc, the dense (111) face is not prone to reconstruction. The more open (110) faces, however, undergo CIR quite readily with strong chemisorption agents such as S atoms. In this case we can extend this rule to CIR and add rule 1 to predict penetration of the adsorbate. In figure 3.27 S atoms are seen to penetrate part way into the open spaces in the Ir(110) surface to form a depressed (2 × 2) superlattice. The penetration is consistent with what would be expected for sulfide type species with strong divalent bonding that pulls the S atoms into the open spaces in the substrate surface lattice.

Following the examples given in figure 3.30, as we progress from the O/Ni(100) result to the O/Cu(110) and N/Ti(0001) ones, notice that adsorbate atoms go deeper and deeper into the surface regions until finally for N/Ti(0001) the N atom is buried

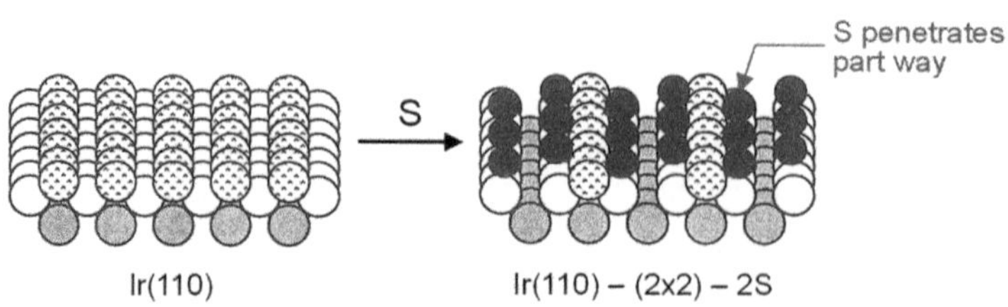

Figure 3.27. Chemisorption of S on a Ir(110) surface to form a (2 × 2) overlayer which penetrates vertically into the Ir surface lattice.

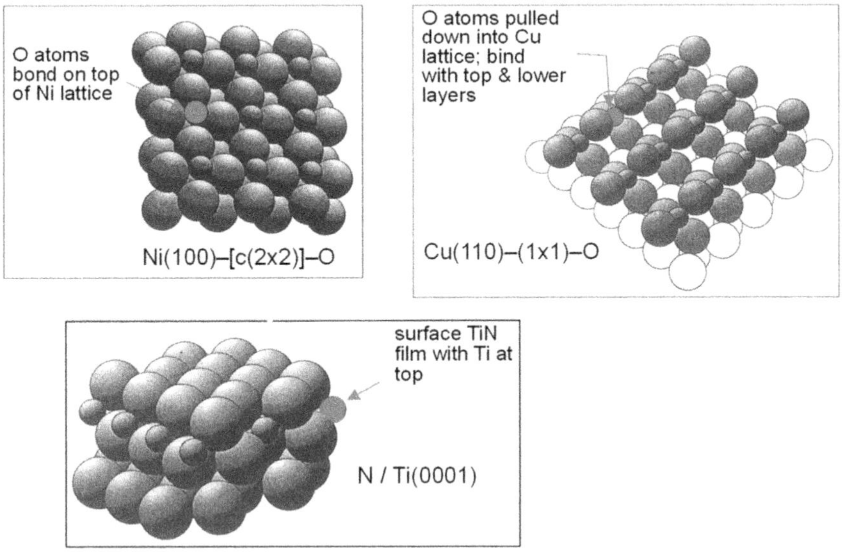

Figure 3.28. Chemisorption of S on a Ir(110) surface to form a (2 × 2) overlayer which penetrates vertically into the Ir surface lattice.

one layer deep into an interstitial site which allows multiple 3D bonding to the N atom. In the latter case the system is approaching the limit in which a new chemical compound is being formed. In each of these systems the energetics of the chemisorption process are moving closer to the limit of forming a new bulk chemical compound.

Now we consider a few generalizations about adsorbate atom, or more generally, heteroatom (different than the atoms of the substrate material) penetration into atomic solids via chemical forces. It follows from above that deep penetration results in formation of interphases or thin films of alloys and new compounds. Penetration is a balance of the chemical reactivity of the heteroatom, the space available in the unperturbed substrate lattice, and the ability of the lattice to be strained. Thus for dense, hard metals with strong lattices only small atoms such as H can penetrate, usually into interstitial sites where space is available. Semiconductors are an important class of technological materials and deposition of heteroatoms on the surface can be important for altering the electronic characteristics. Whereas low atomic number materials, e.g. Si, have strongly bound, dense lattices which make heteroatom penetration difficult, higher atomic number semiconductors such as Ge, GaAs and InAs have softer lattices with more space available. A good example is GaAs, a very important material for high speed electronics and optoelectronic applications, where Al atoms can penetrate and replace Ga atoms to form insulating layers of AlGaAs that allow certain types of valuable device structures to be made.

As a final example we consider the case of large molecule, octadecanethiol [$CH_3(CH_2)_{17}SH$, or ODT], chemisorption on a square GaAs(100) surface terminated by a layer of As atoms to form an alkanethiolate adsorbate overlayer attached to the surface by covalent S–As bonds. This is an interesting case of a readily deformed semiconductor crystal. GaAs exhibits a diamond type of lattice in which the Ga and As atoms alternate, bonded to each other in a tetrahedral geometry, as in Si and C (diamond) solids. Unlike these materials, however, which are mechanically hard and exhibit extremely strong lattices (in line with their high vaporization temperatures), GaAs is easily scratched and broken, and thermally decomposes above several hundred degrees C to release As vapor. Thus when stress is put upon the surface lattice it can easily strain, allowing reconstruction. The ODT adsorbates are too large to form a (1×1) superlattice on the unreconstructed (100) substrate but can fit on a square $c(2 \times 2)$ (NNN) pattern. In spite of the ideal square pattern, the overlayer actually forms a hexagonal lattice, incommensurate with the unreconstructed (100) substrate, with the alkyl chains extended and nearly vertical to the substrate surface, as shown in figure 3.29. Extended alkyl chains are rod-like structures which prefer to pack with a symmetry approaching a hexagonal pattern to maximize attractive forces (like squeezing a bunch of parallel pencils together). The mismatch in adsorbate spacings and symmetry with the underlying (100) substrate, shown in figure 3.29 (right), allows the chains to create a driving force that stresses the underlying As lattice (and lower Ga/As layers), which are relatively easily relieved by distorting the soft substrate over short distances of ~5–10 nm. This creates small domains of hexagonally packed chains on a partially disordered GaAs substrate surface region. This is a good example of CIR in which the driving force is not the adsorbate–substrate chemical bonding but rather the adsorbate–adsorbate interactions which are coupled into the substrate by a strong adsorbate bond.

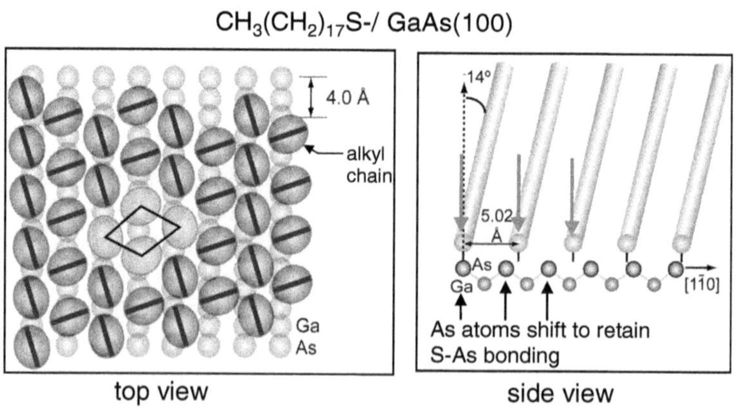

Figure 3.29. Substrate reconstruction induced by the chemisorption of octadecanethiol molecules onto GaAs (100). Left: Top view of the observed pattern of ODT adsorbate molecules with a hexagonal symmetry (shown by the yellow unit cell) and a herringbone pattern of the C–C–C alkyl planes overlaid on an unreconstructed square GaAs(100) substrate with the As atoms on top of the underlying Ga atoms. Right: Side view of a schematic showing the slightly tilted alky chains with their observed spacings on top of the unreconstructed As/Ga surface layers. Note the mismatch in adsorbate and As spacings. (Left panel adapted with permission from [3]. Copyright 2010 American Chemical Society.)

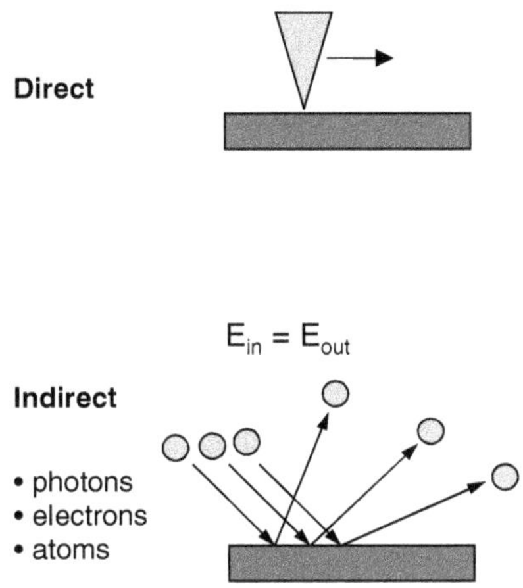

Figure 3.30. Schematic illustration of a proximal probe (top) and particle scattering (bottom) for determining surface structure.

3.4 Common methods for determining surface structure

The common methods used to determine the structure of surface fall into two general classes, proximal probes and particle scattering probes. In the first approach a tip with a radius of curvature approaching the size scale of the surface features of interest is brought to within Ångstrom and sub-Ångstrom distances of the surface and changes in the tip–surface interactions across the surface are interpreted in terms of the local surface features. These are direct measurements in that the instrumentation is used interactively to directly control the approach of the tip to the surface and to scan across the surface to produce the image of the features. Thus the term microscopy is often applied, although the images are on the nano- not micro-scale. In contrast, particle scattering involves impinging beams of fundamental particles, photons (x-rays), atoms or electrons, on the surface and measuring the the angles of the beams which scatter from the surface due to diffraction between ordered surface features at the atomic scale. In this case macroscopic areas of the surface are probed using mm-scale diameter beams and the surface features interpreted from the scattered beams are averages over the probed areas, as opposed to microscope images. The two classes of probes are thus very different in terms of the instrumentation and the theory underlying the interpretations so we will discuss them separately. But overall, they are highly complementary, each providing certain types of information that the other cannot easily do, so for the most informative characterizations multiple probes are often used. The two classes of techniques are illustrated schematically in figure 3.30.

3.4.1 Direct—proximal probes: AFM, STM

Proximal probes are divided into two dominant subclasses: atomic force microscopy (AFM) and scanning tunneling microscopy (STM). The basic differences between the two involve the method used to determine the height of the tip above the surface and the physical mechanism of the tip–surface interaction which gives information on the surface features. In both techniques a sharp nanoscale tip is brought in proximity to the surface of interest by using precision piezoelectric drives, capable of controlled sub-Å motion. Once the tip is near the surface the method for making the measurements differs for the two techniques.

3.4.1.1 Atomic force microscopy (AFM)

In AFM, the tip, typically of a radius of curvature of 2–10 nm (sometimes larger for lower resolution measurements) and located at the end of a stiff cantilever of some material such as silicon, is brought towards the surface until at nm scale distances tip–surface forces begin to operate, bending the cantilever. The deflection is measured, typically by following the movement of a laser beam reflected off the back of the cantilever onto a position sensitive detector, as illustrated in figure 3.31 (left). As the tip approaches the surface attractive forces operate, such as long range van der Waals forces, but at very close approach steeply changing repulsive forces come into play. As the tip moves across the surface with a constant height of a fixed piezo drive, the cantilever bends up and down driven by the local surface forces and a deflection map of the surface is obtained. There are many variations of the technique but the key factor is that tip–surface forces are used to sense where the tip is relative to the surface and thus provides a feedback mechanism to control the tip above the surface with the z-piezo drive.

An example of a standard AFM measurement is given in figure 3.32 where an image obtained from a crystalline graphite sample is shown. Graphite consists of stacks of ~3.5 Å thick (0001) planes with hexagonal symmetry. The measurement is done by pushing the tip onto the surface until repulsive forces operate (usually in the 5–20 nanonewton, nN, range) and keeping the downward force constant as the tip is drawn across the surface while the deflection is measured. The image shows features in a hexagonal pattern, as expected, with a spacing of 2.46 Å. While this gives a good demonstration of the precision of the measurement, note that the actual spacings of the C atoms in graphite are only 1.42 Å, as shown in the figure, thus the observed distances are those between the graphite six-membered rings. The discrepancy points out an important aspect of AFM measurements. The tips are not sharp to the scale

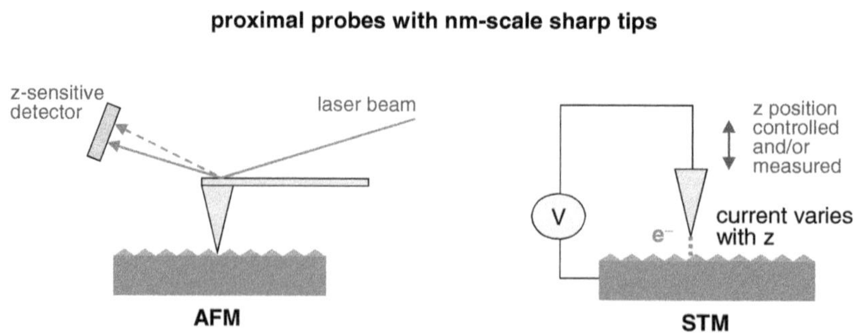

Figure 3.31. Schematic illustration of an AFM (left) and an STM set-up, showing the tip near the surface and the detection method for the tip–surface interaction.

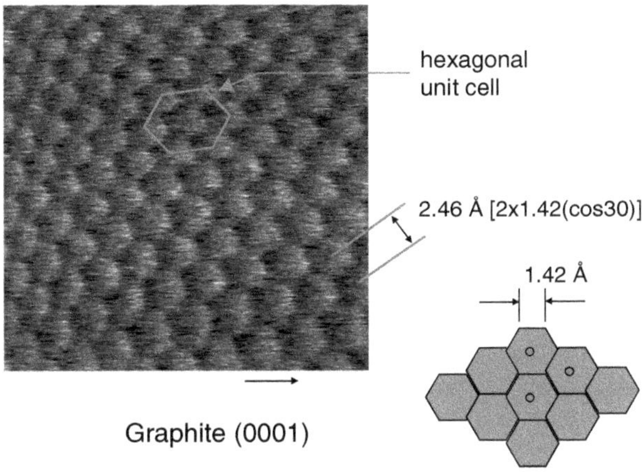

Figure 3.32. AFM contact mode image of graphite (0001) surface. The nearest neighbor C atom spacing is 1.42 Å but the AFM image shows closest spacings of 2.46 Å, the distances between the centers of the rings.

of single atoms but rather are rounded at many nanometers and the deflection of the cantilever arises from tip interactions with a local force field from a group of nearby surface atoms. Thus the mechanism is loosely akin to rolling a bowling ball across a cobbled street with the cobble stones much smaller than the ball. The ball will move up and down but will not be sampling each stone. One important aspect of this is that the surface image may not show a local defect since the force field is an average of the local atoms, not much perturbed by a single missing one. We will note shortly how different this is from STM.

Since this measurement is purely opto-mechanical it can be done in vacuum or liquids and on any substrate, conducting or insulating. On the other hand, for soft materials such as polymers and biological entities the pressure applied to the tip in contact with the sample must be kept very low or damage can result from the tip irreversibly deforming and tearing the surface region. This can be overcome in some cases by doing measurements using very low tip pressures (which may give a loss in resolution) or using the attractive forces from the surface while the tip is pulled back slightly to keep from actually touching the surface (so called noncontact mode). For conducting substrates, metal-coated tips with electrical bias can be used to carry out electrical measurements (conducting probe) on the surface as well, though the resolution is not atomic and this method is not the used for detailed measurement of the physical surface structure. Finally also note that AFM is purely a surface probe of the very top layer of atoms and cannot sense structure below the surface which cannot be physically touched. This is different from STM, as will be discussed below.

3.4.1.2 Scanning tunneling microscopy

In scanning tunneling microscopy (STM) measurements, an electrically biased metal tip is brought in approach to the surface of a conducting substrate. As the tip–surface distance begins to decrease and the electron wavefunctions of the tip and the substrate start to overlap, electrons start to tunnel creating a current across the gap between the tip and the substrate. As the gap decreases the tunneling current (I) goes up exponentially, as given in equation (3.1):

$$I = c \cdot V \exp\left\{-\frac{2\left[2m_e\varphi_{\text{eff}}(V)\right]^{1/2}d}{\hbar}\right\}, \tag{3.1}$$

where V = applied tip–ground voltage bias, m_e = electron mass, φ_{eff} = effective tip–sample potential energy barrier for e^- transfer (a function of V), d = tunneling gap distance, c = a constant for the specific tip–sample combination and $\hbar$ = Planck's constant/2π. The tunneling current provides a mechanism to both control the tip–surface (z) distance and to record the presence of local features of differing conductances and/or heights. Note the generally linear dependence on bias voltage is given in the pre-exponential term, but a further nonlinear dependence in the exponential term. Since the tunneling current is exponential in distance, tunneling from the tip occurs dominantly from the outermost atom on the tip and to the nearest protruding atom on the surface, as suggested by the illustration in figure 3.33. This nonlinear behavior allows atomic scale lateral imaging and subatomic scale z-resolution of surface features. The most common mode of operation is to keep the tip delivering constant current at some bias such as 1 V while recording the height of the tip, controlled by the piezo drive set-up, as the tip is scanned across the surface. Other modes of operation are also used, such as constant height while measuring current variations.

An example of the ability of STM to see isolated molecules is given in figure 3.34 (top) for the adsorption of benzene molecules on a Cu(111) surface at 4 K. The figure shows the large terraces, several steps and a low coverage of individual benzene molecules scattered across the surface. While this shows the ability to see individual molecules, STM also can image at even smaller resolution as shown in figure 3.34 (bottom) where individual benzene molecules adsorbed on Pt(111) are imaged across each molecule. Note the clearly defined pattern of current variation

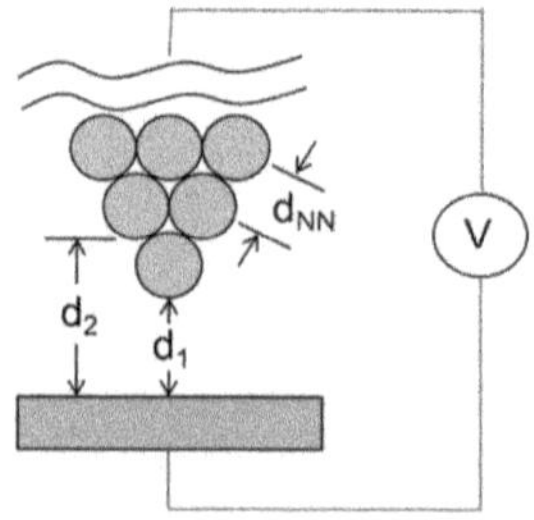

Figure 3.33. Simple representation of an STM tip positioned above a solid surface. The outermost atom on the tip is distance d_1 above the surface plane with a nearest neighbor distance d_{NN} between tip atoms. The tip is positioned by the STM instrument using piezoelectric motors that can move the tip to controlled positions. Application of a controlled voltage bias between the tip and the sample surface allows a tunneling current to flow. Measurement of the current with electronics forms the basis of the surface image. Electronics include a low noise current amplifier and associated items to create a digital file of the tunneling currents as a function of position on the surface.

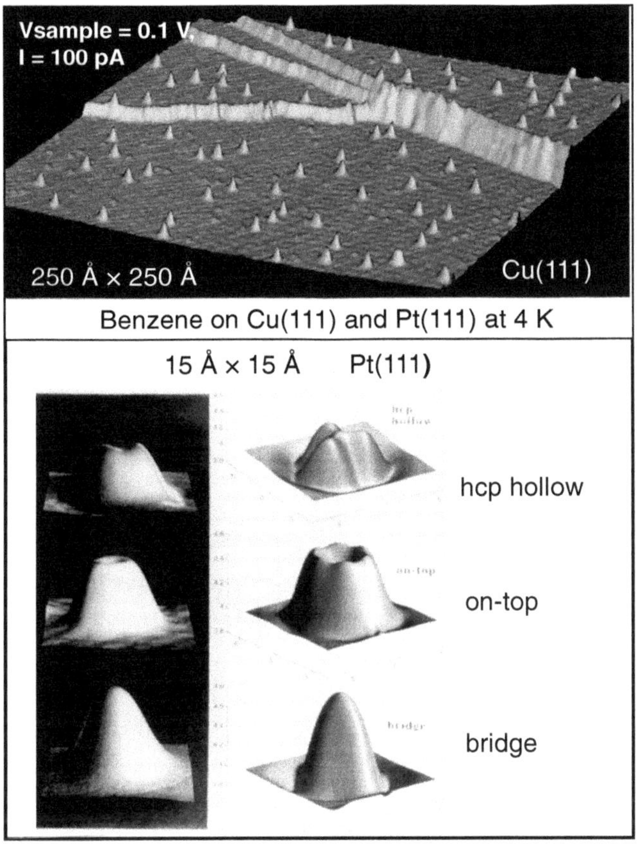

Figure 3.34. STM images of benzene adsorbed on Cu(111) and Pt(111) surfaces at 4 K. The Cu(111) image shows individual molecules across the surface [4], while the Pt(111) images show of the molecules at high resolution revealing patterns with different symmetries [5]. (Image credit, Paul Weiss, UCLA.)

across the molecules depending on the exact adsorption site on the substrate surface, directly on top of a Pt atom (on top), bridging two Pt atoms (bridge site) or in a hollow between 3 Pt atoms (hcp hollow). Each of these images shows different patterns. Thinking in the simplest terms one might have expected the images to have a six-fold symmetry, showing each C atom of the benzene rings (based on the observation that the rings do lay parallel to the surface). This unexpected behavior points out an important aspect of STM—the images arise from a convolution of electronic (tunneling) effects and physical structure. The tunneling process involves interactions between the wavefunctions of the electronic states in the tip and those of the surface. So unlike AFM which senses the physical structure, STM also involves electronic effects. On this basis one interprets the benzene/Pt(111) images in terms of the wavefunctions or distribution of electronic states in the molecules that govern tunneling to the substrate.

Another interesting example of STM is given by images of the adsorption of H atoms on the Pd(111) surface, shown in figure 3.35. This represents a case in which

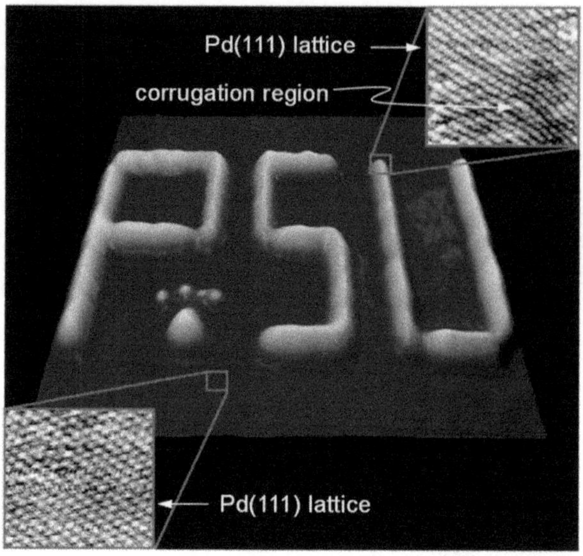

Figure 3.35. STM image of a pattern of H atoms adsorbed below a Pd(111) surface. The atomic resolution lattices show a smooth (111) terrace (lower left) and a warped terrace in the region of the buried H atoms. (Image credit, Paul Weiss, UCLA.)

the very small H atoms are able to adsorb into interstitial sites below the Pd surface, in the process altering the electronic and physical characteristics of the surface lattice in the local region of the adsorbate atoms. The large, low resolution image shows a PSU pattern, written by driving H atoms into the lattice below by applying voltage to the tip. The atomic resolution images show the individual Pd atoms at the surface with a flat, unblemished (111) terrace at the lower left and a warped terrace at the upper right in the region with buried H atoms. The characteristics of the warping must be interpreted in terms of both electronic (wavefunctions) and physical height perturbations.

An example of an STM image of large molecules in a self-assembled monolayer (SAM) is given in figure 3.36 (left), which shows the positions of individual decanethiolate [$CH_3(CH_2)_9S$–] adsorbates on Au(111) ordered in a superlattice. The figure clearly shows the rows of atoms in a hexagonal symmetry and, in addition, shows a line defect where two domains of molecules with apparently different tilt angles come together and a pit defect which has been ascribed to a missing Au substrate atom in the vicinity (typically giving an ~1 Å depression in the local area).

STM involves very low noise measurements of electrical currents, down to the ~0.1 picoampere (pA) level, which requires extremely quiet electronics, especially for samples with molecular overlayers which are not very conductive and lead to low tunneling currents. For example, in the case of the alkanethiolates on Au(111) SAMs above, as the alkyl chains become longer the tunneling current through the monolayer decreases exponentially until at ~12 carbons the noise from even

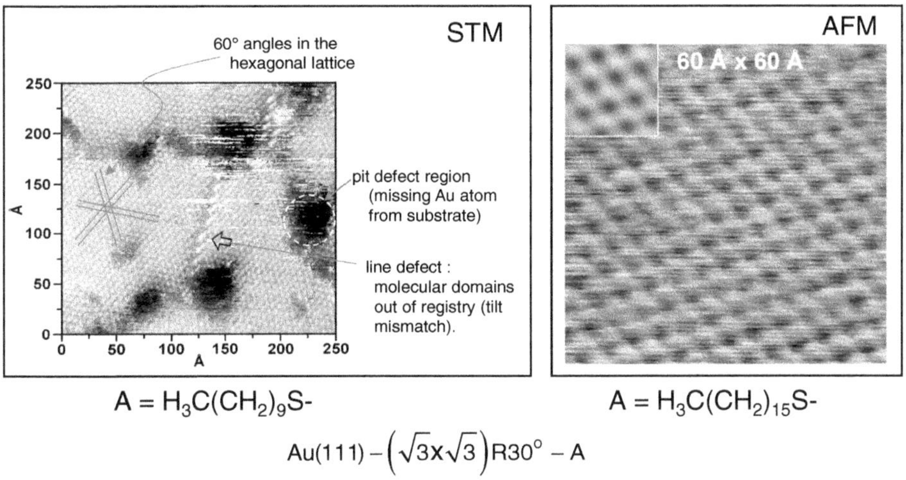

Figure 3.36. STM (left) and AFM molecular resolution images of two alkanethiolate/Au(111) SAMs. STM: the images show hexagonal symmetry marked by parallel red lines at 60° angles, a line defect where two molecular domains meet out of registry and a pit defect due to a missing Au substrate atom. AFM: the images shows the hexagonal molecular lattice (magnified in upper left corner. (Image credit, Paul Weiss, UCLA.)

extremely good electronics overcomes the feature signals across the surface, blurring the molecular resolution images. At this point it is possible to switch to AFM since no electrical conduction is needed. Figure 3.36 (right) shows an AFM contact image of an alkanethiolate/Au(111) SAM for a 16-carbon chain, which is too long to allow STM images to be obtained. Though molecular resolution is obtained with the AFM, the sharpness does not approach what STM can do for the shorter chains.

3.4.1.3 Comparisons of proximal probe techniques

When learning about any characterization technique it is very useful to list the advantages and disadvantages in order to set some perspective on when the technique might be useful, as shown in table 3.3 for the cases of AFM and STM.

3.4.2 Indirect probes: surface diffraction by light and particles

Diffraction is a phenomena in which a wave traveling in a specific direction enters a confined space of a size comparable to the wavelength but upon exiting can assume a number of different directions, each with a specific probability as dictated by the wavelength, confinement dimensions and initial direction. Once the confined space becomes large compared to the wavelength the wave passes straight through (transmission). At the other limit of a vanishingly small space, think of a wall with tiny holes, the wave ignores the hole and treats the wall as a uniform material. The specific behavior depends upon constructive and destructive interference between different parts of the confined wave. This phenomenon can be used to great advantage for precise measurement of spaces between atoms in a crystalline solid or in an ordered surface region using any type of wave with wavelengths at the scale of the atomic spacings. For light, this means x-rays with typical wavelengths

Table 3.3. Some advantages and disadvantages of AFM and STM proximal probes for determining surface structure.

Probe	Advantages	Disadvantages
AFM	Direct view of surface topography	Cannot see individual atoms one at a time (only larger area patterns)
	Conducting or insulating samples	Can damage soft samples (e.g. polymers)
	Lattice imaging possible	
	In-liquid operation	
STM	Atomic resolution	
	Resolve individual atoms	Typically limited to conducting samples (~1 pA limit)
	In-liquid operation (e.g. E-chem) possible	Image is convolution of physical topography and electronic effects
		Difficult for rough surfaces

from ~0.1 to 10 Å. Since particles exhibit wave-like properties according to the de Broglie relationship (equations (3.2a) and (3.2b)) they also can be used to generate diffraction patterns:

$$\lambda = \frac{h}{p} = \frac{h}{mv}, \tag{3.2a}$$

$$\mathbf{p} = h\mathbf{k} = h\left(\frac{1}{\lambda}\right)\mathbf{e}_{\lambda}. \tag{3.2b}$$

The de Broglie equation is shown in two forms (3.2a), where λ = wavelength, p = the magnitude of the linear momentum, h = Planck's constant = 6.63×10^{-34} J · s, and m and v are the mass and speed of the particle; and (3.2b), where $\mathbf{p}$ = the momentum vector and $\mathbf{k}$ = the wavevector defined to be in the direction of the wave propagation given by the unit propagation vector $\mathbf{e}_{\lambda}$. In order to match λ of the particle to the crystal spacings an appropriate particle speed must be selected for the given particle mass to obtain the necessary momentum. Particle diffraction is done using fundamental particles such as small atoms, particularly He; electrons, and even neutrons, each offering unique characteristics for probing spacings in confined geometries. The major difference between x-rays and particles is that while x-rays penetrate deeply into solids, particles generally do not penetrate past the top few layers at the speeds needed for diffraction. This difference is reflected in the very different types of sample–beam geometries used for x-ray versus particle diffraction from surfaces.

There are two general modes of diffraction used for surfaces, slit diffraction and Bragg reflection (BR), as illustrated in figure 3.37 for the simple cases of a single slit and two parallel planes. Both obey exactly the same principles of constructive and destructive interference of waves confined in small spaces. Slit diffraction involves

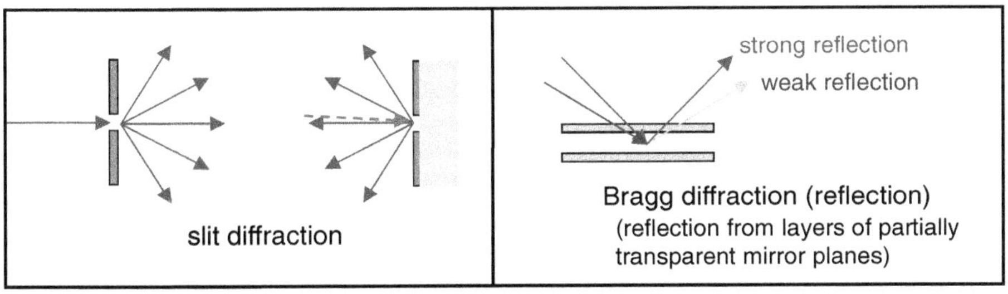

Figure 3.37. Schematic illustration of slit diffraction (left) and Bragg reflection (right). In slit diffraction an incoming wave exits in different directions to give a diffraction pattern. In Bragg reflection an incoming beam to a stack of partially reflecting planes partitions between reflection and transmission at each plane. The combined reflection beams have strong or weak intensities depending on the geometry and the wavelength.

sending a wave (or particle wave) through a small slit, or in 2D a window. For surface diffraction the windows consist of a surface plane of ordered atoms with spaces between and the diffracted beams are reflected back by the material below the surface. BR involves sending a wave at some angle of incidence (θ) through a partially reflecting surface plane with a stack of partially reflecting parallel planes (like partially transparent mirrors) below, each spaced by some distance δz. In a crystal the planes are just the crystal planes and the confinement region for the waves is just the interplanar region with δz equal to the interplanar distance. Since the planes are partially reflective the beam reflects partially of the top surface, transmits to the second surface where it partly reflects, and so on if there are more planes below, depending on how far the waves penetrate. Because of constructive and destructive interference effects between the reflected waves returning from each successive layer, the net result is that for a given λ and δz only certain specific values of θ lead to non-zero reflected beam intensity. For surface diffraction the incoming beams skim across the surface and reflect off parallel rows of surface atoms creating outgoing beams of varying intensity depending on the angle of incidence. BR for measuring surface atom spacings can only be done with x-rays, as will be explained in the next section.

3.4.2.1 Photons—x-rays—grazing incidence x-ray diffraction (GIXRD)

X-rays ranging from ~0.1 to several Å wavelengths are typically employed for diffraction analysis of atom spacings. Since x-rays penetrate deeply into solids (think of dental x-rays) it is not possible to use the slit mode with the incoming beam impinging vertically upon the surface for surface characterization. This geometry will result in diffraction beams exiting from deep within the sample thus swamping any weak diffraction pattern from the surface layer. Use of the BR mode with the incoming beam impinging at easily arranged angles to the crystal surface, will give exit beams from the atoms in sets of parallel planes from deep within the bulk, the standard method of generating structures of bulk crystals, but again, the surface diffraction patterns will be obscured by the bulk patterns. By using a barely skimming beam (grazing incidence) off the surface, however, the surface region is

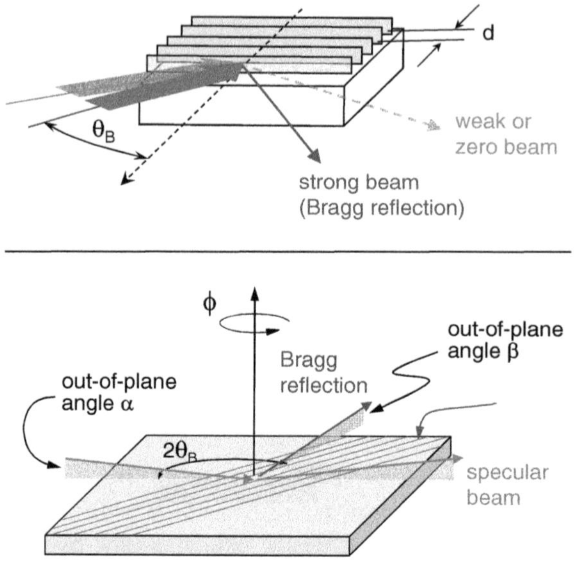

Figure 3.38. Schematics of surface GIXRD. Top: incoming beams skim the surface with a set of parallel surface planes created by parallel rows of atoms. The blue and red beams show strong and weak Bragg reflections, respectively. Bottom: The geometry of the set-up showing the in-plane (θ_B) and out of plane (α, β) angles.

dominantly sampled and the contributions of the reflection beams from parallel rows of surface atoms can be emphasized. This technique, termed grazing incidence x-ray diffraction (GIXRD), works particularly well if the top surface atoms are different than those of the substrate and/or exhibit a distinctly different diffraction pattern, thereby improving the ability to pick out the surface contribution from the total pattern. Having larger atomic number (z) atoms at the surface than in the substrate is helpful since the efficiency of x-ray reflection depends strongly on local electron density, which increases with increasing z. The illustration in figure 3.38 (top) shows a schematic of a surface with rows of atoms, constituting a set of parallel surface planes, with incoming x-ray beams. The x-ray beam penetrates through the planes with partial reflection at each one giving a Bragg beam at the correct angle. The incidence angle is varied (by turning the sample) to provide a pattern of strong and weak beams. Although one plane is shown in the figure, the atoms also form planes at other angles, but with different packing densities (numbers of atoms per unit distance), just as the surface densities of different crystal faces differ. Thus as the beam angle changes Bragg reflections will appear for these other planes in turn. Since it is impossible to arrange to have the beam skim at exactly zero degrees there always is a slight out-of-plane angle, shown in figure 3.40 (bottom), and in fact if the surface species are not just simple atoms but have some out-of-plane feature, for example with a molecular layer with tilted molecules, the x-rays can reflect off this feature at an angle to the surface and the angle can be used to characterize the tilt of this feature.

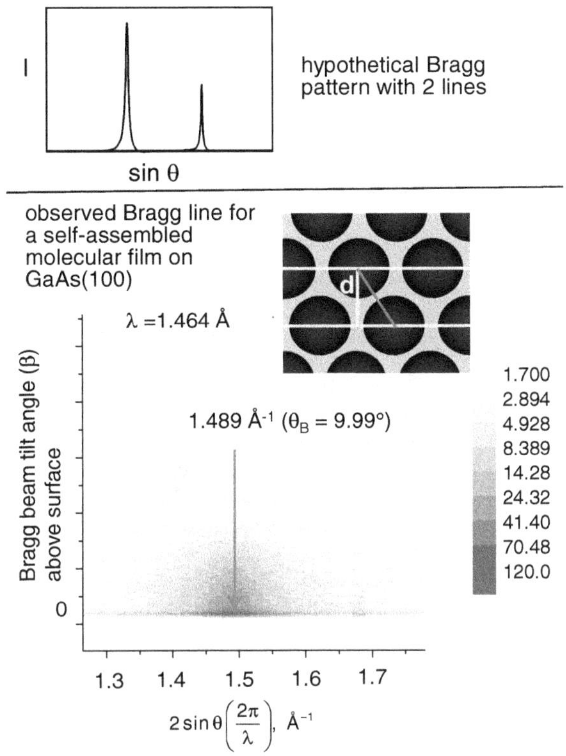

Figure 3.39. GIXRD plots. Top: Hypothetical plot of Bragg reflection intensity versus sine of the incidence angle. Bottom: Point density graph of observed reflection beam intensity shown within a plot reflection beam elevation angle above the surface versus a function of the reflection beam angle. The sample is a monolayer of octadecanethiol [$CH_3(CH_2)_{17}SH$] chemisorbed on a GaAs(001) As atom-terminated surface.

A hypothetical diffraction plot of reflection intensity versus incidence angle with two Bragg peaks is shown in figure 3.39 (top). The bottom panel shows and experimental result, obtained with 1.464 Å x-rays generated at a synchrotron facility, for a SAM of $CH_3(CH_2)_{17}S$– adsorbates on GaAs(100). The molecular adsorbates have their long alkyl chains extended and oriented near vertical to the substrate (14° tilt) with the S atoms bonded to As atoms in the crystal surface. The reflected intensities were taken with a 2D array detector so that both in-plane Bragg reflection and the out-of-plane (elevation of the reflected beam) angles were measured. The plot shows a map of intensities, shown as points densities, for different Bragg angles as a function of elevation angle. The point density for purely in-plane reflection beams peaks at 9.99°, which corresponds to a 4.84 Å atom NN molecule spacing. In the plot note how the out-of-plane beam intensities die off rapidly with increasing elevation angle, consistent with almost no tilt to the chains. The quantity plotted along the x-axis follows the standard Bragg equation relating the angle of incidence θ_B to the conditions for appearance of strong reflections:

$$\sin\theta_B = \frac{n\lambda}{2d}, \tag{3.3}$$

where n is an integer which gives the order of the diffraction beam ranging from 0 (reflection directly back from the surface at normal incidence), 1, 2, etc, d is the interplanar spacing and λ the wavelength.

3.4.2.2 Electrons: ~10–20 eV energies—low energy electron diffraction (LEED)
Whereas x-rays in the ~1 Å wavelength range, equivalent to ~12 keV energies, penetrate deeply into materials, electrons with ~1 Å wavelengths have relatively low energies of ~150 eV, thus the term low energy electron diffraction (LEED). But because of their charge and mass a large fraction of the incoming electrons readily scatter off the electrons in a solid, losing energy so they no longer contribute to the diffraction output. Thus only a small fraction of electrons travel more than a few layers into the solid, typically limiting the diffraction information to just the top ~1 nm surface region. This is ideal for diffraction directly from the top surface and the simple condition such as normal incidence can be used to generate diffraction beams coming out at different angles, essentially a reflection mode of a slit type of diffraction, as depicted in figure 3.40 (top). Note, however, that the use of charged particles to bombard a surface has two drawbacks: (i) if the sample is not conducting it will charge up immediately and repel additional incoming charge, stopping the experiment and (ii) most molecules are readily damaged ('fried') by electron beams. On the other hand, the experiment is relatively easy to run. Electrons are relatively easily generated from hot metal filaments with voltage applied to cause electron emission, the emitted electrons can be precisely manipulated by electrically charged grids to provide controlled monoenergetic beams and the scattered electrons easily detected by standard array charge detectors to give an immediate pattern without scanning. So the experiment is relatively simple, compared to GIXRD which requires a large facility (synchrotron) to generate the high intensity x-ray beam. The penetration of the electrons down several layers into the sample can present a challenge in working up the data since the diffracted beams from the top layer which exit moving towards the second and lower layers can again diffract creating complex patterns for the combined beams exiting from the surface. The multiple scattering, however, is actually an enormous advantage in some cases since the structure of the top few layers can be obtained with patient data analysis. This is particularly an advantage for example in cases such as adsorbate induced reconstruction which extends several layers into the bulk. The interpretations are done using scattering theory to generate simulations of possible structures with their diffraction patterns and then fitting the data to obtain the best proposed structure.

An example of a LEED experiment is shown in figure 3.40 (bottom) for diffraction from an Au(111) – $2(\sqrt{3} \times \sqrt{3})R_{30}$–$C_{60}$ adlayer. The schematic (left) shows the C_{60} molecules ('Buckyballs') placed on the Au(111) hexagonal surface lattice. The C_{60} molecules are too large to fit on NN or NNN sites so are spaced even further apart. Since there is no chemical bonding to the substrate and the molecules are symmetrical the only interactions to the surface are vdW forces and the arrangement appears to

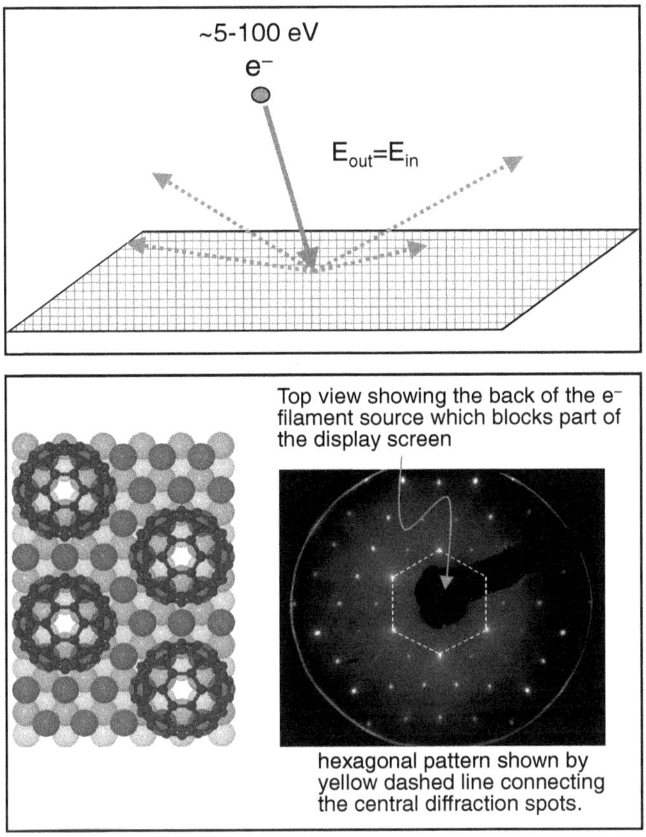

Figure 3.40. Top: Schematic of an ordered surface with an impinging monoenergetic electron beam scattering to produce a LEED diffraction pattern. Bottom: experimental results from a monolayer of adsorbed C_{60} molecules on a Au(111) surface. The observed diffraction pattern shows hexagonal symmetry spots (right). The interpretation of the pattern leads to the superlattice structure shown on the left.

'snug in' with the molecules placed on top of Au atoms. These kinds of details are obtained by careful analysis of the LEED pattern, shown in figure 3.37 (right). In the figure also note the object blocking the central part of the diffraction pattern is the back of the filament source that generates the electron beam. Finally, although electron beams are often quite damaging to molecules, in this case the C_{60} molecules, a form of elemental carbon, are very stable towards electron irradiation. Current LEED instrumentation uses very sensitive detectors for imaging the diffracted electrons so very low electron beam currents can be used to generate clearly observable diffraction patterns and avoid beam damage for fairly stable molecules.

3.4.2.3 Atoms: He atoms at ~thermal speeds—low energy atom diffraction (LEAD)

Atoms can be used to create diffraction patterns but one needs to avoid chemical reactions with the surfaces with the incoming beams so inert gas atoms are typically used with He by far the common choice. According to the de Broglie equation (equation (3.2)) the wavelength of He atoms is at the Å scale at the speeds of gas molecules at ambient temperatures, termed thermal velocities or speeds, as determined from relationship for an ideal gas:

$$KE = ½\, mv^2 = 3/2\, RT, \tag{3.4}$$

where v_2 = mean square speed, R = gas constant (or Boltzmann constant on a per mol basis), m = particle mass and T = temperature. Consistent with the meV scale low energies, the technique is called low energy atom diffraction (LEAD). The experiment is set up by letting a stream of He atoms leak from pinholes, arranged to give a straight beam, into a vacuum chamber where they strike a sample surface placed at chosen angles to the beam. A mechanism at the atom source must be arranged to only select those atoms moving with the same speeds, or equivalently with the same wavelength, to emerge into the vacuum chamber. As fresh gas continues to enter the vacuum chamber pumps are used to maintain low pressure so gas–gas collisions are minimized in the beam–sample–detector region. The diffracted atoms are detected by a mass spectrometer which supplies a signal proportional to the number of He atoms entering the spectrometer inlet area. Since the He source and the mass spectrometer detector are fixed in place rigidly as part of the vacuum chamber, the diffraction pattern can only be generated by sweeping the sample surface orientation through different angles, as illustrated in figure 3.41. This makes the experiment much more arduous than LEED, which detects the diffraction pattern in one shot. This method depends upon the diffraction pattern always being centered around the specular reflected beam, which occurs at $\theta_{in} = \theta_{out} = 45°$, so changing θ_{in} just shifts which part of the diffraction patterns hits the fixed detector.

The figure also shows an experimental result for LEAD from a NaCl crystal oriented to expose the (001) surface to a beam of 10.8 meV He atoms reflecting parallel to the direction of the (110) plane (the [110] direction), as illustrated in the top-right schematic. The experimental result shows the scattering pattern of the He atoms for incidence angles swept from 15°–75° with the presence of three sharp peaks, the 45° peak represents the specular (mirror) reflection while the other peaks represent diffraction from the between the (11) surface planes. When a monolayer of H_2 is adsorbed at low temperature the diffraction peak positions do not change showing the the overlayer has a (1 × 1) structure.

Note that H_2 molecules on a NaCl surface are adsorbed at the surface by extremely weak vdW forces, which points out an advantage of LEAD: the low energy (meV scale) neutral atoms impact surface atoms with very low momentum transfer, leaving adsorbates undamaged. Contrast this to GIXRD with ~1 keV energy photon beams, which are damaging for molecules, and LEED with disruptive ~100 eV beams.

3.4.2.4 Comparisons of proximal probe techniques

A comparison of the advantages and disadvantages of GIXRD, LEED and LEAD are given in table 3.4.

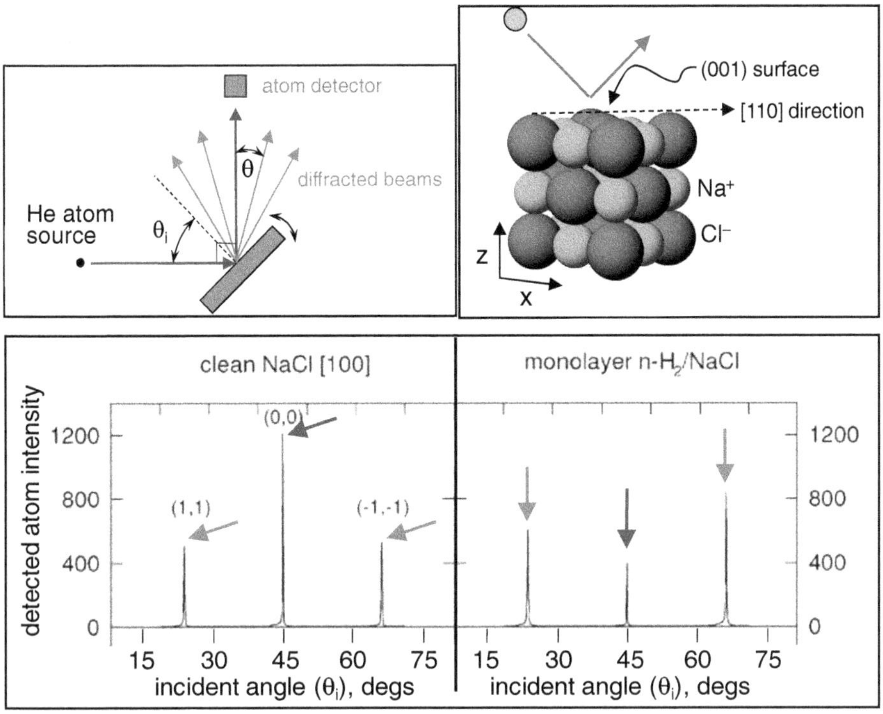

Figure 3.41. Top-left: Schematic of the geometry of an LEAD set-up showing the source, sample and detector. In a measurement the sample angle is swept to allow each diffraction beam to hit the detector in turn. The sample also can be rotated to different angles around the axis perpendicular to the surface to expose different surface planes at different angles to the incoming beam. Top-right: Illustration of a NaCl crystal unit cell with the (001) surface exposed to an atom beam with the angle of incidence parallel to the direction of the [110] plane. Bottom: Experimental results of diffracted 10.8 meV He atom flux versus incidence angle from the (100) surface of NaCl, bare (left) and with a monolayer of adsorbed H_2 molecules (right) at low temperature. The diffraction peaks for both samples are in identical positions consistent with a (1×1) H_2 overlayer.

Table 3.4. Advantages and disadvantages of the three main surface diffraction probes.

Probe	Advantages	Disadvantages
GIXRD	Conducting or insulating samples	Needs powerful beam source at facility (synchrotron)
	3D information possible (e.g. molecular tilts)	Beam damage possible (especially molecules)
	Detailed lattice structure	Needs extremely flat surface
	Liquids possible	
LEAD	Conducting or insulating samples	Difficult experiment (atom sources and mass detectors)
	Low damage for unstable adsorbates	UHV environment
		Needs extremely flat surface
		Long range ordering required for good signal

(Continued)

Table 3.4. (*Continued*)

Probe	Advantages	Disadvantages
LEED	Simple experiment	Only conducting samples
	Reveals detail about multiple top surface layers	UHV environment
		Beam damage possible (especially molecules)
		Patterns can be complicated to interpret

Chapter 3 Problems

1. Copper has a face-centered cubic crystal structure and an interatomic spacing of 0.255 nm. Calculate the surface concentration of atoms (atoms/cm^2) in the (111), (100), and (110) crystal faces.
2. Calculate the energies of electrons, helium atoms, hydrogen molecules, and x-rays which correspond to a wavelength of 0.15 nm.
3. Semiconductor surfaces exhibit several ordered surface structures that are stable in well-defined temperature ranges. It has been reported [6] that the adsorption of oxygen converts the Ge(111)–(2 × 1) surface structure to the bulk-like Ge(111)–(1 × 1) structure. There are reports that some of the surface structures appear only in the presence of trace impurities [7]. Review and discuss those papers that put forward the various suggestions concerning the formation of semiconductor surface structures: [8], [9], etc).

Further reading

General sources for surface structure and reconstruction

1. Somorjai G A and Li Y 2010 *Introduction to Surface Chemistry and Catalysis* 2nd edn (John Wiley). This text provides general advanced undergraduate and beginning graduate level material and chapter 2 is focused primarily on surface structure, including the cases of adsorbates.
2. Oura K, Katayama M, Zotov A V, Lifshits V G and Saranin A A 2003 *Surface Science: An Introduction* 1st edn (Springer). A general advanced undergraduate/beginning graduate level text. Throughout the text the focus is primarily on surface structure.
3. Wise H and Oudar J 2001 *Material Concepts in Surface Reactivity and Catalysis* (Dover). Chapters 1 and 2 cover fundamental concepts and principles in surface structure.

Detailed texts on characterization methods

There are a number of texts that delve into considerable detail on the characterization methods and applications relevant to the current chapter material. These texts generally include characterization of both bare surfaces and surfaces with adsorbates and include chemical characterization of the surfaces as well, a subject that falls outside the scope of the current chapter on physical structure. Here are a few:

1. Vickerman J C and Gilmore I S 2009 *Surface Analysis: The Principal Techniques* 2nd edn (Wiley). A reference text covering the principal surface analysis techniques ranging across ion scattering, electron and x-ray probes and scanning probes.
2. Alford T L, Feldman L C and Mayer J M 2007 *Fundamentals of Nanoscale Film Analysis* 1st edn (Springer). A graduate level text on common surface analysis techniques
3. Meyer E, Hug H J and Bennewitz R 2004 *Scanning Probe Microscopy, The Lab on a Tip* 1st edn (Springer). A thorough review of scanning probe techniques and applications.
4. Woodruff D P and Delchar T A 1994 *Modern Techniques of Surface Science* 2nd edn (Cambridge University Press). A graduate level reference text covering common surface analysis techniques including a variety of spectroscopies, electrical measurements and scattering.

References

[1] Somorjai G A and Rioux R M 2005 High technology catalysts towards 100% selectivity: fabrication, characterization and reaction studies *Catal. Today* **100** 201–15

[2] Stocker W, Beckmann J, Stafler R and Rabe J P 1996 Surface reconstruction of the lamellar morphology in a symmetric poly(styrene-block-butadiene-block-methyl methacrylate) triblock copolymer: a tapping mode scanning force microscope study *Macromolecules* **29** 7502–7

[3] McGuiness C L, Diehl G A, Blasini D, Smilgies D-M, Zhu M, Samarth N, Weidner T, Ballav N, Zharnikov M and Allara D L 2010 Molecular self-assembly at bare semiconductor surfaces: cooperative substrate–molecule effects in octadecanethiolate monolayer assemblies on GaAs(111), (110), and (100) *ACS Nano* **4** 3447–65

[4] Stranick S J, Kamna M M and Weiss P S 1995 Interactions and dynamics of benzene on Cu {111} at low temperature *Surf. Sci.* **338** 41–59

[5] Weiss P S and Eigler D M 1993 Site dependence of the apparent shape of a molecule in scanning tunneling microscope images: benzene on Pt{111} *Phys. Rev. Lett.* **71** 3139

[6] Henzler M 1969 Correlation between surface structure and surface states at the clean germanium (111) surface *J. Appl. Phys.* **40** 3758

[7] Van Hardeveld R and Hartog F 1969 The statistics of surface atoms and surface sites on metal crystals *Surf. Sci.* **15** 189

[8] Heron D L and Haneman D 1970 Two-centre theory of large surface unit cells on semiconductors *Surf. Sci.* **21** 12

[9] Palmberg P W 1968 Structure transformations on cleaved and annealed Ge(111) surfaces *Surf. Sci.* **11** 153

IOP Publishing

Surface and Interface Science

David L Allara and Robert L Opila

Chapter 4

Surface and interface thermodynamics

Interfacial phenomena play a crucial role in many daily and industrial processes. These processes are related to the adsorption on solid and fluid interfaces, wettability, nucleation, adhesion, and other phenomena that strongly impact the quality and cost of products and processes. Understanding interfacial phenomena at equilibrium requires the description of surface thermodynamics.

4.1 Thermodynamic considerations for surfaces

The application of quantitative thermodynamics to any system at equilibrium requires that we define the system in terms of its associated thermodynamic variables. For typical systems the core variables include pressure (P), temperature (T) and concentrations (c_j) or mole fractions (x_j) of the components, heat capacity (C_P or C_V), the containing volume (V) and the internal energy and entropy (U and S). Other variables may be defined on the basis of the core variables, e.g. Gibbs free energy (G). The variables P, T, c_j, x_j, and C_P, C_V are intensive quantities, meaning they do not change with system size and represent the intensity of some condition imposed on the system. In contrast, V, U, S and G are extensive variables since their values scale (or extend) with system size, i.e. if we double the system size without disturbing its inner relationships all the extensive variables double their value. The value of thermodynamics is that the first and second laws impose two firm constraints on the relationships between the core variables. This in turn, along with obvious constraints such as conservation of mass, provides us with two knowns to help determine what happens when a system is perturbed to change its condition by a process that keeps the system at equilibrium.

What do we need to add to our standard description of a system in thermodynamic terms to accommodate the presence of a surface? The simple answer is just to include the area (A) of the surface that defines the termination of an object, similar

doi:10.1088/978-0-7503-4916-1ch4

to including the volume containing the system, and to include the variables associated with the adjoining phase. If the adjoining phase is a vacuum or dilute vapor this simplifies things a bit. Let's first look at volume change. Changing V is a dimensional change in which all parts of the system work together to shrink or expand, which usually means work of compression or expansion, as well as changes in various extensive thermodynamic variables. Similarly for surfaces, when all parts of the system work together to change A, there will be work involved and changes in extensive variables. So our first task will be to examine the work of surface expansion (or contraction).

4.2 A simple, but powerful coordination picture for understanding and predicting the unfavorable energetics of surfaces—the missing molecule (MM) model

Before we begin the thermodynamic analysis of surface work, it is very useful to discuss a simple molecular (or atomic) scale model for understanding what is involved in the expansion (or contraction) process for a simple condensed phase, typically pictured as a liquid since we expect liquids to rapidly adjust their internal structure to mechanical perturbations of surface area. For example, consider a nearly spherical drop of a liquid (mercury is a good choice) which readily deforms under a simple force, such as pressing, to form more surface at the expense of shrinking volume, then spontaneously springs back when the force is released, as depicted in figure 4.1 (left). Why and how does this happen in terms of the internal structure of the medium?

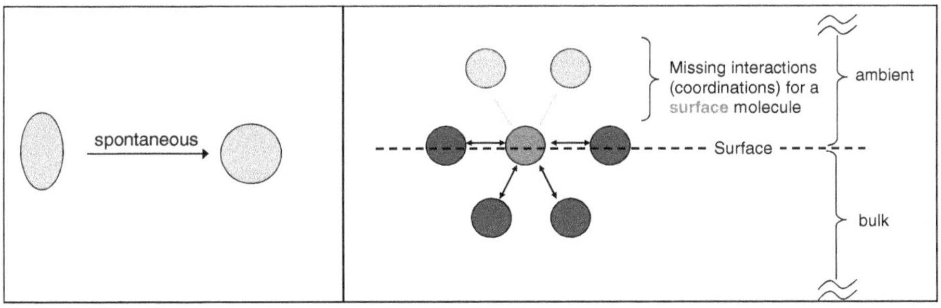

Figure 4.1. Illustrations of the ideal processes involved in expansion and contraction of the surface area of a condensed phase. Left: A drop of liquid which as been squeezed into some non-spherical shape will tend to spring back spontaneously to a spherical shape when the squeezing force is removed. Right: The missing atom model for understanding surface forces and the work of expansion and contraction of surface area. The figure shows the surface of a condensed phase with an ambient phase, treated as a vacuum, above it. At the surface one atom (red) is selected and some coordinating NN atoms depicted (dark blue). In the ambient phase the atoms that would have been present if the red atom were place in the bulk are shown as gray circles. It is clear that moving the red atom into the bulk to reduce the surface area is accompanied by a gain in energy from the extra coordination not available at the surfaces, thus making the reduction in surface area a spontaneous process.

As seen on the right-hand side of figure 4.1, we can reduce the surface area in a condensed phase (let's imagine a liquid to make it easier) by pushing the red molecule (or atom) back into the liquid where it is reunited with all the possible coordination neighbors that were missing in the surface (gray molecules). This is a favorable process energetically, all other factors being equal, since CN increases, and thus attraction or bonding energy increases. So this missing molecule model (let's call it m^3) immediately gives us a way to estimate the energy or work change in expanding or shrinking surface area.

As we go along, we will point out how the surface coordination picture allows us to make a number of useful predictions of surface and interface properties, including, in a later section, chemical reactivity. We can extend the simple m^3 picture to handle things more exactly and here are two important points about what you would need to do. But do not let them deter you from getting the simple picture to work to your advantage in understanding general behavior of surfaces.

1. For ordered solids exact pictures of coordination numbers and energies are possible but for liquids and amorphous solids average structures and average interaction energies must be considered.
2. The surface coordination picture and associated concepts and calculations will apply to any aspect of the surface energy that stems from the interaction energies between the atoms (or molecules). It is essentially a model that considers the positions of the nuclei and their interactions through the outer shell electrons. A separate model of the surface must be used to consider the effects on the electronic properties such as surface dipole moments, energy for ionization of an electron from the surface and the electronic conduction properties at the surface. (We will consider this aspect of surface in chapter 6 on the surface electrical properties and develop another useful picture based on delocalized electrons in the solid.)

4.3 Single component systems

To keep things simple, we start our study of surface thermodynamics by considering objects composed of a single pure substance and look at the surface properties of these single component systems. Later we will consider multicomponent systems such as solutions with surfactants, which is the basis for treating many biological systems, and metal alloys. We break up the single component section into two main parts. In sections 4.3.1 and 4.3.2 we treat the surface as an interface between a pure liquid (or solid) in equilibrium with its vapor and look at the interface from a mechanical point of view in which expansion or contraction of a surface at rest requires force or pressure and we investigate pressure in each phase and in the surface layer. Thus, the only thermodynamic aspect is simple PV work. In section 4.3.3 we add another thermodynamic variable, the chemical potential and look at how changes in bubble and drop shapes can affect properties such as vapor pressure of a pure substance. Just before section 4.3 we pause to briefly review the relevant fundamentals of equilibrium thermodynamics for PV work and surface expansion work, so we have a suitable background to combine these two effects.

4.3.1 Surface tension

4.3.1.1 Mechanical definition in terms of surface forces

Historically the subject of liquid surfaces viewed the energy or force required to increase the surface area in terms of mechanical force, viewing the surface as a skin under tension that resisted stretching. Thus, the term surface tension arises, to which we assign the symbol γ with units of force/area, newtons per meter or $N \cdot m^{-1}$ in SI units. Liquids with high surface tensions such as mercury resist being pushed around (e.g. squeezed through small holes) while those with low surface tensions are very cooperative in such manipulations (such as low molecular weight oils). Also, historically, the units of γ were specified in dynes per centimeter (dyn cm^{-1}), which are equal to millinewtons m^{-1}, or mN m^{-1}.

Here is a simple hypothetical experiment to show what is going on. The apparatus is shown in figure 4.2 and is sort of a tricky bathtub with a back side (red) that can slide left or right with no friction to increase or decrease the surface area of the liquid exposed at the top (blue). Note this experiment in practice is devilishly hard to do since we need frictionless walls, etc, so it is strictly a *gedanken* experiment (meaning one just for thinking, not doing, or equivalently means you get a zero on a test question that asks how to make a real measurement of γ). We will soon see there are simple ways to actually measure γ (which result in better test question scores).

When the wall, initially in complete force balance so the liquid does not push it away by hydrostatic pressure, has some infinitesimal force **F** applied perpendicular to the wall, the wall moves a distance **dx** in a direction parallel to the applied force to create a change in surface area dA. Since the force and distance moved are both in the same direction we can drop the vector notation. If the liquid has a large surface tension, then more force is needed to 'stretch' the surface than for a low-tension liquid such as a hydrocarbon. Now we figure out the force per area change and make sure the process is done in infinitely small steps that are easily reversible, thus

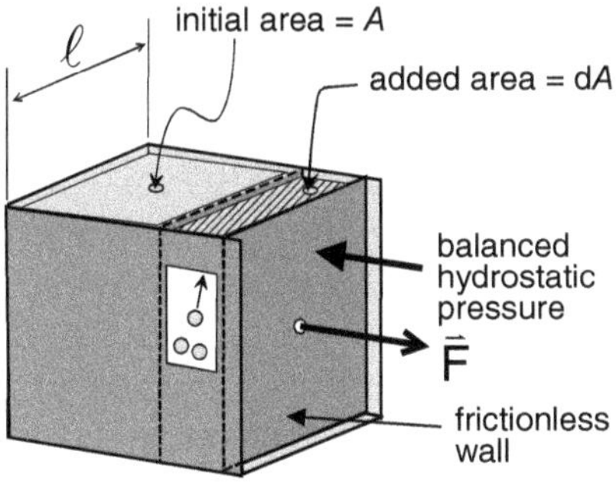

Figure 4.2. A *gedanken* bathtub experiment to show what is involved in surface tension in terms of applying a force to create or remove surface area of a liquid. The tub, filled with the liquid of choice (blue color) has a movable side wall (red) starts out with a static force that keeps it in place but can be pushed back and forth (with no friction on the sides) by an extra applied force (noted as the vector **F**). The tub width is ℓ, the total surface area of the liquid on top of the tub at any time is A with incremental change dA (indicated by the shaded area on the right) when a force is applied.

keeping the system at equilibrium during the process (another reason this is a *gedanken* experiment). Here are the equations:

$$\begin{aligned} \mathrm{d}W &= \text{reversible work of creating new surface} \\ &= \mathbf{F}\cdot \mathrm{d}\mathbf{x} = F\mathrm{d}x = \left(\frac{\ell}{\ell}\right)F\mathrm{d}x = \left(\frac{F}{\ell}\right)\mathrm{d}A \equiv \gamma \mathrm{d}A. \end{aligned}$$

From this mechanical experiment we have a new quantity, the modulus for the force per unit length of surface required to pull open new surface area, which we call the surface tension defined as

$$\boxed{\gamma \equiv \left(\frac{F}{\ell}\right)}. \tag{4.1}$$

In equation (4.1) the quantity ℓ defines the length of the *contact line* across the edge of the liquid which is put under tension by force F. As the contact line increases in length for a given substance, the force required to create surface area scales directly. This makes sense because there is proportionally more edge of the liquid to pull on (essentially a wider bathtub). The idea of a contact line is very important in treating phenomena such as liquid rise in a capillary and the shape of drops on a surface. More about that later.

What happened during the process that was responsible for the liquid to resist the applied force that was trying to move the wall? Look in the little window on the side of the tub and you can see molecules that move to the surface as required to fill the opened surface space to create more area. The cost of stripping off part of their solvation shells to get to the surface is given by the reversible work. Since the work is done reversibly, we can equate this to a change in free energy. More about that soon.

Finally, let's look at surface tension in terms of the work expended to create an infinitesimal amount of area. We keep all the quantities in SI units which gives

$$\gamma = \frac{F}{\ell}\left(\frac{\text{mN}}{\text{m}^2}\right) = \frac{F\cdot \ell}{\ell\cdot \ell} = \frac{\mathrm{d}W}{\mathrm{d}A}\left(\frac{\text{millijoules}}{\text{m}^2} \text{ or } \frac{\text{mJ}}{\text{m}^2}\right).$$

Thus, mN m^{-1} = mJ m^{-2}, which in centimeter per gram per second cgs units (historical) is dyn cm^{-1} = (dyn.cm) cm^{-2} = ergs cm^{-2}. Note that work (energy) per area and force per length are interchangeable dimensions. These conversions are useful in looking at data from various sources.

4.3.1.2 The surface membrane concept

So far we are thinking about surface tension in terms of stretching a surface membrane. But let's be careful not to treat this like blowing up a balloon and stretching an elastomer membrane! When a force is applied to increase surface area of a condensed phase under reversible conditions, no matter how large the area becomes, the force per unit area is constant; we can bring molecules to the surface forever, all other factors constant, with no change in the work cost. In contrast, stretching membranes is very nonlinear and eventually the membrane irreversibly

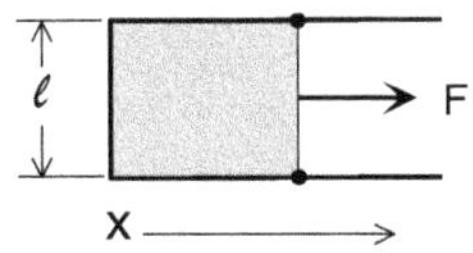

Figure 4.3. Schematic of a soap film stretched in a wire frame with a movable front wire of length ℓ connected by eyelets to the main frame so it can slide. A force F pulls on the front wire in the x-direction to stretch the soap film.

breaks. You can only stretch so far. The surface membrane has a constant work needed to pump molecules to or from the surface according to the value of γ. Soon we will look at this in terms of energy per area.

First let's look at a simple, common example of the work to stretch a soap film. You could do this experiment yourself and make a measurement without any fancy equipment. We refer to figure 4.3 which shows a simple wire frame with a soap film trapped by the perimeter and a movable wire on the right of length ℓ (a contact line put into action) to which we can apply a known force. What force is required to stretch the film? This is a complicated liquid since it is not single component but a dilute aqueous soap solution. But we will consider it as a simple pure substance for now and figure out details later. The liquid has a surface tension γ. As we stretch equation (4.1) holds. Since there are two surfaces involved, we have

$$F = 2\gamma\ell, \tag{4.2}$$

where γ is the value for the soap solution, not pure water (there is a considerable difference, as you might guess from the considerable changes that soap added to water can bring about). Now go ahead and make a measurement. Do you think you could get your hands on a light enough weight to make an accurate measurement? By childhood experience with soap bubbles and practice in using a good balance you might guess this is not a problem. So, this could be your first measurement of surface tension?

Here is a thought problem. What happens at the surface area is increased? Something moves to the surface for sure and coordination is lost, but what moves? The surface tension of pure water is ~73 mN m^{-1} and a typical soap solution is ~25 mN m^{-1}. So which molecules are moving to the surface, water or soap? This brings up the question of what a soap molecule is. Here is a common one: $H_3C(CH_2)_{16}CO_2Na$ (sodium stearate). Notice that it has a hydrocarbon end, which 'hates' water, and a polar end (a salt in this case) which 'hates' hydrocarbon and 'loves' water. The molecule is built to have part of it in water and part in an oil type of phase or at least out of water. Is it easy to imagine that this molecule prefers to go to the surface of water over a water molecule?

Questions for thought:

1. If the soap (or better surfactant) molecule above did move to the air/water interface which end would point outwards into ambient air?
2. Does this solve the problem of why the surface tension of a soap solution surface is lower than that pure water?

3. Looking back in chapter 2 on the types of adsorbate interactions and surface structures and considering the soap at the surface as an adsorbate, what might a progression of structures from low soap concentration in the solution to high concentration look like? Could we approach what looks like a self-assembled monolayer this way?

Since we are getting a feeling for what surface tension means physically, let's look at values for a variety of substances, as shown in table 4.1. If the substances are solids just imagine that if one were patient enough in expanding the surface area bit by bit as the atoms adjust positions (maybe even over years or centuries) you would get these values.

The first task to do in looking over a new table of data is to relate the trends to the basic principles with which you are familiar. See if the trends make sense or if you could have guessed them from what you already know. With this in mind is it reasonable that tungsten (W) has a surface tension greater than nearly two orders of magnitude of CCl_4? This probably seems ok since CCl_4 is a pretty 'squishy' liquid (please forgive the touchy-feely approach) compared to the metal which drives machinists crazy if they have to cut into it, breaking their tool bits. In fact, in looking through the list it should seem sensible in very general terms as viewed from the perspective of how hard it is for the material to release some coordination of the bulk molecules or atoms to put them in the surface. In machining (typically with a

Table 4.1. Surface tension, γ, of selected solids and liquids.

Material	State	Reference	γ (J m^{-2})	T (°C)
W	Solid	[1]	2.9	1737
Au	Solid	[1]	2.100	1027
Ag	Solid	[1]	1.140	907
Ag	Liquid	[2]	0.879	1100
Fe	Solid	[1]	2.150	1400
Fe	Liquid	[2]	1.880	1535
Pt	Solid	[1]	2.340	1311
Cu	Solid	[1]	1.670	1047
Cu	Liquid	[1]	1.300	1535
NaCl	Solid	[3]	0.227	25
MgO	Solid	[3]	1.200	25
KCl	Solid	[3]	0.110	25
N_2	Liquid	[2]	0.009 71	−195
Water	Liquid	[2]	0.072 75	20
Benzene	Liquid	[2]	0.028 88	20
n-octane	Liquid	[2]	0.021 80	20
Benzaldehyde	Liquid	[2]	0.025 20	20
Ethanol	Liquid	[2]	0.022 75	20
CCl_4	Liquid	[2]	0.026 95	20

hardened steel tool bit) it is necessary to expose fresh surface. So, the principle of surface tension is involved, although certainly not in terms of a reversible process as smoke is emitted from the lathe, milling machine or drill press! The trends look OK in general based on our chemical intuition, but we can dramatically sharpen them up by putting everything on a per mole basis, which is really the way we have learned to treat chemical systems. We will do this shortly.

4.3.1.3 Work of cohesion—making new surfaces by fracture or disjoining

The process of fracturing a material into two pieces, as shown in figure 4.4 for the tensile rupture of a solid rod, creates two new surfaces. Since the process involves the sudden removal of the neighboring molecules or atoms to expose two new surfaces it follows that the ideal work δW per unit area of the tensile rupture process, when carried out reversibly, is given by

$$\frac{\delta W}{A} = 2\gamma. \tag{4.3}$$

The ideal case would involve reversible rupture under a vacuum where adsorbates would absent in order to avoid perturbations of the surfaces. In practice when experiments are done to apply controlled force to the point of tensile failure for pure materials, particularly engineering materials such as metals, ceramics or polymers, the work required is always far less than the theoretical value from equation (4.3). There are many reasons for this including the fact that the actual physical mechanism of tensile fracture (and other kinds of fracture as well) are very complex and there is much discussion among experts about the idea of what a reversible fracture might entail. These aspects are involved in the questions below.

This aspect of surfaces seems very engineering oriented in terms of materials performance and deviates from the previous emphasis on liquids. It gives us a chance to put to use some of the material on surface structure that we have learned in chapter 3.

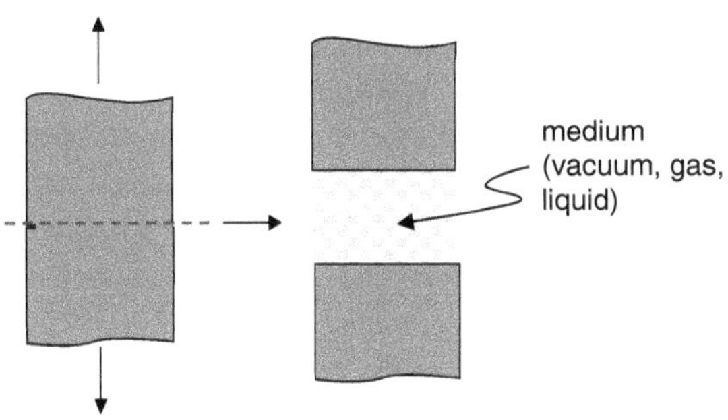

Figure 4.4. Schematic illustration of a solid rod or bar of a pure material which is placed under tensile stress and fractures cleanly to produce two new surfaces. To be general, the process can take place under vacuum or in a medium of gas or liquid. The ideal process takes place under vacuum where the new surfaces are pristine.

Here are some questions for thought:

1. Assuming the forces pulling on the two ends of a rod are very uniform, how and where in the structures might the fracture initiate for a typical crystalline metal, e.g. iron or copper?
2. Would single crystal rods with the crystal planes oriented in different directions to the applied forces have different tensile strengths? Why? How might you estimate the differences?
3. Would surface reconstruction affect the force required for ideal behavior? Would this increase or decrease the force required for rupture? Why?
4. Why would the rupture force required vary with the presence of gases?
5. Based on 3 and 4, what would the mechanism of a reversible rupture require? To help think about this imagine what the process would be for a brittle crystalline material (say a single crystal rod of silicon) which is ruptured by bending.

4.3.1.4 Pressures in the surface region

In-plane surface pressure: From the previous sections we learned there are forces unique to the surface region. If there are forces pushing or pulling on something, then these can be expressed as equivalent terms as pressures. To do this we need to think more about the surface region as a real layer of material. Up to this point we have considered the surface as an infinitely sharp boundary line, though there have been hints that it might very well be 1, 2 or even 3 molecules (atoms) thick and thus be loosely considered as a substance itself with slightly different properties than the adjacent bulk. With this in mind let's do a simple calculation on the forces on the sides of this thin layer. Obviously, the surface molecules have substantial forces trying to drive them into the bulk so they can gain coordination energy. But they are geometrically trapped since the surface must exist and cannot simply go away. What is the lateral pressure on these surface layer molecules that is trying to push them back into the bulk? To solve this problem we consider a small slab of surface of length ℓ, as shown in figure 4.5, guess the surface layer thickness δz is the diameter of one molecule, and look at the pressure perpendicular to the ℓ, δz wall that is necessary to just balance this internal pressure and keep the surface at rest. The balance can be dealt with in terms of the force F that is required to create new surface per area of side wall exactly balancing the lateral, in-plane surface pressure P^{s}_{xy}, which is a negative or internal pressure since the associated force goes inwards to decrease the volume of the slab. Since P^{s}_{xy} must be equal on all four surface walls we just need to look at one of them and figure out the value. Here are the relationships based on the definition of the surface tension as force/area:

$$P^{\mathrm{s}}_{xy} = \frac{F_{xy}}{A_{\delta z,\,\ell}} = \frac{\gamma\ell}{A_{\delta z,\,\ell}} = \frac{\gamma\ell}{\delta z \cdot \ell} = \frac{\gamma}{\delta z}, \tag{4.4}$$

where $A_{\delta z,\ell}$ = the area of the side wall of the surface slab of thickness δz and width ℓ along the force contact line. For convenience in doing related calculations some information on pressure units is given in table 4.2.

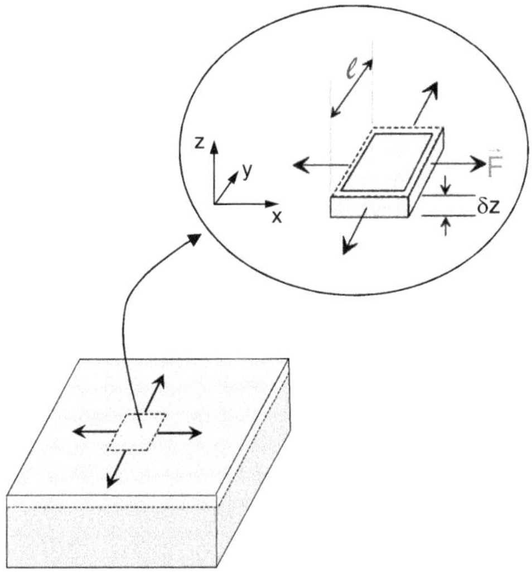

Figure 4.5. A schematic of a slab of a surface layer within the surface of a liquid. The magnified region on the upper right shows the surface slab in detail with a surface layer thickness is δz. When the liquid is at rest there is a balance of forces on all sides of the slab. The force F outward along the x-direction perpendicular to the side of length ℓ (running parallel to the y-direction) balances the internal forces that pull the area smaller to remove surface. For a slab thickness δz the pressures on the walls (red) can be calculated from the forces. When the surface expands F exceeds the internal inward force and new area is created, as suggested by the dashed lines.

Table 4.2. Notes on pressure units.

SI units:
pascal (Pa) = $\text{kg} \cdot \text{m}^{-1} \cdot \text{s}^{-2}$ (recall $P = F/\text{unit area} = \text{N} \cdot \text{m}^{-2}$)
from this it follows:
$P = \text{J} \cdot \text{m}^{-1} = (\text{kg} \cdot \text{m}^2 \cdot \text{s}^{-2}/\text{m}^{-3}) = \text{J} \cdot \text{m}^{-3}$ = energy/unit volume

---

Interconversions:
Bars and atmospheres (common units):
1 Pa = 10^{-5} bar $\sim 1.06 \times 10^{-5}$ atm;
Typical room pressure $\sim$ 1 atm (14.6 psi)
Torr (after Torrcelli; used for variations of room pressure and in vacuum work)
1 atm = 760 Torr = 760 mm Hg (in a column on Earth)
1 bar = 750 Torr

Putting some values in equation (4.4), we set $\gamma \sim 30$ mN m^{-1} for a typical organic liquid (e.g. benzene or toluene) and $\delta z \sim 0.5$ nm (5 Å, the diameter for a typical small organic molecule) which gives $P^{s}_{xy} \sim 60 \times 10^6$ N m^{-2}. Converting with 1 pascal = 1 N m^{-2} and 1 bar = 1×10^5 Pa we obtain P^{s}_{xy}~600 bar or ~600 atm. This is a huge pressure! For a refractory metal with strong lattice energies, $P^{s}_{xy} \sim 40$ kbar, or ~40 000 atm.

These pressures represent the forces that are trying to suck the molecules back into the bulk to increase their coordination at the expense of decreasing the surface area. Think of this sort of like a frustrated bathtub drain where the gravity is unsuccessfully trying to pull the water laterally over into the drain to decrease the gravitational potential energy of the disappearing water molecules. The only way that molecules can be taken back into the bulk in a liquid surface, however, is for the shape of the object to change such that the surface occupies less area (an S/V change). If the object is a sphere with no other external forces, there is no hope but if the object is misshapen and can become spherical P_{xy}^{s} is the pressure that will push to make this spontaneous.

Let's finish this topic by writing what is happening as a simple chemical equation:

$$\mathrm{M_B^{CN_B}} \rightarrow \mathrm{M_S^{CN_S}} + (\mathrm{CN_B} - \mathrm{CN_S}), \tag{4.5}$$

where M = molecule, $\mathrm{M_B^{CN_B}}$ = a molecule in the bulk with a surrounding coordination shell of $\mathrm{CN_B}$ bulk molecules, $\mathrm{M_S^{CN_S}}$ = a molecule in the surface region with a surrounding coordination shell of $\mathrm{CN_S}$ surface region molecules. In this expression with the direction as written, fully coordinated molecules are transferred to the surface region with the release of ($\mathrm{CN_B}$–$\mathrm{CN_S}$) molecules which are left back in the bulk. In the process surface area is created according to the size of the molecules as they push into the surface. The energetics of the process depend on the bonding energy per coordination. We will use equation (4.6) later to estimate surface energies and γ values from energies of vaporization according to the chemical equation

$$\mathrm{M(B)} \rightarrow \mathrm{M(vapor)}. \tag{4.6}$$

In this process all coordination of the molecules is stripped and isolated vapor molecules formed. Next, we will discuss the hydrostatic pressure above and below the surface layer and look at surface layer density. There are some interesting things going on to learn. But first a few questions to ponder.

Some questions to think about at this point:

1. How might the surface pressure concept explain how a water skate bug stays on top of a pond?
2. What is the physical mechanism that underlies P_{xy}^{s} and how does it relate to the forces that hold condensed substances together?
3. How thick exactly is the surface region (δz)? Have we learned enough at this point to make good guesses for different substances? What kinds of measurements can be made?
4. Is there a hydrostatic pressure difference across the surface region in the z-direction, or equivalently is the pressure in the z-direction within the surface region different than in the bulk or exterior phases? Why or why not?

Hydrostatic pressures across the interface: From the above discussion we decided that there is a huge negative pressure P_{xy}^{s} in the xy surface plane which is ready to pull molecules inward to the bulk whenever the surface area can be decreased. Now we look in the z-direction and investigate the hydrostatic pressure. It is useful to make the simple assumption that the liquid depths of interest are small, and gravity has no effect on pressure. In figure 4.6, a container of liquid is shown with the surface

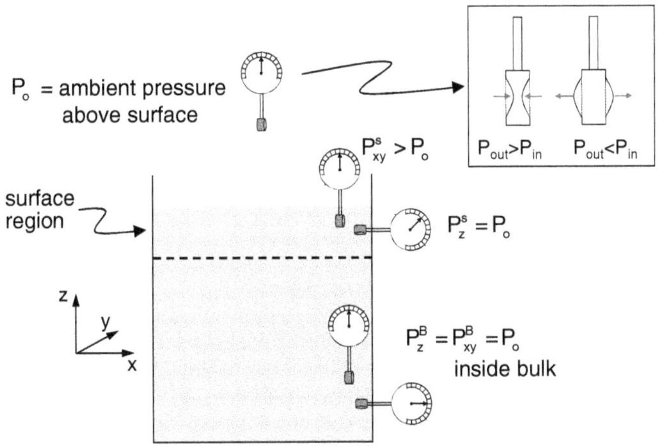

Figure 4.6. Schematic of a hypothetical experiment in which a nanoscale pressure gauge is inserted into regions within and above a container of liquid to test the pressure in different directions. The gauge senses pressure along a specific axis direction based on a diaphragm set-up (shown in the upper right inset). P_o, P^s_{xy}, P^s_z, P^B_{xy}, P^B_z are the ambient, surface xy, surface z, bulk xy and bulk z pressures, respectively.

region (perhaps a layer of molecules thick or so) above the dashed line and the coordinate directions shown as used in the previous figure. Imagine we are clever and make a nano-pressure gauge which measures pressures along one axis by use of some sort of nano-diaphragm set-up with fancy gadgets inside and which is only a molecule long, as shown in the upper right-hand insert of the figure. We set the pressure inside of the diaphragm volume P_{in}, at pressure P_o, the ambient pressure above the liquid so that when the gauge is inserted above the liquid it reads zero. Next the gauge is inserted in the bulk liquid, no matter which way the diaphragm axis is oriented, the gauge reads zero since $P^B_z = P^B_{xy} = P_o$, where P^B_z and P^B_{xy} = the bulk pressures in the z- and xy-directions. Thus, as you would expect the hydrostatic pressure in the bulk equals the ambient pressure (recall we are neglecting gravity effects).

Now go to the surface. In the z-direction we should obtain $P^s_z = P_o$. Why? What happens if the pressure in the z-direction is greater on one side of the interface layer than the other? Since the surface is nice and quiet and lying flat we can agree on the above equality. How about pressures in the xy-plane? We obtain $P^s_{xy} \ll P_o$. Why? As the liquid contacts the diaphragm walls they can respond by moving outwards to allow the surface area of the liquid to reduce, as suggested in the upper right insert of the figure, and this process will only stop when the pressure in the gauge equals the expansion pressure P^s_{xy}. So now we can make a plot of pressure versus direction, shown in figure 4.7.

The P versus z plot through the interface region is purely hypothetical plot of course, since in the real world there is no way to actually measure pressures in the surface region. Furthermore, the surface region itself is a gradual change of molecular characteristics and environment across a width of a few molecules so the surface region boundaries are very fuzzy. Most of what we know about the

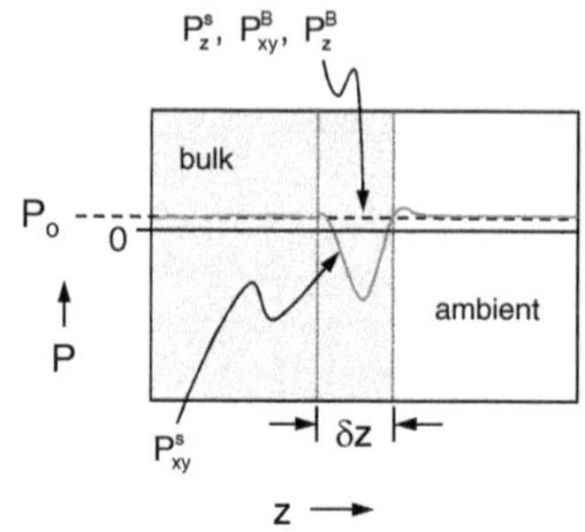

Figure 4.7. Plots illustrating the general direction-dependent pressure profiles as a function of distance (z) in moving from the bulk region of a liquid through the surface region and into the adjacent ambient atmosphere. The pressure is measured by the hypothetical diaphragm gauge in figure 4.6. In the bulk and ambient regions, the pressure is the hydrostatic pressure P_o which is equal on all sides in the measurement region. In the surface region the z-direction pressure is P_o but in the xy-plane the pressure is highly negative because of the forces which pull the diaphragm outward to reduce the surface area of the liquid.

internal pressures in the surface region come from theory approaches such as molecular dynamics simulations on simple molecules, e.g. argon. But here is one relationship that has to be correct:

$$\gamma = \int_{-z}^{+z} (P_o - P_{xy})\mathrm{d}z = \int_{-z}^{+z} (P_o - P^s_{xy})\mathrm{d}z.$$

All the effect of surface tension depends on the internal pressure created in the interface region by the forces exerted on the molecules to drive them back into the bulk. Only in the interface region is $P_{xy} \neq P_o$ and the entire effect of the surface imbalance is expressed as the integrated change in xy pressure times distance over the thickness of the interface region, which up to this point we have defined as δz. But now see that it is not a sharply defined region, but rather one that can gradually change over some small distance range. Finally, while we are looking at the direction dependence of pressures, you should note that these directional force effects in a medium are handled mathematically by the use of pressure tensors, something familiar to mechanical engineers and those that work with the response (e.g. deformation) of materials stresses in different directions.

4.3.2 Work is required to change the curvature of surfaces

4.3.2.1 Simple spherical bubbles

Any change in the shape of an object (all other factors equal) requires a change in the S/V ratio. In geometric terms this requires some change in surface curvature, most easily seen by starting from a stable spherical drop of liquid (put it in outer space if you wish to avoid gravity distortions) and misshaping it. The creation of more sharply curved regions (changing the local radii of curvature) on the surface brings an energy penalty due to the creation of more surface area. A simple illustration of this is given for the simple system of a spherical bubble sustained inside of a liquid by a pressure imbalance, as illustrated in figure 4.8. The hydrostatic pressure in the

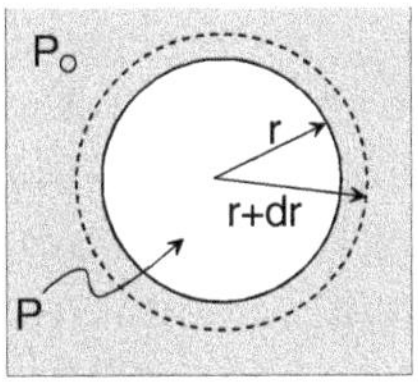

Figure 4.8. A bubble of radius r in a liquid is sustained by a pressure P relative to the outside pressure of P_o. The bubble undergoes an infinitesimal expansion to radius $r + \mathrm{d}r$ at constant pressure P.

medium (in equilibrium with the outside atmosphere) is P_o and the pressure inside that sustains the bubble is P. The hydrostatic pressure across the interface is no longer equal since the surface is wrapped around on itself and exerts a compressive pressure that to tries to force surface layer molecules into the bulk. But since the drop is spherical no more change in surface area is possible and the system is at balance with extra pressure ΔP inside the bubble to keep the interior liquid surface from collapsing; thus $P > P_o$. This excess pressure in the bubble that arises from surface tension is also called the Laplace pressure.

How exactly is ΔP related to the surface tension? To make things simple we will assume the liquid is non-volatile (for example a liquid polymer) and the gas inside the bubble is not soluble in the liquid so we do not have to concern ourselves with troublesome vapor pressures and disappearing gas. The simplest approach is to directly calculate the compressive force by running a circle of radius r around the bubble which gives $F = \gamma\ell = \gamma(2\pi r)$. This compressive ring is like tightening a belt. It also can be referred to a line tension. Since this force is equal around all circumference lines everywhere on the sphere with surface area $4\pi r^2$ we have

$$\Delta P = \frac{F}{A} = \frac{2\gamma}{r}. \tag{4.7}$$

We could also do this a fancier way that looks more like thermodynamics by making a small perturbation dr of the radius and impose the conditions needed to keep the system in balance or at equilibrium. An expansion by dr is favorable in terms of PV work since $P > P_o$ but unfavorable in terms of creating more area. So dW/dr for each process must be equal and opposite as the drop is perturbed keeping equilibrium, as given in the equations below:

$$\mathrm{d}W_\gamma = \gamma \mathrm{d}A = \gamma \mathrm{d}(4\pi r^2) = 8\pi\gamma r \mathrm{d}r \;\text{ or }\; \frac{\mathrm{d}W_\gamma}{\mathrm{d}r} = 8\pi r\gamma$$

and

$$\mathrm{d}W_{\Delta PV} = -P\mathrm{d}V - P_o\mathrm{d}V = -(\Delta P)\mathrm{d}V = -(\Delta P)\mathrm{d}[4/3\pi r^3] = -(\Delta P)\mathrm{d}(4\pi r^2 \mathrm{d}r),$$

or

$$\frac{\mathrm{d}W_{\Delta PV}}{\mathrm{d}r} = -4\pi r^2(\Delta P).$$

This gives $8\pi r\gamma + [-4\pi r^2(\Delta P)] = 0$ or $\Delta P = \frac{2\gamma}{r}$ just as we obtained above.

We will do this derivation an even fancier way with differential geometry in sections 4.3.2–4.4 where we derive the Young–Laplace equation with the reward of being able to handle non-spherical surface shapes.

Equation (4.7) gives a very useful and fundamental behavior of bubbles and provides a basis to understand many common phenomena including *bumping* of liquids heated above their boiling point temperatures (superheated liquids) and foaming of liquids when soaps (surfactants) are added. To approach these problems properly we need to introduce the effects of chemical potential and mass transport and the temperature dependence of the surface tension, which we shall do in a few sections.

4.3.2.2 Measurement of surface tension from maximum bubble pressure

With equation (4.7) in hand we apply the behavior of a bubble with changes in applied pressure to demonstrate a simple, practical way to measure surface tension, as illustrated in figure 4.9, where a bubble, formed by gas injection into a liquid from a capillary of radius r, has a pinning or contact line circumscribing the outer perimeter of the capillary end. From the figure it is obvious that as pressure increases from the lowest value necessary to start a bubble to high pressures, the center point of the bubble curvature continually shifts outward whereas the radius first decreases, goes through a minimum value at r and then increases thereafter. At the minimum radius r the curvature becomes spherical. Applying equation (4.7) with an inverse ΔP versus radius dependence shows that the applied pressure ΔP must be a maximum at radius r. Thus, we have $\gamma = \frac{r \cdot \Delta P_{max}}{2}$ and all that is needed to obtain γ is to slowly increase the gas pressure until a maximum value is reached after which less pressure is needed to keep growing the bubble. At the ΔP_{max}, the bubble shape is

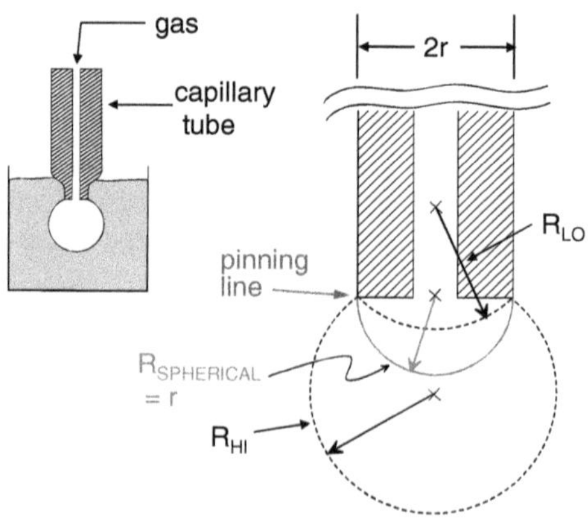

Figure 4.9. Schematic of an apparatus for measuring the surface tension of a liquid by creating a bubble in a liquid by injecting gas under pressure. Left: Gas is injected into a liquid through the tip of a capillary tube. Right: Magnified region at the tip of the capillary tube which shows the bubble formed at the outer edges of a capillary of radius r. At low pressure the bubble radius is R_{LO}, at high pressure, R_{HI} and at the appropriate intermediate pressure, r, where the bubble assumes a spherical curvature. In each case the bubble contact line is constant but the center point defining the radii shifts.

automatically exactly spherical with radius equal to the capillary radius *r*, so in principle we do not have to measure the radius, although one might do this from time to time to make sure things are working correctly. Since it is easier just to note the applied gas pressure by reading a gauge, as opposed to making a measurement of the bubble radius, this method is quite simple.

Some questions to think about:

1. What possible errors in the measurement might arise, other than faulty gauges, etc?
2. Why is the contact or pinning line at the outer perimeter of the capillary rather than the interior opening? What factors might cause the pinning line to shift from the outer to the inner edges, particularly for aqueous or polar liquids?

4.3.2.3 The inverse of simple spherical bubbles—simple spherical liquid drops

Now we look at the inverse problem of a simple liquid drop suspended in a gas environment, as shown in figure 4.10, and continue the assumptions of a non-volatile liquid and an insoluble gas (as the surrounding atmosphere) along with adding the absence of gravity effects so as not to distort the spherical shape. The drop is terminated by a surface of radius *r* and the inside has an extra pressure increment ΔP arising from the surface compressive forces. If you are sharp thinking, you will see that the equation relating ΔP, γ and *r* is just equation (4.7). It is the same liquid/gas interface layer trying to get rid of itself, squeezing on whatever is trying to impede this! So we can use the same equation for two different cases of spherical objects. Notice the direction of the curvature with respect to the liquid and the pressure differential in figures 4.8 and 4.10. In the case of the bubble $P_{\text{out}} > P_{\text{liq}}$, the surface of the liquid is a depression in the liquid, and the surface curvature is negative; whereas in the drop case with $P_{\text{liq}} > P_{\text{out}}$, the curvature is positive. So, we have a general principle that a pressure differential exists across any curved interface of pure liquid and the sign of the differential depends on the sign of the curvature. With this principle in hand we can now look at more complicated drop or bubble shapes and generate a more general relationship.

4.3.2.4 ΔP across non-spherically curved surfaces: the Young–Laplace equation

We saw from the gas bubble example that changes in the curvature of a surface of an object arise from a change in the pressure differential change across the surface and this change is proportional to the surface tension. For example, for a planar liquid

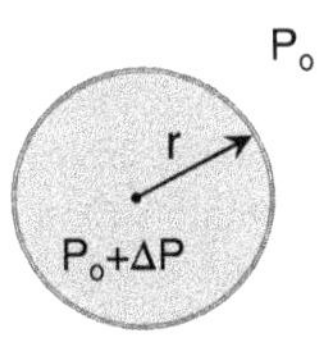

Figure 4.10. Illustration of a spherical liquid drop of radius *r* surrounded by a gas at pressure P_o. The inside of the drop has an extra pressure increment of ΔP above P_o due to surface tension compressive forces. The drop follows the same ΔP versus *r* equation as the bubble since the same gas/solid interface is involved.

surface (like an undisturbed pond) the pressures on both sides of the ambient/liquid interface are equal. But if the interface is disturbed by introduction of local pressure differences (throw a rock in the pond) then local curvature will arise in response. The shapes of liquid surfaces stressed by non-uniform pressure differentials is typically treated analytically in terms of a differential geometry problem.

First let's set up how to think about a non-uniform curved surface. Consider a small section of a curved surface of area A, as depicted in the left side of figure 4.11 where the section curvature is characterized by the two principal radii R_1 and R_2, located at origins o_1 and o_2 along the z-axis, which sweep across the surface in the x- and y-directions, respectively. The principal radii define the shape, such as the major and minor radii of an ellipsoid. As an example, if o_1 were located infinitely far away along z from the surface then $R_1 \to \infty$ and the surface section would have a cylindrical shape due to the remaining curvature along the y-direction. Typically, the curvature of a surface is defined in terms of a parameter which is the reciprocal radii $1/R_1$ and $1/R_2$.

Now we need an equation which relates the total pressure difference between the sides of the interface to the surface tension and the two radii of curvature. Before we do a somewhat detailed solution from differential geometry let's have a go at some simple guess at to what the answer must be.

Since we have two curvatures let's suppose that each has a ΔP necessary to sustain that specific curvature. Thus applying equation (4.7) in two parts, each contributing one half of the curving of the bubble, we have

$$\Delta P_1 = \frac{\gamma}{R_1} \text{ and } \Delta P_2 = \frac{\gamma}{R_2}$$

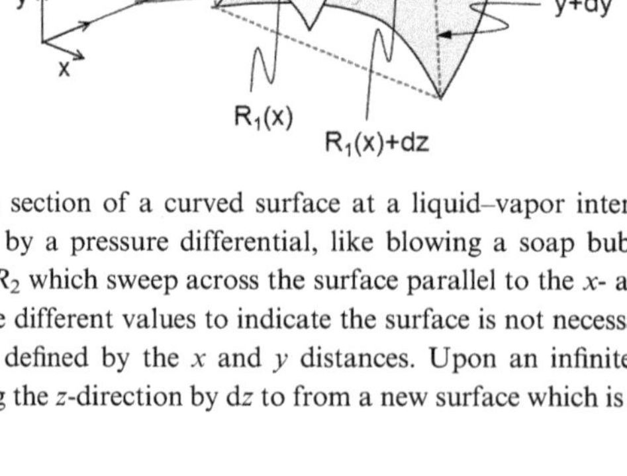

Figure 4.11. Schematic of a section of a curved surface at a liquid–vapor interface which is infinitesimally stretched to a larger radius by a pressure differential, like blowing a soap bubble. The initial curvature is defined by the radii R_1 and R_2 which sweep across the surface parallel to the x- and y-directions, respectively. To be general, R_1 and R_2 are different values to indicate the surface is not necessarily spherically shaped. The section of the membrane is defined by the x and y distances. Upon an infinitesimal expansion the bubble surface moves outward along the z-direction by dz to from a new surface which is now defined by the distances x + dx and y + dy.

so the total pressure would be

$$\Delta P = \gamma\left(\frac{1}{R_1} + \frac{1}{R_2}\right),$$

which turns out to be the correct answer! It is not clear, however, that this simple approach should actually be right since our assumption based on some hybrid of two spherical bubbles could be seriously questioned. It certainly gives the correct answer for a spherical surface where $R_1 = R_2$ so we can use this derivation as a mnemonic device (a trick for remembering).

Next, to set up the problem on hand in a sophisticated way, referring back to figure 4.11, we see that the corners of the section are connected by parallel lines across the x- and y-directions to define a rectangle of area xy which frames the section as a projection of the surface along the z-axis onto the xy-plane. The specific shape of this section is defined by the ratios of the sides of the projected rectangle to the respective radii, $\rho_x = x/R_1$ and $\rho_y = y/R_2$.

Now to obtain the Y–L equation, infinitesimally expand the section by scaling the dimensions and area while preserving the shape, calculate the $P\mathrm{d}V$ work required and set this equal to the $\gamma\mathrm{d}A$ work. Use these steps:

1. Increase R_1 and R_2 by $\mathrm{d}z$ so they now sweep across the inner surface of the expanded section. This also expands the framing rectangle sides to $x + \mathrm{d}x$ and $y + \mathrm{d}y$.
2. Maintain the rectangle side to radius proportions ρ_1 and ρ_2 which preserves the shape:

$$\frac{(x + \mathrm{d}x)}{(R_1 + \mathrm{d}z)} = \frac{x}{R_1};\ \frac{(y + \mathrm{d}y)}{(R_2 + \mathrm{d}z)} = \frac{y}{R_2} \Rightarrow \mathrm{d}x = \frac{x\mathrm{d}z}{R_1},\ \mathrm{d}y = \frac{y\mathrm{d}z}{R_2}.$$

3. Since we are only interested in the changes in shape and the bubble surface shape is uniform over the entire bubble section, we can take smaller and smaller sections of the surface to simplify the math by letting let the radii sweep angles $\delta\theta_x$ and $\delta\theta_y \rightarrow 0$, which also shrinks the size of the projection rectangle (this is essentially the same as approximating a circle as an infinite polygon). In this limit

$$R_1\delta\theta_1 \rightarrow x,\ R_2\delta\theta_2 \rightarrow y \text{ and the surface area} \rightarrow \text{projection rectangle area}$$

$$A \rightarrow x \cdot y,\ A + \mathrm{d}A \rightarrow (x + \mathrm{d}x)(y + \mathrm{d}y), \text{ which gives}$$

$$\mathrm{d}A \rightarrow (x + \mathrm{d}x)(y + \mathrm{d}y) - xy = (x\mathrm{d}y + y\mathrm{d}x) = \mathrm{d}(xy).$$

4. Calculate the work of surface expansion for the infinitesimal area expanded by the infinitesimal radii increases:

$$\mathrm{d}W_{\gamma A} = \gamma\mathrm{d}A = \gamma\mathrm{d}(xy).$$

5. Calculate the associated work of volume expansion for the over pressure:

$$\mathrm{d}W_{PV} = \Delta P \cdot \mathrm{d}V = \Delta P(xy \cdot \mathrm{d}z).$$

6. Solve for ΔP:

$$\Delta P = \gamma \frac{\mathrm{d}(xy)}{xy \cdot \mathrm{d}z}.$$

7. Apply the preservation requirement (step 2) for the expanded area section to have the same shape as the original section to step 6 to obtain the Y–L equation:

$$\boxed{\Delta P = \gamma\left(\frac{1}{R_1} + \frac{1}{R_2}\right)}. \tag{4.8}$$

It looks like we were correct with our simple first guess based on equation (4.7). We can get back to a spherical surface with the case $R_1 = R_2$ for which the Y–L equation reduces to equation (4.8). For a planar surface the radii $\rightarrow \infty$ and the surface curvature, defined as the reciprocal radii, $\rightarrow 0$ with $\Delta P \rightarrow 0$.

The Y–L equation is used to handle cases of non-spherical objects such ellipsoidal or elongated drops or bubbles and for analysing capillary phenomena. In the case of complex surfaces such as ones with bumps, protrusions, depressions and other similar features, the surface often can be modeled by looking at each local feature, estimating the local radii of curvature and then applying the Y–L or a similar equation to the local feature. Pressure gradients within objects are caused by local forces that cause surface roughening or misshaping and are typically not stable with time; eventually the object will relax back to a spherical shape if at all possible, for example a white capped, rough sea will relax back to a smooth one once the wind stops. In such examples the local surface features are caused local momentum fluctuations within the liquid. If we treat the liquid as an elastic medium, it is easy to see that there should be certain frequencies of internal motion that will become resonant with some driving force, thus resulting in phenomena such as periodic surface waves. The amplitude and wavelength will be a function of the surface tension and the liquid density. In fact, this is one way to measure surface tension. The subject of surface dynamics will be discussed later in chapter 5.

Part T
Brief review of equilibrium thermodynamics for surface applications

T.1 Review of fundamentals of the thermodynamics of reversible work

Before moving on it will be very useful to stop and do a quick review of equilibrium thermodynamics. In the last sections we discussed forces, pressures and surface work under equilibrium conditions. This gets us right into thermodynamic principles since the work done under equilibrium conditions means reversible work and can be directly equated to free energy by the first and second laws of thermodynamics. So it will be a great advantage to use these laws in our development of surface phenomena and it is worth a bit of time to develop what is needed to move ahead. It is assumed that the reader is already familiar with basic chemical and materials thermodynamics as provided in physical chemistry, materials science or physics courses, so the development here will be brief and designed for the purposes at hand. The main topics are:

- Definition of a system and surroundings and associated conventions.
- Statement of the first and second laws for equilibrium conditions and reversible processes.
- Equilibrium thermodynamics from the point of view of work processes.
- Fundamental constraints on intensive variables from the first and second laws.
- *T* and *P* effects on reversible mass transport work for a pure substance.
- Gibbs free energy for constant *T*, *P* conditions.

T.2 Definition of system and surroundings and associated conventions

Thermodynamic processes are analysed in terms of the interchange of work and heat between the system of interest and its surroundings, as illustrated in the schematic in figure T1. The system is chosen by drawing a boundary around the parts of the process that take place in some physical region where the observer cannot enter without disturbing the process, e.g. around a solution of chemical species which are in equilibrium with the surroundings as the laboratory and beyond. Inside the system

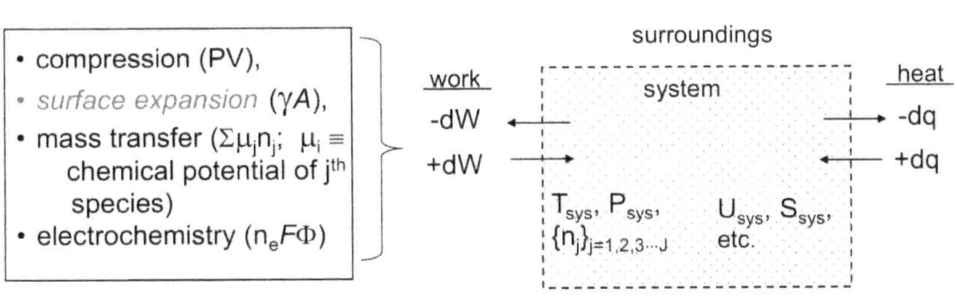

Figure T1. A schematic defining the system and surroundings for analysis of thermodynamic balances and relationships.

there are specific values of termperature and pressure as well as some collection of chemical species, written as the list $\{n_j\}_{j\,=\,1,2,3,\ldots,J}$, where n_j = number of moles (or molecules) of the *j*th species with *J* species total. If the system is uniform in conditions there is only one value of *T* and of *P* specified as T_{sys} and P_{sys}, otherwise these values may differ from local region to local region. Transfer of an infinitesimal increment of energy as work or heat transferred into the system from the surroundings (e.g. a gas is compressed or the solution in a beaker is heated on a hot plate) is assigned a positive sign by the convention we will use ($+dW$ or $+dq$; the *greedy* system convention). The surroundings has its own temperature (T_{sys}), pressure (P_{sys}), chemical species, energy (U_{sys}), entropy (S_{sys}) as well as other thermodynamic variables, which may or may not be uniformly distributed according to the fluctuations in the surroundings.

Let's discuss work processes briefly since this is where we are headed. As seen in the figure on the left inset, the four common types of works are listed: PV work of volume expansion/compression, mass transport (governed by chemical potential), surface expansion/contraction (governed by surface tension), and charge transport (governed by both electrical and chemical potential). While our primary interest is in surface work, for systems where added chemical species affect the surface condition (e.g. add a pinch of soap from the surroundings) we also will be interested in mass transport work and the chemical potentials of the species involved.

Here is the key rule for this set-up—the observer cannot directly know (or directly measure) values of the system variables but can only infer them from whatever information is available in the surroundings. The observer can know the volume of the system since that represents space lost to the surroundings. Now we go to the first and second laws of thermodynamics.

T.3 Statement of the first and second laws for equilibrium conditions and reversible processes

Referring to chart T1 for this discussion, we see that the first law is straightforward; energy flow between the system and surroundings, which consists of heat and work,

1st Law: $dU = dq + dW$

2nd Law: $dS_{sys} + dS_{surr} = dS_{\infty} = + dS_{sys} + dq_{surr}/T_{surr}$

$dS_{\infty} > 0$ for irreversible process or non-equilibrium

$dS_{\infty} \rightarrow 0$ for reversible process or equilibrium

for *reversible* process: $T_{sys}=T_{surr}=T$, $P_{sys}=P_{surr}=P$ } fundamental relationship for reversible changes

thus, $dS_{\infty} = dS_{sys} + dq_{surr}/T = dS_{sys} - dq_{sys}/T = 0$

Combining with Laws 1 & 2 for equilibrium or reversible process:

$$dU - TdS - dW = 0$$

Chart T1. Summary of the combination of the first and second laws of thermodynamics for reversible processes and equilibrium conditions.

is conserved. It does not matter how the process is carried out; whatever form and amount of energy leaves one side of the boundary ends up on the other side.

The second law is a bit more subtle since it involves the quantity entropy (S). We define entropy in terms of the thermodynamic variables as $dS_R \equiv \left(\frac{dq}{T}\right)_R$, where the subscript R refers to the region in which the change occurs, either in the system or surrounding. Thus to know an entropy change we need to know the sign and magnitude of the heat flow and the temperature in the region of interest. This definition does not tell us much about a fundamental interpretation of entropy but for now we put that aside for later discussion. Just keep in mind that entropy change is defined entirely in terms of the heat change at some local temperature.

The second law for our purposes states that an infinitesimal change in the sum of the entropy of the system and surroundings, which we call dS_∞, is greater than or equal to zero. If the change is irreversible and out of equilibrium (like an explosion or mixing two liquids, for which in either case you cannot possibly undo what you just did without an enormous cost of work) $dS_\infty > 0$ and we have no information on what is the actual value. If the systems and surroundings are not at equilibrium there could be all kinds of crazy T_{sys} and P_{sys} gradients within the system and there is absolutely no way to know what these are without sticking probes everywhere in the system and disturbing it. Even then you would be inferring the local conditions based on some kind of transfer mechanism from the system into the probe, which is part of the surroundings where you are. Further, if some work is transferred into the system from the surroundings there is no way to really know what happens in the system without disturbing it and this changes the boundary of system and surroundings. The major problem is we do not know T_{sys} and P_{sys} and cannot do thermodynamic balances and calculations that involve T and P. So how we can know what is going on?

The solution is to control the processes to take place in infinitesimal steps, each of which are fully reversible without extra work required to put things back together again. This results in $dS_\infty = 0$ and we gain another fixed constraint to analyse the thermodynamic balance between the system and surroundings. You cannot really do processes in infinitesimally small, reversible steps but you can approach the limit in many cases and be satisfied that the errors involved are small enough to be neglected. Here is what you really gain. Since systems and surroundings come into equilibrium and will maintain the same condition forever unless a change is deliberately made you can now assert with great certainty that T and P are equal across the system/surrounding boundary and since you are in the surroundings and know T_{surr} and P_{surrr} you then know T_{sys} and P_{sys}. With that the system is much better defined. Combining the first and second laws under reversible changes/ equilibrium conditions, as shown in chart T1, gives: $dU - TdS + dW = 0$.

T.4 Equilibrium thermodynamics from the point of view of work processes

To apply the combined law, one needs to break dW down into the specific work of interest and then proceed to analyse the processes and equilibrium conditions involving that work. The most common forms of work are PV, surface, mass transport and charge transport work expressed as the sum $\mathrm{d}W = -P\mathrm{d}V + \gamma\mathrm{d}A + \sum_{j=1}^{J}\mu_j\mathrm{d}n_j + F\Phi\mathrm{d}n_e = -P\mathrm{d}V + \mathrm{d}W_{\text{non-}PV}$, where μ_j and n_j are the chemical potential and number of the jth chemical species, respectively, $F =$ the Faraday constant (coulombs mol^{-1}), $\Phi =$ electrical potential and $n_e =$ number of electrons (or charges) transferred and the latter three collected into the term $\mathrm{d}W_{\text{non-}PV}$ to separate from PV work. Putting this form of $\mathrm{d}W$ into the general first and second law expression gives

$$\boxed{\mathrm{d}U - T\mathrm{d}S + P\mathrm{d}V = \mathrm{d}W_{\text{non-}PV}}. \tag{T1}$$

This is the basic equation from which we will start to develop approaches for different types of systems and problems. Once we specify the type of work of interest the equation immediately takes the shape needed to apply the constraints of the first and second laws of thermodynamics. For shorthand we will call equation (T1) the 1 + 2 law. And, while you are looking at the terms for the different types of work you should think about why the signs are + and − in the different cases.

T.5 Fundamental constraints on intensive variables from the first and second laws

If we scale the size of the system without making any internal changes in the system configuration those variables that scale with system size, e.g. V, U, S, etc, are called extensive, since their values extend with system size. In contrast, those variables that remain constant, e.g. P, T, are called intensive since they represent some type of intensity. The 1 + 2 law puts a constraint on the interrelationship between the intensive variables. We can see what that is by scaling the system, leaving these variables unaffected and then looking at an infinitesimal change in the scaled system with 1 + 2 law applied. Here is how that works. For simplicity we will only consider PV, $\gamma\mathrm{d}A$ and $\mu\mathrm{d}n$ work.

1. Scale the system to a macro system, rewriting the 1 + 2 law:

$$U - TS + PV - \left(\gamma A + \sum_{j=1}^{m}\mu_j n_j\right) + c = 0,$$

   where we add *a* constant of integration c just to be general. Essentially, think of this as adding a huge number of infinitesimally tiny systems each with energy $\mathrm{d}U$, entropy $\mathrm{d}S$, volume $\mathrm{d}V$, etc, to obtain U, S, V, etc, total with any additional change in entropy due to the expanded volume accounted for by c.
2. Sum all the ways to make an infinitesimal change in the system, namely, generate a total derivative (notice the integration constant is gone now):

$$\mathrm{d}U - T\mathrm{d}S - S\mathrm{d}T + P\mathrm{d}V + V\mathrm{d}P - \left(\gamma\mathrm{d}A + A\mathrm{d}\gamma + \sum_{j=1}^{J}\mu_j\mathrm{d}n_j + \sum_{j=1}^{J}n_j\mathrm{d}\mu_j + \right) = 0.$$

3. Now subtract the 1 + 2 law, which constrains the sum, to obtain

$$\boxed{-SdT + VdP - \left(\sum_{j=l}^{m} n_j d\mu_j + Ad\gamma\right) = 0}. \tag{T2}$$

Equation (T2) constrains the relationship between the intensive variables T, P, μ_j and γ when a system undergoes some change. Here we apply the 1 + 2 law to see the restriction on the way that the intensive variables can change with an infinitesimal, reversible perturbation. This is an important form of the 1 + 2 law which we will use from time to time. The process of converting a function of several variables $f(x_1, x_2, x_3, \cdots)$ to another function $g\left[\left(\frac{\partial f}{\partial x_1}\right), \left(\frac{\partial f}{\partial x_2}\right), \left(\frac{\partial f}{\partial x_3}\right), \cdots\right]$ which expresses the same constraints but in terms of the corresponding partial derivatives of f is an example of the Legendre transform.

T.6 *T* and *P* effects on reversible mass transport work for a pure substance

Let's apply equation (T2) to the specific case of mass transport work and the chemical potential for the case of a pure substance which is perturbed by changes in P and T. Here are the steps.

1. Start with equation (T2) restricted to changes in the chemical potential for a pure substance, which we write per mole by dividing through by n_j, the number of moles of species j and rearrange terms:

$$-\bar{S}_j dT + \bar{V}_j dP = d\mu_j,$$

where the super bars mean per mole.

2. Now setting constant P and T in turn we obtain

$$\boxed{\left(\frac{\partial \mu_j}{\partial T}\right)_P = -\bar{S}_j} \tag{T3a}$$

$$\boxed{\left(\frac{\partial \mu_j}{\partial P}\right)_T = \bar{V}_j}. \tag{T3b}$$

Let's pay attention to equation (T3b) since that is the one of the most interest for future reference. As we squeeze on the substance by increasing pressure we see that the chemical potential increases. This means that the substance is resisting the pressure change, becoming relatively less stable, and will immediately return itself to lower pressure to stabilize if allowed. Most importantly, the rate of increase of chemical potential with P is proportional the volume per mole (essentially the density) of the substance. That makes sense. Really densely packed substances have much less available internal space to lose by compression. We will use this equation later when we look at the effects of reducing the size of drops, thereby increasing their curvature which in turns forces their vapor pressure to increase (the Kelvin equation).

T.7 Helmholtz and Gibbs free energy for constant *V*, *T* or *T*, *P* conditions

At this point we already have all the thermodynamic tools to move ahead on a variety of surface and interface topics but we pause just to make two definitions of convenience—the Helmholtz and Gibbs free energies. All this does is allow us to use shorthand notations for the terms $dU - TdS + PdV$ which appear in the 1 + 2 law, which is shown at the top of chart T2 for reference. There are no new thermodynamic principles. Once we define the two free energies, we follow the steps shown in the chart to restate the 1 + 2 law in terms of the free energies for fixed conditions appropriate to each type of free energy. The end result is that under constant *V*, *T* the Helmholtz free energy change is equal to the reversible non-PV work and for constant *P*, *T* the Gibbs free energy is equal to the reversible non-PV work as summarized in these equations that represent the 1 + 2 law under specific conditions of *V* or *P* and *T* held constant:

$$(dA)_{V,\,T} = dW_{\text{non-}PV} \tag{T4a}$$

$$\boxed{(dG)_{P,\,T} = dW_{\text{non-}PV}}. \tag{T4b}$$

All we gained is fewer symbols to write, but we also introduced the term free energy which now can represent reversible non-*PV* work in a process under specific conditions. So that is all there is to free energy for our purposes.

From time to time we will use dG and G or dA and A when we do not feel like writing out all the constituent terms of the more fundamental variables and

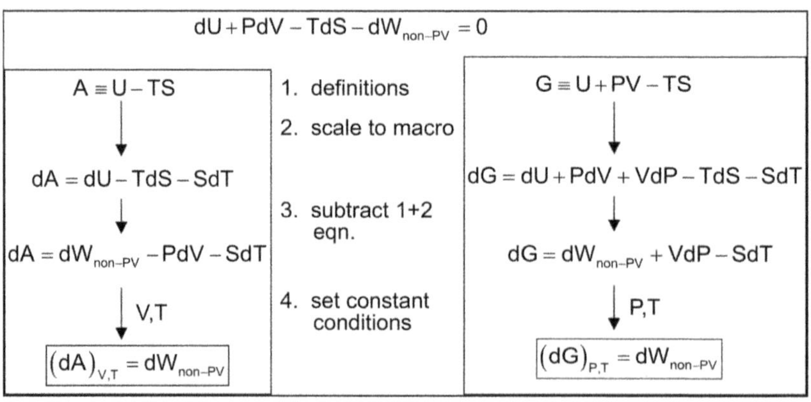

Chart T2. Steps for applying the 1 + 2 law (top of chart) to the definition of the Helmholtz (*A*) and Gibbs (*G*) free energies to equate free energy changes to a reversible non-*PV* work process at either constant *V*, *T* or *P*, *T* conditions. The steps show that the Helmholtz and Gibbs free energies are most useful for constant *V*, *T* and *P*, *T* conditions, respectively.

whenever we want we can recover the U, TS and PV terms from the definitions of the free energies. Here is an example for mass transport at constant P, T. This serves to connect G to chemical potential.

1. For only mass transport work we have from the 1 + 2 law:

$$dU + PdV - TdS = dW_{\text{non-}PV} = \sum_{j=1}^{J} \mu_j dn_j.$$

2. Using equation (T4b) for the Gibbs free energy we have

$$(dG)_{T,\,P} = \sum_{j=1}^{J} \mu_j dn_j.$$

3. Taking the partial derivative with respect to the number of moles of the jth species gives

$$\boxed{\left(\frac{\partial G}{\partial n_j}\right)_{T,P} = \mu_j = \left(\bar{G}_j\right)_{T,P}}. \tag{T5}$$

Thus at constant T, P the chemical potential of a chemical species is equal to the Gibbs free energy per mole of that species.

Gibbs free energy is so commonly used in thermodynamics of surfaces that it is important to know how and where it can be used. The Helmholtz free energy is less commonly used. Shortly we will see that since most surface phenomena or processes take place under constant T and P and since there is negligible change in V that the two free energies are essentially interchangeable (as long as the error in doing so are tolerable for the experiments considered).

T.8 Entropy defined in terms of numbers of ways to arrange a system—the Boltzmann equation

In section T.2 entropy was defined formally in differential terms of as the increment of heat transferred into a region per unit temperature of the region, $dS_R \equiv \left(\frac{dq}{T}\right)_R$. There is another more fundamental way to define entropy in terms of the Boltzmann equation:

$$\boxed{S = k_B \ln W \text{ or } \bar{S} = R \ln W} \tag{T6a}$$

$$\boxed{dS = k_B d(\ln W) \text{ or } d\bar{S} = Rd(\ln W)} \tag{T6b}$$

,

where k_B = the Boltzmann constant (1.381×10^{-23} J K^{-1}), R^{-1} = the molar gas constant (8.314 J K^{-1} · mol), $\bar{S}$ = entropy per mole, W = the number of choices available for the system to arrange itself (degrees of freedom) at constant energy, and the two equations refer to the integral and differential forms of the equation. If a small increment of heat dq is transferred into a system at some temperature T then the entropy change is defined to be dq/T. This form contains no information directly at the molecular level. Equivalently, however, the entropy change also can be treated in molecular terms by the Boltzmann equation which looks at changes in the number of ways the system can arrange itself, e.g. different energy levels, molecular orientations and conformations, and changes in energy level populations, etc. In fact, this fundamental approach includes the well-known third law of thermodynamics which states that $S = 0$ for a perfect crystal at 0 K, since such a system has only way to arrange itself and $\ln(1) = 0$. As an example, for a system of one molecule which has two possible conformations, then relative to this same system but with just one available conformation we would have $\Delta S = S_2 - S_1 = k_B \ln\left(\frac{W_2}{W_1}\right) = k_B \ln 2$. Clearly equations ((T6a), (T6b)) offer a very powerful tool for analysing the underlying molecular (or atomic) level basis of entropy changes in systems, in stark contrast to the equivalent but uninformative heat flow definition, and can be usefully applied to the behavior of surfaces and interfaces.

End of thermodynamics review

4.3.3 Surface excess thermodynamic quantities

4.3.3.1 Basic definition of surface excess quantities

With the basic toolbox of thermodynamic definitions and relationships based on the 1 + 2 law in hand, let's figure out how to apply it to surfaces and interfaces. If we follow the logic embedded in the missing molecule model, then we need a strategy that focuses on the difference between the values of bulk and surface quantities. Suppose we have an infinitely precise (and accurate) set of measuring tools to obtain thermodynamic quantities such as the energy of vaporization, heat capacity or the entropy of a substance relative to a perfect crystalline form at absolute zero. One would expect to find that the results depend on the size of the object since the S/V ratio would vary and we know that the molecules in the surface will have different values due to their different environments compared to molecules buried in the bulk. Thus, we would make errors in assigning the measured quantity to a bulk property. So, a good strategy would involve defining surface quantities in terms of these errors. This is illustrated in the picture in figure 4.12 which shows an object with a surface layer of some appropriate thickness to fully capture the surface environment and a layer of exactly the same area and thickness deep in the bulk, which becomes a reference. Then for any thermodynamic quantity X for a substance or material; e.g. U, S, G, n, C_P, etc; we define a surface excess quantity as $X^S = X^{\text{surf layer}} - X^{\text{bulk layer}}$, where $X^{\text{surf layer}}$ = total value of X per area in the surface layer, $X^{\text{bulk layer}}$ = total value of X per area in the bulk layer, with both layers of equal thickness, and thus X^S = surface excess value of X per unit area. It is clear from the several examples in chapter 1 that this error can be substantial as objects approach the nanometer size scale. For intensive quantities involving composition in multicomponent

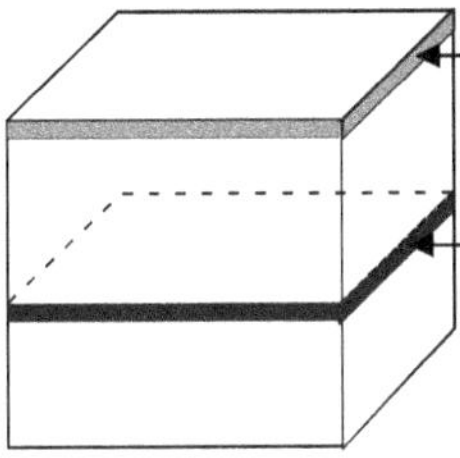

Figure 4.12. Illustration of an object with a surface layer of specific thickness and a layer deep in the bulk with the exact same thickness. The bulk layer is used as a reference for obtaining the difference in some thermodynamic quantity between surface and bulk.

objects, e.g. mole fractions, the surface excess of a component is an indication of preferential enrichment or depletion in the surface region.

4.3.3.2 Connecting thermodynamic variables to the surface tension—merging of the mechanical and thermodynamic description of surface expansion

Surface tension in the previous sections was treated as a mechanical phenomenon in terms of forces and pressures in the surface (interface) region. The definition of surface tension (equation (4.1)) was based on the equation for the reversible work of creating an infinitesimal patch of surface area: $\mathrm{d}W = \gamma \mathrm{d}A$. Since this involves a reversible process and if no other non-PV work is being performed in the perturbation of the system, then from the 1 + 2 law (equation (T1)) we have

$$(\mathrm{d}U - T\mathrm{d}S + P\mathrm{d}V) = \mathrm{d}W_{\text{non-}PV} = \gamma \mathrm{d}A.$$

Although it was not stated explicitly in the mechanical definition of γ, it would make sense that T and P were held constant. With that we see that the above equation can now be written in equivalent terms of the Gibbs free energy change. Using equation (T4b) for the specific case of $\mathrm{d}W_{\text{non-}PV} = \gamma \mathrm{d}A$, we have $(\mathrm{d}G)_{T,\,P} = \gamma \mathrm{d}A$, which then gives

$$\left(\frac{\mathrm{d}G}{\mathrm{d}A}\right)_{T,\,P} = \gamma \equiv G^{\mathrm{S}}. \tag{4.9a}$$

Since the area derivative of G gives the extra value of the Gibbs free energy added to the object per increment of new surface error, it then must represent the excess free energy, which we write as G^{S}, following our previous convention for surface excess quantities. Scaling an infinitesimal piece of the macro system up to an entire macro system to obtain $\int_{\to 0}^{G} \mathrm{d}G = G = \gamma \int_{\to 0}^{A} \mathrm{d}A = \gamma A$, then gives

$$\frac{G}{A} = \gamma = G^{\mathrm{S}}. \tag{4.9b}$$

Equations (4.9a) and (4.9b) represent two forms of expressing the excess surface free energy, either as the rate of change in free energy with surface area or as the integrated free energy per unit area of the entire object. These equations merge the thermodynamic and mechanical treatments of surface expansion.

Since by definition $G \equiv U + PV - TS$, it follows that equation (4.9a) (remembering to hold T, P constant) can be expanded in terms of the fundamental thermodynamic variables to give

$$G^{\mathrm{S}} = \left(\frac{\mathrm{d}G}{\mathrm{d}A}\right)_{T,\,P} = \left(\frac{\partial(H - TS)}{\partial A}\right)_{T,\,P} = \left(\frac{\partial H}{\partial A}\right)_{T,\,P} - T\left(\frac{\partial S}{\partial A}\right)_{T,\,P} = H^{\mathrm{S}} - TS^{\mathrm{S}}$$

and

$$H^{\mathrm{S}} = \left(\frac{\partial(U + PV)}{\partial A}\right)_{T,\,P} = U^{\mathrm{S}} + P\left(\frac{\partial V}{\partial A}\right)_{T,\,P} \sim U^{\mathrm{S}},$$

where the approximation holds generally since $V \sim$ constant for small surface expansions of typical size objects. Note that with $V \sim$ constant the Helmholtz free energy, $A = U - TS$, could also be used in place of G via equation (T4a).

4.3.4 Temperature dependence of surface thermodynamic quantities

What is the relationship between temperature and surface tension? How can we apply the 1 + 2 law to determine this? These are two intensive variables which do not depend ideally on the system size. So, the simplest way is to go back to the form of the law which is expressed as a constraint on the changes of intensive variables, given as equation (T2). If we restrict the work to only surface work, then equation (T2) simplifies to the relationship between T, P and γ:

$$-S\mathrm{d}T + V\mathrm{d}P - A\mathrm{d}\gamma = 0,$$

which when recast and imposing constant pressure (which one normally does in most surface measurements in which temperature is varied) gives

$$\left(\frac{\partial\gamma}{\partial T}\right)_P = -\frac{S}{A} = -S^{\mathrm{S}} = \left(\frac{\partial S}{\partial A}\right)_P. \tag{4.10}$$

Thus, at some particular temperature for a substance we look at small perturbation $\mathrm{d}T$ of the temperature and see that the change in γ is determined by the negative value of the excess surface entropy S^{S}. You will see in the next section that both γ and S^{S} change with temperature so the derivative $\left(\frac{\partial\gamma}{\partial T}\right)_P$ will eventually change, and in a way that you would expect as you raise T towards the vaporization point. Using the equivalence of γ and G^{S}, the definition of $G^{\mathrm{S}} = H^{\mathrm{S}} - TS^{\mathrm{S}} \sim U^{\mathrm{S}} - TS^{\mathrm{S}}$ and the continued assumption of negligible volume change with surface area perturbations, we have for the other standard thermodynamic variables,

$$T\left(\frac{\partial\gamma}{\partial T}\right)_P = -TS^{\mathrm{S}} \sim G^{\mathrm{S}} - U^{\mathrm{S}} = \gamma - U^{\mathrm{S}},$$

which we rewrite as

$$U^{\mathrm{S}} \sim \gamma + T\left(\frac{\partial\gamma}{\partial T}\right)_P = \gamma - TS^{\mathrm{S}}. \tag{4.11}$$

Using the above relationships, we could also derive other relationships such as surface excess heat capacity:

$$\left(\frac{\partial U^{S}}{\partial T}\right)_P \equiv C_V \sim \left(\frac{\partial H^{S}}{\partial T}\right)_P \equiv C_P.$$

Equation (4.11) is very important since it gives us a way to determine the excess surface energy U^S of a pure substance by measuring the temperature dependence of γ, a rather simple measurement in many cases. The excess surface energy should relate directly to the missing molecule model since it represents the energy change that arises from loss of coordination and bonding with neighbors when a molecule moves from the bulk to the surface. The related value of the entropy change S^S, which we obtain from equation (4.11), also contains important information about the changes in internal structure and arrangements in in creating a surface molecule. This is a bit more complicated to discuss and we come back to this shortly when we convert the thermodynamic excess values from a per area to a per mole basis.

For now, let's go on and examine the way their surface and thermodynamic properties actually behave with temperature changes and see what we can learn about the surface layer.

4.3.4.1 Surface tension variation with T

Table 4.3 shows selected data on the temperature dependence of surface tension for a variety of substances along with the values of the slopes, which from equation (4.10) give the values of $-S^S$. Here are several things to notice:

1. γ decreases with T in all cases, so this is a very general observation and a clue that something fundamental is happening as T increases.

Table 4.3. Temperature dependence of the surface tension and surface energies for selected substances.

Liquid	T (K)	T (°C)	$\gamma = G^S$ (mJ m^{-2})	$(d\gamma/dT) = -S^S$ (mJ m$^{-2}\cdot$ K)	U^S (mJ m^{-2})
He	2.5	−270.7	0.308	−0.07	0.5
N_2	75	−198.2	9.71	−0.23	27
Ethanol	293.2	20	22.75	−0.086	48.0
Water	293.2	20	72.88	−0.138	113.3
Diethyl ether	293	20	17	−0.116	51.0
CCl_4	293	20	26.9	−0.092	53.9
Toluene	293	20	28.5	−0.081	52.3
C_7H_{14}	293.2	20	15.7	−0.1	45.0
Benzene	293.2	20	28.88	−0.13	67.0
n-octane	293.2	20	21.8	−0.1	51.1
$NaNO_3$ (mp)	581.2	308	116.6	−0.05	145.7
Hg	293	20	484	−0.22	548.5
Cu (mp)	1358	1085	1550	−0.176	1789.0
Ag (mp)	1234	961	910	−0.164	1112.4
Fe (mp)	1808	1535	1880	−0.43	2657.5

2. In all cases, $S^S > 0$ and there seems to be no correlation between the general properties of substance and the S^S values. For example, the metals copper and silver have values of 0.176 and 0.164 which are quite similar to water at 0.138 and even smaller than liquid nitrogen at 0.23. Nothing seems to pop out here in terms of our typical way of gauging a material property on the basis of the type of material.
3. In all cases, $U^S > G^S$ ($= \gamma$), which from the equation. $G^S = U^S - TS^S$, with $S^S > 0$, means that the reversible work of bringing a molecule to the surface has to overcome the unfavorable $U^S > 0$ value but is assisted by a favorable entropy gain, as reflected in the $-TS^S < 0$ value.

Analysis of U^S and S^S can give us information on the energetics of bringing a molecule to the surface layer and the structural changes, respectively. It will be helpful to convert the data from per area quantities to molar quantities to compare with familiar values of energies and entropies of the pure liquids and for processes such as vaporization. We will delve into this soon.

Below are a few questions to ponder. The last question gives us an introduction to this next section.

Questions for thought:

1. Think of a common example in everyday life (around the kitchen works well) in which you easily observe that $d\gamma/dT < 0$.
2. Treating surface expansion as a chemical reaction (following equation (4.5)) is the reaction endothermic or exothermic? How does Le Chatelier's principle apply in terms of explaining the change in γ with increasing T?
3. Using the missing atom (molecule) model, why would $S^S > 0$?
4. If $S^S > 0$ then is this a stabilizing factor for the surface layer? If so, can we combine the Boltzmann equations (T6a) and (T6b) with the missing molecule model to understand this effect?
5. $d\gamma/dT$ looks like a small but real effect. Would you expect it to continue as T continually increases? What would eventually happen?

4.3.4.2 The vanishing of the interface at the critical point

What happens to the surface of a liquid as the temperature of a liquid is continually increased? An easy question—it vanishes when the liquid vaporizes, and naturally the characteristic of surface tension vanishes. An experimental demonstration of this behavior is shown in figure 4.13 for the example of CCl_4. In this experiment the pressure was continually increased to suppress vaporization of the liquid phase. You can see that as $T\uparrow$, $\gamma\downarrow$ and vanishes at T_C, the critical point temperature of CCl_4, above which there is no liquid/vapor interface; rather, only a single fluid phase exists with a liquid-like density. According to equation (4.10) the slope of the plot at each point gives the value of $-S^S$. The plot is fairly linear in the lower T regime, so this gives a fairly constant value of the excess surface entropy. Having both G^S ($= \gamma$) and S^S in hand we can straightforwardly calculate the excess surface energy U^S (equation (4.11) works well). The values of U^S, shown on the right-hand vertical axis, exhibit a sharp drop somewhat above 200 °C and crash to zero at T_C where the interface

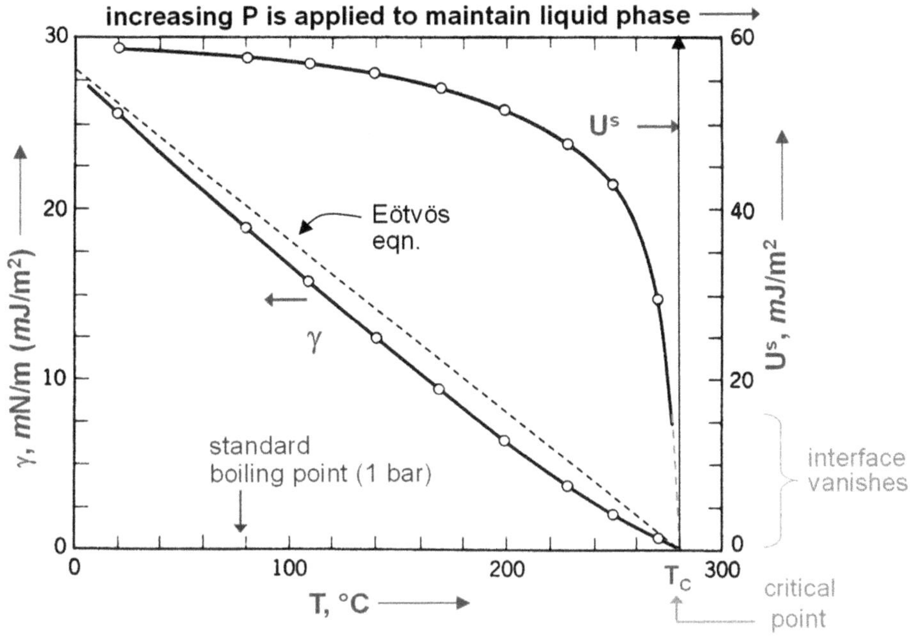

Figure 4.13. A plot of surface tension (left vertical axis) versus temperature for CCl_4 liquid. To prevent the liquid from vaporizing as T increases, the pressure of the system is continually increased. At the critical temperature T_C the vapor/liquid interface vanishes. Also shown (right vertical axis) are the values of the excess surface energy. The red dashed line represents an estimated extrapolation of U^S to zero, where the interface vanishes. The dashed black line of γ versus T is a prediction from the empirical Eötvös equation.

vanishes. The solid–vapor interface temperature behavior for sublimation processes of solids is similar except that the solid does not melt before it vaporizes, and the solid/vapor interface vanishes directly.

Because of the usefulness of these plots in providing values of S^S and U^S, there have been attempts to use empirical models to predict the how surface tension varies with temperature. For example, a simple empirical relationship is given by the Eötvös equation

$$\gamma = \frac{k}{\left(\bar{V}_{\text{liq}}\right)^{2/3}}(T_C - T), \tag{4.12}$$

where k = an empirical constant which is usually chosen as a value that fits a large number of similar types of liquids, $\bar{V}_{\text{liq}}$ = liquid molar volume and T_C = the critical temperature of the substance in kelvins. The dashed γ versus T line in figure 4.13 is a prediction from the Eötvös equation with $k = 2.1 \times 10^{-7}$ J $K^{-1} \cdot mol^{-2/3}$, a value that fits for many organic substances. Since this empirical equation is a linear correlation, the prediction cannot capture the curvature of the actual data near T_C. Other more sophisticated equations have been developed to capture this behavior.

One should note that the character of these γ versus T plots has continued to be of interest over the years because of the deep dependence on important fundamental properties of the substance. A number of semi-empirical and theory equations have been developed that correlate the behavior as functions of properties such as the molar densities of the vapor and liquid phases, T_C, boiling temperature, energies of vaporization and attractive energies between the molecules. Since many these properties are given on a per mole basis, in order to make such correlations it is necessary to convert surface properties from a per area basis to a per mole basis, which is the subject of the next section. At that time, we can see what might fundamentally underlie the simple Eötvös equation.

4.3.5 Excess quantities on a per mole basis—comparisons with bulk thermodynamic properties

If we describe the surface of a substance in terms of a thin slab of molecules (or atoms) of thickness δz we can convert surface excess quantities such as U^S and S^S from a per area basis to a per mole or per molecule basis. The task at hand for a given substance or material is to identify the thickness of the surface region and the number of molecules in the volume defined by unit area and thickness δz. Once we have the per mole basis then we can compare the quantities directly with standard per mole quantities for bulk materials. The comparison can give us insight into the energetics and structure of the surface layer.

4.3.5.1 Conversion of surface excess quantities to per mole quantities

The area to mole conversion is given by

$$X^{S\prime} = X^S \cdot \bar{A}_m, \tag{4.13}$$

where, $X^{S\prime}$ = the quantity X on a per mole basis, X^S = the usual surface excess of X on a per area basis and $\bar{A}_m$ = the area per mole of molecules or atoms in the surface region that carries all the surface properties.

For planar surfaces, the area per molecule (or atom) is given by the general equation

$$\bar{A}_m\ (\mathrm{m}^2 \times \mathrm{mol}^{-1}) = \frac{\bar{V}_m\ (\mathrm{m}^3 \cdot \mathrm{mol}^{-1})}{\delta z\ (\mathrm{nm}) \cdot 10^{-9}}, \tag{4.14a}$$

where $\bar{V}_m$ = the molar volume of the material and δz = the thickness of the surface layer which carries the surface excess properties. The molar volume can be obtained from the bulk density ρ_b as needed by the equation

$$\bar{V}_m\ (\mathrm{m}^3 \cdot \mathrm{mol}^{-1}) = \frac{M\ (\mathrm{g} \cdot \mathrm{mol}^{-1})}{\rho_b\ (\mathrm{g} \cdot \mathrm{cm}^{-3})} \times 10^{-6}.$$

It is often convenient to use molecular-scale units, which give on a per molecule basis

$$a_{\mathrm{molecule}}\ (\mathrm{nm}^2) = \frac{\bar{A}_m}{N_{Av}} \times 10^{18},$$

where N_{Av} = Avogadro's number and a_{molecule} = area per molecule. Naturally, if the molecule is not spherical then one must assume an orientation at the surface to get a more accurate value. In general, in the absence of other information, δz is taken to be the diameter d of a single molecule or atom of the material and we shall assume this henceforth, unless otherwise stated. For estimates of δz in terms of the number of layers of molecules or atoms, an average size of the molecule can be estimated from the molar volume:

$$d\ (\mathrm{nm}) \sim \left[\frac{\bar{V}_\mathrm{m}\left(\mathrm{m}^3 \cdot \mathrm{mol}^{-1}\right)}{N_\mathrm{Av}}\right]^{\frac{1}{3}} \times 10^9.$$

We can combine this equation with equation (4.14a) to obtain

$$\bar{A}_\mathrm{m} = \bar{V}_\mathrm{m}{}^{2/3} \cdot N_\mathrm{Av}{}^{1/3}, \tag{4.14b}$$

where the value of $\bar{A}_\mathrm{m}$ is based on the average molecular diameter as the surface layer thickness and the packing density as derived from the molar volume of the substance.

For crystals of known structure with a specific face exposed, a more exact relationship (setting δz = atomic diameter d) is

$$\bar{A}_\mathrm{m} = \frac{A_\mathrm{cell}}{N_\mathrm{cell} \cdot d} N_\mathrm{Av}, \tag{4.14c}$$

where A_cell = the area per unit surface cell determined from the lattice constant and cell geometry and N_cell = the number of atoms per unit cell. The picture for a crystal face is shown in figure 4.14. In cases where $\delta z > d$ the bulk properties change to surface properties over multiple layers and exact calculations of the conversion of surface excess properties to a per mole basis require more information from both theory and experiment.

Using equations (4.13) and (4.14b) along with handbook values for density, molar volume and molar mass, values of γ (= G^S), S^S and U^S (mJ m^{-2}) for selected liquids in table 4.4 have been converted to a per mole basis to give $G^{\mathrm{S}\prime}$ (kJ mol^{-1}), $S^{\mathrm{S}\prime}$ (J K^{-1}·mol) and $U^{\mathrm{S}\prime}$ (kJ mol^{-1}), as summarized in table 4.4. In addition, the values of $-TS^{\mathrm{S}\prime}$ are included since these give the entropic contributions to the work of promoting a molecule to the surface layer. Notice that the entries in the table have been grouped into classes of substances: simple nonpolar organic molecules, an alkane, hydrogen-bonded molecules and liquid metals. In principle, $U^{\mathrm{S}\prime}$ should be a direct measure of the missing coordination energy for a molecule at the surface versus in the bulk so this list gives us a chance to think about this effect and related changes in the molecular or atomic environment for different types of substances.

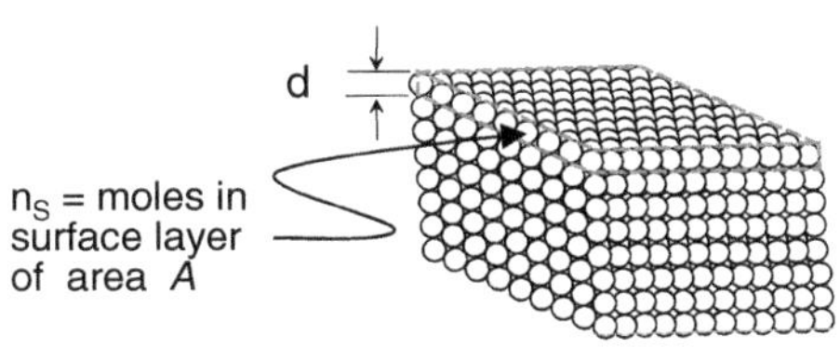

Figure 4.14. Representation of a crystal surface with a specific plane exposed to the ambient. The top layer of atoms of thickness d contains n_S moles per unit area.

Table 4.4. Comparisons of estimated surface excess entropies and energies on a per mole basis of selected liquids with their corresponding values for vaporization of the bulk phase. The surface tension values for liquids which are normally solids at ambient temperatures were determined at the melting point. The values of ΔH_{vap} were determined at the boiling temperature (T_{bp}) of the substance and ΔS_{vap} was calculated for this temperature from $\Delta H_{vap}/T_{bp}$. The energy of vaporization ΔU_{vap} is calculated from ΔH_{vap} by subtracting the correction RT_{bp}. The values of the per mole quantities are only stated to one decimal point given the rough assumptions in the calculations.

Liquid	T (K)	$\gamma = G^S$ (mJ m^{-2})	$G^{S\prime}$ (kJ mol^{-1})	$S^{S\prime}$ (J mol · K^{-1})	$-TS^{S\prime}$ (J mol^{-1})	ΔS_{vap} (J K · mol^{-1})	$U^{S\prime}$ (kJ mol^{-1})	ΔH_{vap} (kJ mol^{-1})	ΔU_{vap} (kJ mol^{-1})
Diethyl ether	293	17	3.1	21.2	−3.13	96.6	9.3	27.5	23.9
CCl_4	293	26.9	4.7	16.0	−2.30	85.9	9.4	32.5	26.9
Toluene	293	28.5	5.3	15.0	−4.40	87.3	9.7	38.1	30.0
Benzene	293	28.88	4.8	21.4	−6.27	87.2	11.0	33.9	27.8
n-octane	293	21.8	5.4	24.9	−7.29	100	12.7	39.9	36.6
Ethanol	293.2	22.75	2.8	10.7	−3.13	109.7	6.0	38.6	35.7
Water	293.2	72.88	4.1	7.8	−2.30	109.1	6.4	40.7	37.6
Hg	293	484	24.1	11.0	−3.21	84.2	27.3	59.2	53.9
Al	933	1050	43.8	11.4	−10.7	107	55	294	271.2
Ag	1234	910	38.5	6.9	−8.55	103	47	251	230.3
Cu	1358	1550	51.1	5.8	−7.89	106	59	300	276.7
Fe	1808	1880	62.1	14.2	−25.7	116	88	350	324.5

4.3.5.2 Comparisons of per mole surface quantities with analogous per mole bulk quantities

There are several important aspects of the thermodynamic data to consider, chemical potentials, entropies and energies.

Chemical potentials (μ)

First, the values of $G^{S\prime}$ represent the excess Gibbs free energy per mole of surface layer molecules. Looking back at equation (T5) in section T.7 of the thermodynamics review, we see that the Gibbs free energy per mole at fixed T and P is equal to the chemical potential μ for these conditions, so this gives us an opportunity to look again at the imbalance between bulk and surface but now in terms of chemical potential. It follows from the definition of surface excess quantities that $G^{S\prime} = \mu_s - \mu_b$, where μ_s and μ_b = the surface layer and bulk chemical potentials of the pure substance, and since $G^{S\prime} > 0$ then $\mu_s > \mu_b$ and the chemical potential is higher in the surface layer than the bulk. This, of course, must be the case since the molecules in the surface layer are driven to escape into the bulk, and which fits with the usual thinking of the chemical potential in terms of an 'escaping' tendency. In general, when there are two regions of a system (or system and surroundings) with different chemical potentials there is a driving force to undergo mass transport to stabilize the system (e.g. remove a concentration gradient). But in this case there is

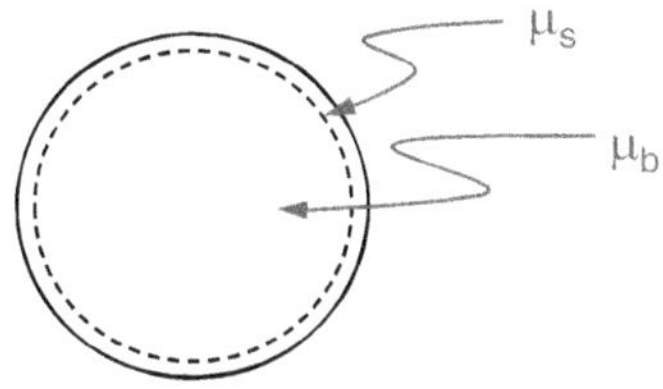

Figure 4.15. A schematic of a stable spherical drop of a pure liquid in which the intrinsic chemical potential imbalance between the surface and bulk region is frozen in by the inability of the drop to decrease the surface area.

no *allowed path* to move molecules from the surface into the bulk if the shape of the object cannot change to decrease the surface area. This is depicted for a spherical drop in figure 4.15 in which, all other factors being equal, the spherical shape has the minimum possible surface curvature thus stabilizing the shape in terms of surface forces but the imbalance in chemical potential remains, frozen in by the lack of a path for mass transport relaxation. This analysis seems trivial since we already knew about the intrinsic instability of the surface but we will see later that when multicomponent solutions are considered it becomes entirely possible for mass transport processes of the component species to occur at constant surface area to vary mole fractions in the bulk and surface layer so as to minimize surface tension and enrich or deplete concentrations in the surface region. So at that time we will need to consider both surface and mass transport work.

Now let's push comparisons of the per mole surface thermodynamic quantities with analogous bulk quantities and see what can be learned about the surface layer properties. The basis of this approach is that $G^{S\prime}$ represents the reversible work per mole required to strip coordinations from a bulk molecule as we push it into the surface layer and thus it seems logical to compare $U^{S\prime}$ and $S^{S\prime}$, the component values of the free energy, with ΔU_{vap} and ΔS_{vap}, the energies and entropies of vaporization, which represent the changes per mole for stripping *all* the bulk coordinations to push the molecule into the vapor phase. We will start with entropy since this will be a short discussion (entropy is always a bit complex to understand on a simple molecular basis without introducing statistical thermodynamics concepts and equations). Then we will go to energy in the next section to develop some ways to estimate surface tension from bulk data.

Entropies (S)

In table 4.5 notice that the values of $S^{S\prime}$ for small organic molecules cluster around $\sim 18\ \text{J} \cdot \text{K}^{-1} \cdot \text{mol}^{-1}$, excluding n-octane at $\sim 25\ \text{J} \cdot \text{K}^{-1} \cdot \text{mol}^{-1}$, which is likely different since it is a long floppy molecule and definitely will configure differently at the surface compared to more symmetrical molecules such as benzene or CCl_4. The larger entropy suggests there is a greater release of degrees of freedom in going to the surface for a hydrocarbon chain compared to a smaller molecule, which fits with the hydrocarbon chain extending somewhat at the surface with more degrees of freedom compared to the bulk where the chain is fully interacting with neighbors.

Next, the hydrogen bonding molecules tend to have low values near $\sim$9 J $\cdot$ K^{-1} $\cdot$ mol^{-1}, which suggests that the H-bonding restrictions likely hold to a good extent in moving to the surface. Finally, the liquid metals have lower values as well, though iron appears at the very high end, approaching that of molecules. It would appear that the metal atoms reconstruct to find as many NN interactions as possible.

Now let's compare these values with entropies of vaporization. In table 4.4 you can see that the values of ΔS_{vap} fall in the general range of $\sim$85–110 J K^{-1} $\cdot$ mol. This follows from the well-known Trouton rule which states that the entropies of vaporization of pure substances is $\sim$86–89 J K^{-1} $\cdot$ mol, with exceptions at higher values mostly for hydrogen bonding substances and a few at lower values for small molecule gases which condense at cryogenic temperatures, especially He. The Trouton rule is a rough and ready guide but does have a reasonable physical basis for many common molecules in that the process of vaporization is quite similar in promoting the molecules from a dense liquid environment in which they are confined by attractive forces within molar volumes which are several hundred or so times less than the molar volume of the vapor where the molecules freely move in all directions. If we estimate ΔS_{vap} purely on this basis as the entropy of expansion for an ideal gas [$R\ln(V_{vapor}/V_{liquid})$] we will come up with values $\sim$ ½ of the Trouton rule values, which means there are other more subtle factors to consider, e.g. unlocking of frozen rotational degrees of freedom in the bulk in going to the gas phase, but this is not a bad guess anyway.

Looking at the analogous process of releasing some of the attractive interactions in going from the bulk to the surface we see from table 4.4 that roughly 1/10 to 1/4 of the bulk $\rightarrow$ vapor entropy change is reflected in the bulk $\rightarrow$ surface change and for typical non-hydrogen bonding organic molecules the range seems to be narrower, $\sim$1/6 to 1/4. For example, for benzene $S^{S/}/\Delta S_{vap}$ = 21.4/87.2 $\sim$ 0.25. This comparison suggests that we can think of the surface environment in terms of an expanded local volume for motion of the molecules compared to the bulk, and applying the missing molecule model, particularly in the z-direction.

We can put the Boltzmann equation (equations (T6a) and (T6b)), using the per mole basis) to work here to obtain another perspective. Let's calculate the difference in the ways to arrange the molecules in the surface relative to the bulk (at T, P) as follows:

$$S^{S/} = \Delta S_{bulk}^{surf} = R\left[\ln\left(\frac{W_{surf}}{W_{bulk}}\right)\right]$$

or

$$\frac{W_{surf}}{W_{bulk}} = \exp\left(\frac{S^{S/}}{R}\right).$$

Using an average value of $S^{S/}\sim$ 18 J K^{-1} $\cdot$ mol for a typical small organic molecule with $R = 8.314$ J K^{-1} $\cdot$ mol, we obtain $\frac{W_{surf}}{W_{bulk}}\sim$ 8, which is equivalent to stating that there are $\sim$8 times the number of ways to arrange the states of surface molecules relative to a bulk environment. The states can include the energy states and physical positions so we cannot be definite about what these are without serious analysis of the quantized energy levels of the system. Let's also apply the analysis to ΔS_{vap} to see what that gives. Using a typical Trouton value of $\sim$87 J K^{-1} · mol, we obtain $\frac{W_{vapor}}{W_{bulk}} \sim 4 \times 10^4$, so there are $\sim$tens of thousands times the number of states available in the vapor compared to the bulk, of which a good portion will be associated with the translational degrees of freedom of the vapor molecules speeding around in the large volume of the vapor.

At this point keep this factor in mind before pushing any comparisons and number crunching much harder. Our estimate of the thickness of the surface layer was $\delta z \sim d$, the average molecular diameter but then consider that the surface layer that carries all the change from bulk to surface may very well be more than one molecule deep, though likely not more than three with an average closer to one, which would essentially dilute the values of our per mole surface excess quantities (share the changes among more molecules) by some factor less than 2. But overall, the simple comparisons above seem to capture the basic entropy effects quite well and all fits nicely into the missing molecule model. So keep the numbers above in mind for when we move to the chapter on surface dynamics where we look at surface vibrations and motions. Although we will provide a few more details there, realize that we have already learned the important concepts at this point.

Before we leave entropy, let's use some of what we have learned about $S^{S\prime}$ to pick apart the empirical Eötvös equation (equation (4.12), section 4.3.5.2) and see what turns up. Here is the approach, keeping all units in the SI convention.

1. Start with the Eötvös equation, take the derivative with respect to T and set equal to $-S^S$ from equation (4.10) to obtain $\left(\frac{\partial \gamma}{\partial T}\right)_P = -\frac{k}{(\bar{V}_{liq})^{2/3}} = -S^S$, or in terms of the Eötvös constant: $k = -S^S(\bar{V}_{liq})^{2/3}$.
2. Applying the per area to per mole conversion (equation (4.13) based on the molar volume to convert S^S to $S^{S\prime}$ gives $S^{S\prime} = S^S\bar{A}_m = S^S(\bar{V}_m)^{2/3} \cdot N_{Av}{}^{1/3}$.
3. Substituting 2 in the k equation and evaluating N_{Av} gives: $k = 1.19 \times 10^{-8} \cdot S^{S\prime}$.
4. Taking the value $k = 2.1 \times 10^{-7}$ J K^{-1} · $mol^{2/3}$ for a typical organic molecule gives: $S^{S\prime} \sim 17.7$ J · K^{-1} · mol^{-1}.

Thus the Eötvös constant for a typical organic molecule is actually based on an average value of $\sim$18 J · K^{-1} · mol^{-1} for the excess surface entropy, which fits quite well with the average value extracted from the simple, small, non-H-bonding molecules in table 4.5. Thus, we see there is a fundamental basis underlying the approximation and the Eötvös equation is based on a linear dependence on T with a constant value of the excess surface entropy per mole up to T_C.

Energies (U)

Following the strategy developed above for $S^{S\prime}$, we analyse $U^{S\prime}$ in terms of the energy required to strip some coordination from a bulk molecule to push it to the surface and compare with the energy of vaporization ΔU_{vap}, the corresponding energy to strip all the coordinations and vaporize the molecules. Thus we are returning to the associated chemical equations presented earlier and associating them with energy changes:

$$M_B^{CN_B} \rightarrow M_S^{CN_S} + (CN_B - CN_S), \quad \Delta U_{vap} \tag{4.15}$$

$$M(B) \rightarrow M(vapor) \quad U^{S\prime} = \Delta U_{rxn}, \tag{4.16}$$

where ΔU_{vap} and $U^{S\prime} = \Delta U_{rxn}$ represent the energy changes in the corresponding reactions and the values of the coordination numbers (CNs) represent the stoichiometry coefficients.

Table 4.4 shows the values of ΔH_{vap} since these, in comparison to ΔU_{vap}, are readily available in handbooks. Conversion to ΔU_{vap} follows directly from the definition of H to give $\Delta H_{vap} = \Delta U_{vap} + P\Delta V_{vap}$. Using the standard assumption of ~negligible liquid volume relative to the vapor volume together with the ideal gas law for a mole of vapor, gives the estimate $\Delta H_{vap} \sim \Delta U_{vap} + RT$. Since values of ΔH_{vap} are usually given for the respective standard boiling points (bp) these values are used in table 4.4 and thus the correction RT_{bp} will vary. For many liquid metals $T_{bp} \sim 2500$–3000 K so RTbp ~20 kJ mol^{-1}, but for typical organic liquids T_{bp} ~350–400 K so RT_{bp} is considerable smaller correction,~3 kJ mol^{-1}. We note these details since we must be mindful in using the ΔU_{vap} values, each appropriate for some T_{bp}, for interpretations of γ values obtained at lower temperatures, particularly ambient.

Applying to the data in table 4.4 for typical nonpolar molecules we obtain $U^{S\prime} \sim f \cdot \Delta U_{vap}$, where the fraction f runs from ~ 1/3 to 1/5. For H_g, f~ 1/2 and for the polar liquids, ethanol and water, f ~ 1/4 and 1/6, respectively. So the numbers jump around but in general, $U^{S\prime}$ represents some significant fraction of ΔU_{vap}. This suggests that if we apply the missing atom molecule approach and count the lost coordinations we might be able to make some fair estimates of the fraction f. We do this now in the next section.

4.3.6 Estimation of U^S and γ from intermolecular (atomic) interactions

We can refine our analysis of the correlations between $U^{S\prime}$ and ΔU_{vap} (and ΔH_{vap}) by estimating the fractions of coordinations lost for a molecule in moving from bulk to surface following reactions (4.15) and (4.16). Keeping the same assumption as above for the surface region as one molecule (or atom) thick, the analysis depends upon assigning the energy per coordination interaction on the basis of the value of ΔU_{vap} and then estimating the energy change of a bulk molecule coming to the surface as follows:

$$\text{Complete vaporization: } \Delta U_{vap} = -1/2(N_{Av} \cdot \varepsilon_{NN} \cdot CN_{bulk}),$$

where N_{Av} = Avogadro's number, ε_{NN} = the potential energy of coordination or bonding for a NN pair, CN_{bulk} = the coordination number of a molecule in the bulk, the factor of ½ is used to correct for over counting the number of bonds since there are only half as many bonds as total molecules and the negative sign accounts for the attractive potential energy $\varepsilon_{NN} < 0$ whereas $\Delta U_{vap} > 0$:

Bulk to surface transfer: $U^{S\prime} = -1/2(N_{Av} \cdot \varepsilon_{NN})(CN_{bulk} - CN_{surf}) - U_{relax}$,

where CN_{surf} = the coordination number of a molecule in the surface layer and U_{relax} = the relaxation energy of a molecule after coming to the surface (reconstruction) and the minus sign and ½ factor applied as for complete vaporization.

Neglecting the U_{relax} term and combining the equations we obtain

$$\frac{U^{S\prime}}{\Delta U_{vap}} \sim \frac{(CN_{bulk} - CN_{surf})}{CN_{bulk}} = f = \sim \frac{U^{S\prime}}{\Delta H_{vap}}, \tag{4.17}$$

where f = the fraction of coordinations lost in going from bulk to surface. In order to use equation (4.17) we need values for CN_{bulk} and CN_{surf}. Since equation (4.17) is intended primarily for application to liquids where we do not have to deal with the uncertainties of U_{relax}, which can be severe for hard materials, we have to guess the coordination numbers since these are difficult to obtain (usually obtained from various kinds of x-ray and neutron scattering measurements and from theory). Typical liquids have CN_{bulk} ~4–5 and dense liquid metals may be on average ~12, similar to hexagonal close packing but with the loss of ordering. In coming to the surface one would expect to lose ~ 1/4 to 1/3 of the coordinations for molecules and perhaps ~1/4 for liquid metals. You can get some idea of how this might work by looking at the case of crystals.

You might expect a liquid on average to always arrange the molecules or atoms at the surface in the most stable configuration so we will look at monatomic crystals with the most stable face and count coordinations. Referring to the atom coordination structures shown in figure 4.16, for an FCC lattice, which has $CN_{bulk} = 12$, the most stable face is the (111) hexagonal surface with only three missing coordinations so $CN_{surf} = 9$ and thus $(CN_{bulk} - CN_{surf})/CN_{bulk} = f = 1/4$. For a cubic crystal with a (001) face the most stable face has only one missing coordination and $(CN_{bulk} - CN_{surf})/CN_{bulk} = f = (6 - 5)/6 = 1/6$. Based on this from equation (4.18a) we predict

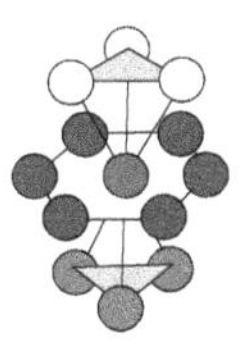

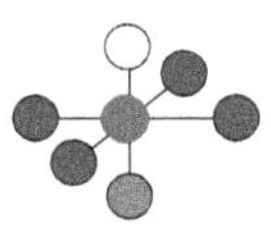

Figure 4.16. Illustration of the atom coordination numbers for selected atom in an FCC and a cubic lattice. The central atom (red) is arranged in a 3–6–3 and 1–4–1 stacking for the two lattices, respectively, such that when the central atom is in the surface plane there will be three and one missing atoms above, respectively.

$$\text{FCC hexagonal face:} \quad U^{S\prime} \sim 1/4\Delta U_{vap}$$

$$\text{Cubic square face:} \quad U^{S\prime} \sim 1/6\Delta U_{vap}.$$

You can see that these predictions actually fit within the empirical range above for typical molecules. The case of mercury with the experimental value of $U^{S\prime} \sim 1/2\Delta U_{vap}$ seems to be off by a factor of 2 or so and we might have expected this simple liquid metal to follow the case of the hexagonal FCC face with $f \sim 1/4$.

Now we tackle the prediction of γ from ΔU_{vap} or ΔH_{vap}. Here are the steps:

1. Estimate $U^{S\prime}$ from ΔU_{vap} or ΔH_{vap}.
2. Obtain or estimate a value of $S^{S\prime}$.
3. Calculate $G^{S\prime}$ from: $G^{S\prime} \sim U^{S\prime} - TS^{S\prime}$.
4. Convert from a per mole to a per area basis: $\gamma = G^{S} \sim U^{S} - TS^{S} = \frac{U^{S\prime}}{\bar{A}_m} - \frac{TS^{S\prime}}{\bar{A}_m}$.

The challenge in this approach is that it is difficult to estimate reasonable values of $S^{S\prime}$ for any arbitrary molecule or atom on a simple basis, as compared to estimating $U^{S\prime}$ from ΔU_{vap} and coordination numbers. But here are some methods used for estimating γ:

1. At low temperatures.

   Looking at figure 4.13 you can see that as T drops $U^{S} \rightarrow \gamma$. Thus, at low T one can directly use the estimate

$$\gamma \sim U^{S} = \frac{U^{S\prime}}{\bar{A}_m} = \frac{1}{\bar{A}_m}\frac{(CN_{bulk} - CN_{surf})}{CN_{bulk}}\Delta H_{vap}. \tag{4.18a}$$

   This is often used as a rough estimate even at higher temperatures where $U^{S} > \gamma$. The dominant error is in the neglect of the entropy of stabilization of the surface layer.

2. For liquid metals.

   It has been found empirically that for many liquid metals this equation holds:

$$\gamma \sim \frac{1}{6}\frac{\Delta H_{vap}}{\bar{A}_m}. \tag{4.18b}$$

3. For many typical molecules.

   Use empirical equations, such as the Eötvös equation with the molecular parameters of $\bar{V}_m$ and T_C which are based on average values of the excess entropy per mole. There are many empirical equations in the literature that have been developed over the years.

4. Looking at table 4.3 you can see that γ (mJ m^{-2}) $\sim$ ½ ΔH_{vap} (kJ mol^{-1}) for organic molecules so you can use this as a really rough guess for this class of molecules.

Overall, in this section we have seen that it is possible to understand the physical basis of surface tension in terms of analysing the constituent changes in the entropy and energy changes in bringing a molecule from the bulk to the surface terms on a per mole or molecule basis. For entropy the complexity of ways to arrange the states of molecules and atoms makes direct estimates difficult for any particular molecule but some estimates can be made by comparing to the entropy of vaporization and the use of Trouton's rule to set a limited range of values. For energy, direct comparisons with the heat of vaporization together with a simple model of the change in coordination number allow formulation of simple relationships for estimates. In general, though, the best we can do overall is to note that surface tension roughly tends to track the energy of vaporization, which fits well with the missing atom/molecule model. But do not be discouraged that you cannot easily estimate the loss in CN in promoting a molecule to the surface. It remains a current and challenging problem in the field of molecular liquids to figure out local coordination so there is often no easy answer and serious experimental and theory approaches are needed to obtain reliable information.

4.3.7 Effects of surface phenomena on the equilibrium between adjoining phases

Until now we have looked primarily at the behavior of the interface of a pure liquid (more generally condensed matter) surface in equilibrium with an adjacent inert gas or the vapor phase of the liquid. The main focus has been on the change in structure and thermodynamic properties in passing from the bulk into the interface and the effects of surface expansion (or contraction) on the pressure differential on the two sides of the interface, as related to small drops or bubbles. In the case of an object in contact with a complementary phase, such as a system with a solid–liquid, solid–vapor or liquid–vapor interface, we might consider how the phase equilibrium is perturbed by the presence of a surface. For example, consider an ice cube floating in water at exactly 0 °C where equilibrium is established. But if the ice cube were at the size scale of a few nanometers, would you expect exactly the same melting temperature? Look back to chapter 1 and you will see examples where such properties change with S/V. So how do we attack this problem?

Phase equilibrium is defined at the point in T, P space at which the different phases all co-exist, which requires that the chemical potentials are all equal in accordance with no net transport from one phase to another that could remove one phase in favor of another. So the key to involving surface effects in phase equilibrium is to look at the effects of the S/V ratio on the chemical potentials. We dispensed with using the S/V ratio in a quantitative way long ago in favor of describing the size and shape an object in terms of two radii of curvature, as demonstrated with the Y–L equation (equation (4.8)). So, let's proceed in this direction. To use the Y–L equation we must start with connecting the pressure differential on the two sides of a condensed matter/vapor interface to the chemical potentials of the two phases in contact. This sets us up perfectly for the problem of looking at the vapor pressure of a liquid drop as a function of drop size. Let's see how that works and when finished we arrive at the famous Kelvin equation and learn why it is often difficult for the skies to open and rain even when there is plenty of humidity and the conditions seem perfect.

4.3.7.1 Surface curvature effects on vapor pressure: the Kelvin equation

Here are the two basic relationships we need:

1. The Y–L equation, which we repeat here just to have it before us for convenience:

$$\Delta P = \gamma\left(\frac{1}{r_1} + \frac{1}{r_2}\right).$$

2. The relationship between chemical potential of a species j in a mixture and the surrounding pressure on the species (equation (T3b) from the mini-thermo review):

$$\left(\frac{\partial \mu_j}{\partial P}\right)_T = \bar{V}_j, \text{ or in differential form, } \mathrm{d}\mu_j = \bar{V}_j \mathrm{d}P,$$

where $\bar{V}_j$ = the molar volume of species j. In the case of a single pure substance j is used to represent a specific phase of the substance, in the present case a liquid or a vapor.

The general schematic of the system is shown in figure 4.17 for three types of surface curvature. In working with mass transport and chemical potentials (such as in equilibrium constants for chemical reactions), absolute values for the species of interest at a specific condition are rarely used and one inevitably chooses some reference state for comparisons. Since we know that a flat surface with $R_1 = R_2 \to \infty$ has no pressure differential with the ambient phase we choose that as a reference state with a saturation vapor pressure $P^{\text{o}}_{\text{vap}}$ at temperature T, which we keep constant. For this surface the vapor and liquid chemical potentials are equal so $\mu^{\text{o}}_{\text{vap}} = \mu^{\text{o}}_{\text{liq}}$. To make the conditions simple we consider only liquid and vapor are present with no ambient gas (e.g. air), though the presence of an ambient pressure will have no effect on the system at ordinary pressures since the liquid always

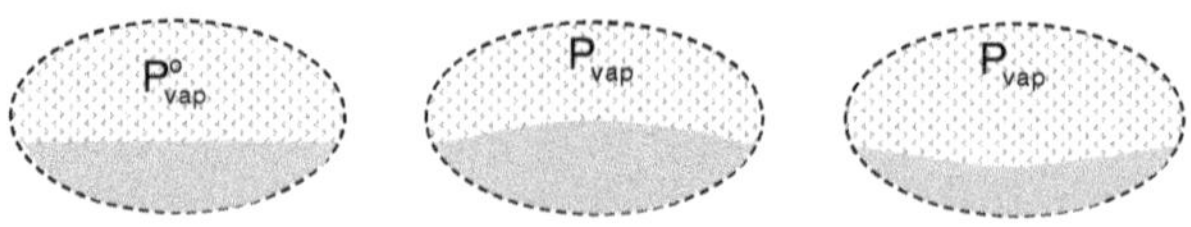

Figure 4.17. Schematic of three states of curvature of the surface of a pure liquid surface in contact with the pure vapor at equilibrium at some temperature. Each picture represents a small section of a larger volume which would include the entire liquid object in the presence of its vapor phase. Left to right: a flat surface in equilibrium with vapor pressure $P^{\text{o}}_{\text{vap}}$, an outward curved surface such as that of a drop and an inward curved surface such as would exist within a drop as an enclosed bubble filled with vapor. The vapor pressures above the curved surfaces will differ from that above the flat surface because of surface tension effects.

responds with an identical balancing pressure. With no ambient inert gas pressure, for a flat surface, $P^{o}_{liq} = P^{o}_{vap}$.

Here are the steps:

1. Calculate the chemical potential of the vapor above a surface with some curvature, as defined by the values of R_1, R_2, by integrating the equation in two above from a flat surface to the curved surface:

$$\mu_{vap} - \mu^{o}_{vap} = \int_{P^{o}_{vap}}^{P_{vap}} \bar{V}_{vap} \mathrm{d}P = \int_{P^{o}_{vap}}^{P_{vap}} \frac{RT}{P} \mathrm{d}P = RT \ln\left(\frac{P_{vap}}{P^{o}_{vap}}\right),$$

where the ideal gas law is applied to convert $\bar{V}_{vap}$ into a pressure term.

2. Calculate the chemical potential of the liquid under a curved surface relative to the flat surface using a similar integration, keeping mind that $\mu^{o}_{liq} = \mu^{o}_{vap}$:

$$\mu_{liq} - \mu^{o}_{vap} = \int_{P^{o}_{vap}}^{P_{liq}} \bar{V}_{liq} \mathrm{d}P = \bar{V}_{liq}\left(P_{liq} - P^{o}_{vap}\right).$$

3. Set the liquid and vapor chemical potentials for the curved surface system equal to obtain

$$\bar{V}_{liq}\left(P_{liq} - P^{o}_{vap}\right) = RT \ln\left(\frac{P_{vap}}{P^{o}_{vap}}\right) \text{ or } P_{liq} = \frac{RT}{\bar{V}_{liq}} \ln\left(\frac{P_{vap}}{P^{o}_{vap}}\right) + P^{o}_{vap}.$$

4. Write the Y–L equation for the conditions, remembering the overpressure is the difference between the liquid and vapor pressures, $P_{liq} - P_{vap}$, for the specific curved surface, both referenced to the same flat surface vapor pressure, which then cancels:

$$\Delta P = \left(P_{liq} - P_{vap}\right) = \gamma\left(\frac{1}{r_1} + \frac{1}{r_2}\right) \text{ or } P_{liq} = \gamma\left(\frac{1}{r_1} + \frac{1}{r_2}\right) + P_{vap}.$$

5. Combine 3 and 4 to obtain the *Kelvin equation*:

$$\boxed{RT \ln\left(\frac{P_{vap}}{P^{o}_{vap}}\right) = \bar{V}_{liq} \cdot \gamma\left(\frac{1}{r_1} + \frac{1}{r_2}\right) + \bar{V}_{liq}\left(P_{vap} - P^{o}_{vap}\right)}. \tag{4.19a}$$

Under typical conditions the last term can be neglected so we write the approximate form

$$\boxed{RT \ln\left(\frac{P_{vap}}{P^{o}_{vap}}\right) \sim \bar{V}_{liq} \cdot \gamma\left(\frac{1}{r_1} + \frac{1}{r_2}\right)}. \tag{4.19b}$$

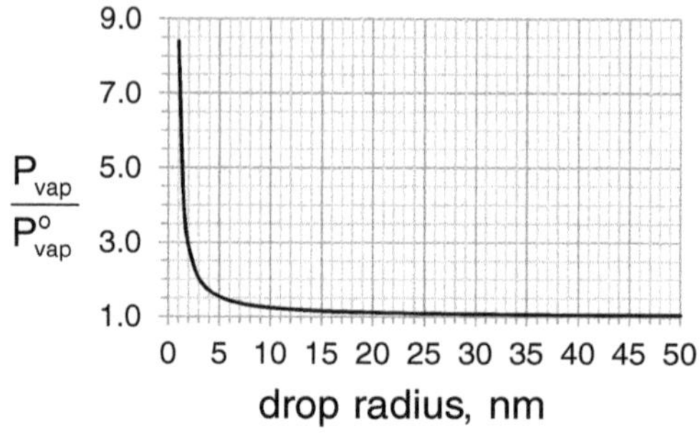

Figure 4.18. Relative vapor pressure of CCl_4 vapor above pure liquid at 298 K calculated as a function of drop radius from equation (4.19b). Notice the sharp upturn in vapor pressure when the drop radius reaches the several nanometer scale and the overpressure factor rises towards an order of magnitude.

We can apply the Kelvin equation. with no loss in significance by using a simple spherical surface with $R_1 = R_2 = r$. So let's start with a liquid drop and look at the vapor pressure versus r, as illustrated by the plot in figure 4.18 of calculated values of $\left(\frac{P_{vap}}{P^{o}_{vap}}\right)$ versus drop radius for CCl_4 at 298 K.

Similarly, we could analyse water vapor pressure above a water drop. For example, using the standard value of surface tension around ambient temperatures, a simple calculation shows that a 20 nm diameter water drop is in equilibrium with a vapor pressure that is ~11% times that of a flat-water surface. The overpressure of small drops has profound impacts in condensation phenomena. Imagine a system in which vapor pressure of some substance is exactly at the saturation value at some specific temperature. Will the vapor instantly condense to liquid? Thermodynamically it should but then if the condensation starts off with nucleation of nanoscale clusters there is a problem—the vapor pressure is not high enough to achieve equilibrium so any small drop likely will explode back to vapor where the molecules have a lower chemical potential. Small drops are often unstable for this reason. We will look at this in detail in the next section and add the parallel behavior of nucleation of nanocrystals from solution.

Let's invert the system to negatively curved surfaces, of which a common example arises for the case of a spherical bubble filled only with the vapor of the liquid. As we learned earlier the surface tension overpressure across a liquid/vapor interface is always exerted on the phase around which the interface is wrapped. So in the case of a drop the liquid is compressed whereas in the case of a bubble the gas is compressed. Now we ask a simple question. A flat puddle of liquid exists which is in equilibrium with its vapor at the equilibrium pressure P^{o}_{vap}. A cavity suddenly opens within a liquid and fills with vapor. What is the vapor pressure in the cavity? Apply the Kelvin equation, and you see that the internal bubble vapor pressure is higher than the outside liquid pressure. So, what happens if you let the system sit around and reach equilibrium? Obviously, the vapor bubble is unstable and will collapse unless you intervene in some way to stabilize it.

Some questions for thought:

1. How can you get a saturated vapor to condense if initially formed droplet clusters are unstable? Let's be specific. Water vapor in the atmosphere reaches saturation but it does not rain. What natural factors might cause rain in spite of the droplet size problem. Man has intervened to solve this problem from time to time by flying a plane through the clouds and dispersing AgBr or similar powders. How does this work? What happens typically in nature that allows rain to form quite efficiently when ideally it should not.
2. Consider a pot of water that is exactly at the boiling temperature but does not boil while you wait and wait. You heat the water to a higher temperature and the water suddenly bumps, splashing on the stove. What might be the problem? We addressed this question earlier but now we have more information at our disposal. What different things could you do to overcome the problem and how do they work? Why would a boiling chip help? Which way do each of these quantities change as T increases: P_{vap} and γ?
3. What do you think would happen in 1 and 2 for the case of H_2O if a bit of surfactant were added?
4. Apply the reasoning from 1 to the case of a saturated solution of an inorganic salt from which you expect large crystals to form, but none do (at least soon enough). What might you do to drive the crystallization?

4.3.7.2 Nucleation phenomena under driven conditions

The Kelvin equation got us thinking about this whole problem of the instability of small objects with respect to continuous adjacent phases under equilibrium conditions and how this underlies difficulties in getting water to boil and rain to form without pushing the conditions away from equilibrium a bit to favor these processes. Now let's use exactly the same logic, equations and approach to look at the issues in getting nanoparticles and clusters to form from solution near equilibrium conditions. We know at this point that clusters of a liquid formed by condensation from an equilibrium saturated vapor will not be stable thermodynamically because of the free energy cost of having a small radius surface wrapped around the cluster. Similarly, a nanoparticle precipitating from solution at exactly equilibrium conditions will be unstable until perhaps at a large enough size that surface effects nearly vanish. The only way to get from the continuous phase to the condensed object phase is to set the system up to be off equilibrium with a higher chemical potential of the 'reactant' continuous phase than the 'product' condensed phase. We need to adjust the chemical potential offset sufficiently to be able to get through the costly high S/V part of the nucleation and growth. Since at constant T and P the chemical potential is equal to the Gibbs free energy we can use these two quantities interchangeably. With that, let's follow the free energy trail of the growth of an incipient cluster or nanoparticle under driven conditions, as illustrated in figure 4.19.

To be general, we consider both nucleation and growth of a nano-object by vapor condensation and a nanoparticle driven by precipitation from solution with the ambient P and T held constant. In the first case the system is driven by an overpressure

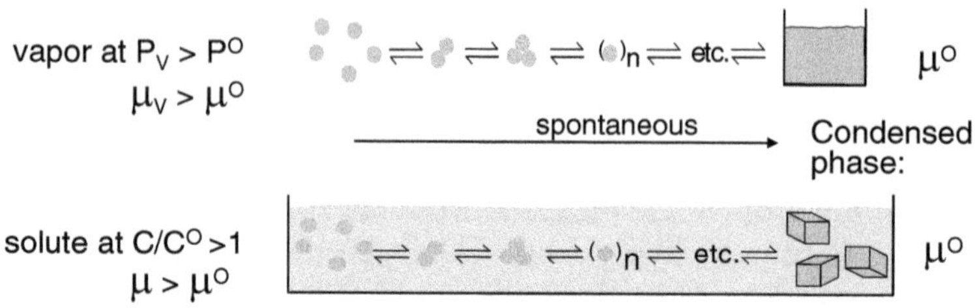

Figure 4.19. Top: Illustration of a vapor molecule progressively forming larger clusters molecule by molecule with eventual formation of a bulk condensed phase. The reactant vapor is held at a higher chemical potential than that of the bulk solid by supersaturating the vapor at the condensation temperature. Bottom: Similar to the vapor case, solute molecules at a supersaturated concentration progressively form clusters with eventual formation of the bulk solid. In both cases the system is pushed off equilibrium by super saturation to drive spontaneous formation of the condensed product.

of the vapor, $P_V > P^o$, where P^o = the equilibrium saturated vapor pressure above a flat liquid surface, which is equivalent to a chemical potential offset $\mu_V > \mu^o$ or equivalent free energy offset $\Delta G > 0$, where ΔG = the free energy difference between reactant (vapor) and product (drop) states. In the second case the system is similarly driven by a super saturation of concentration C/C^o, where C^o = the saturated solution concentration in equilibrium with a large object approaching infinite size, or by the equivalent offset of chemical potentials or free energies, $\mu > \mu^o$ or $\Delta G > 0$.

Now we write the change in free energy as each cluster is formed in turn by adding a molecule. To make things simple we will neglect complicated cluster shapes, which surely will happen at the few molecule (or atom) scale, and assume each cluster is spherical with radius r. This lets us take the surface tension contribution into account without any fancy models. But, keep in mind that the idea of surface tension in terms of a simple missing atom model is not really valid at the smallest sizes where there really is no true bulk structure so our approach will be a crude approximation at that scale, but will get better as the cluster size grows. For the arbitrary case of an n-molecule cluster with radius r forming by vapor condensation we have

$$\Delta G(r) = \left(\frac{n}{N_{Av}}\right)\left(\mu_L^o - \mu_V\right) + 4\pi r^2\gamma = -nRT\,\ln\left(\frac{P_V}{P^o}\right) + 4\pi r^2\gamma,$$

where N_{Av} = Avogadro's number, n = the number of molecules in the cluster, μ_L^o = standard chemical potential for a pure flat surface liquid, μ_V = chemical potential of the vapor and P_V and P_V^o are the actual and the standard vapor pressures. At equilibrium with no surface effects, $\mu_L = \mu_L^o = \mu_V^o = \mu_V$. To set the system up to be driven we arrange to have $P_V > P_V^o$ and $\mu_V > \mu_V^o$, μ_L^o which then makes the sign of $\Delta G(r)$ vary between + and − depending on the contribution of the positive surface work terms. Of course, referring back to the mini-thermo review, $(dG)_{T,\,P} = dW_{non\text{-}PV}$ and if $dG > 0$ then $dW_{non\text{-}PV} > 0$ and the surroundings must do the work on the system, which means that the only way the reaction can go forward is for you to arrange to put work in from the surroundings, thus the reaction is not spontaneous. Conversely, if $dG < 0$ then the system spontaneously does the work for free and you (in the surroundings) can harvest the benefits (the products).

Since we want to use r as the growth variable, we need to connect n to r. We do this by estimating how many molecules are contained in a sphere of radius r using

$$n(r) = \frac{\frac{4}{3}\pi r^3}{\left(\frac{\bar{V}_L}{N_{Av}}\right)}, \quad \text{where } \bar{V}_L = \text{molar volume of the liquid.}$$

Now we combine this into the ΔG equation above to obtain

$$\boxed{\Delta G(r) = -\left(\frac{\frac{4}{3}\pi r^3}{\bar{V}_L}\right) RT \ln\left(\frac{P_V}{P^o}\right) + 4\pi r^2 \gamma = -ar^3 + br^2}, \tag{4.20}$$

where a and b are used as shorthand notations for the coefficients of powers of r. Now we are set up to watch the competition between the volume (a) and surface (b) contributions with increasing r. Naturally the r^3 term will eventually win out over r^2 but let's see at what sizes of drops (or nanoparticles) the race is unmistakably won. This is shown in the hypothetical behavior in the plots in figure 4.20.

Overall, the main ΔG curve starts at zero, increases to a maximum at $r_{critical}$, then decreases, crossing the zero line at r_o and finally becoming negative. The top curve shows the positive br^2 surface contribution to the work of forming the clusters, which mitigates against the process, and the bottom curve (dashed line) shows the negative ar^3 contribution which drives cluster formation. When the overall ΔG value becomes negative the reaction thereafter is completely spontaneous

There are several points to note in the plots. First, the point of the maximum ΔG value is called the critical nucleation point and the associated radius is $r_{critical}$. Before this value cluster growth becomes increasingly uphill and cluster formation is

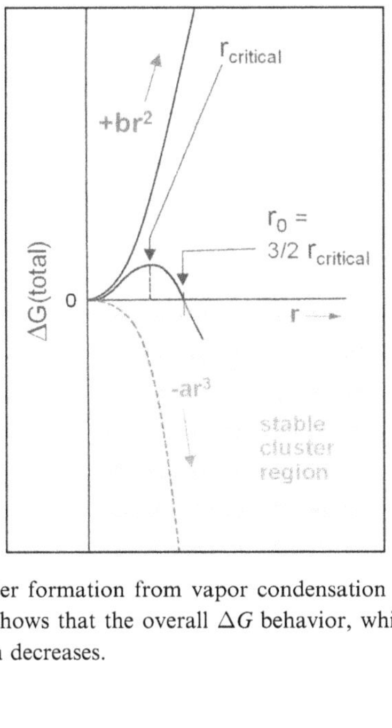

Figure 4.20. Plots of ΔG of cluster formation from vapor condensation at a fixed temperature versus the cluster radius. The middle curve shows that the overall ΔG behavior, which starts at zero, increases until a critical radius is reached, and then decreases.

statistically improbable so only a very small number of clusters form by fluctuations and these are unstable so tend to dissociate rapidly to go back downhill in free energy. At r_{critical} the next evolving clusters have lower free energies than $\Delta G_{\text{critical}}$ (or lower chemical potentials) and in order to dissociate must go uphill in free energy, which now provides a work barrier (remember, this is a free energy or work barrier, not just an energy (U) barrier) and allows these clusters to have a longer lifetime than those with $r < r_{\text{critical}}$. As a consequence, r_{critical} is considered to be the point at which nucleation starts irreversibly. The subsequent point where ΔG crosses the zero line at r_o is the transition into the fully stable cluster region where clusters will continue to grow larger until all the vapor (or reactant to be general) is consumed or something is done to intervene in the process. Note this point since it underlies methods to control the size distribution of clusters.

Here are some useful relationships that can be obtained for the critical point and transition point by respectively setting $\left(\frac{\partial \Delta G}{\partial r}\right) = 0$ and $\Delta G = 0$:

$$r_{\text{critical}} = \frac{2b}{3a} = \frac{2\gamma \bar{V}_L}{RT \ln\left(\frac{P_V}{P^o}\right)}$$

$$r_o = \frac{b}{a} = 3/2 r_{\text{critical}}.$$

Water condensation is a key meteorological process in the atmosphere and has been intensely studied as a function of a wide variety of atmospheric conditions. A simple plot of ΔG versus r of the condensation process at 298 K calculated from equation (4.17) for an over pressure of $P/P^o = 1.10$ is shown in figure 4.21. The critical radius occurs for ~6 nm water droplets containing ~10^4 molecules. So at least according to the ideal behavior in equation (4.17) condensation will not occur until drops of this size

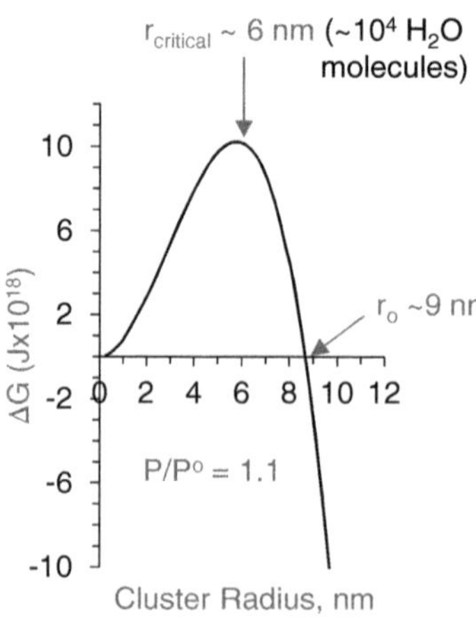

Figure 4.21. The calculated dependence of ΔG versus r for the condensation of water vapor at 298 K into water droplets driven by a vapor overpressure of $PV/P^o = 1.10$. The critical point occurs at a radius of ~6 nm, which is equivalent to a cluster of ~10^4 H_2O molecules. The calculations were done using equation (4.20).

are formed, whatever amount of time that takes. Smaller drops simply explode and return the vapor to the atmosphere a high fraction of the time. But eventually enough reach the critical radius and condensation ensues. This is sort of like trying to roll heavy stones up a hill; eventually you will get a few to the top through persistence (recall the Greek myth about Sisyphus). The kinetics of these processes is a very complex story and requires the application of statistical mechanics and thermodynamics but has been well studied because of the extreme importance in fields ranging from meteorology to nanoparticle synthesis. Unfortunately, the details are outside the scope of equilibrium thermodynamics, but we can return briefly to this issue in the dynamics chapter.

Finally, note again that the above discussion and equations applies equally well with obvious changes to the case of solutes crystallizing from saturated solutions. For example, one can make nanoparticles of CdSe by precipitation from solutions of Cd^{+2} and Se^{-2} ion species. In essence, this would be a highly overdriven system since CdSe is very insoluble with a very low solubility product, so the saturated equilibrium ion concentrations are typically orders of magnitude below the concentrations used to form the nanoparticles. Nonetheless, the principles of nucleation sequence and the critical radius are still operative, though the kinetics can be quite complex because of the highly non-equilibrium condition.

Some more questions for thought:

1. Typical condensation of H_2O vapor in the atmosphere (rain) usually occurs much easier than predictions from the Kelvin equation. How might this happen in terms of temperature gradients and dust particles?
2. In the case of crystallization of solutes from saturated solutions, why might scratching the container inner walls contacting the solution be of help in starting nucleation?
3. Looking at the free energy equations, how might you drive the nucleation of a solid crystal from its pure melt? You cannot supersaturate so what can you change and how does that work?
4. Could you construct an equation to estimate $r_{critical}$ in terms of whatever you are varying to drive the crystallization?
5. Is there some simple way to apply the fundamental relationships in nucleation to the case of boiling a liquid?

4.3.7.3 The equilibrium shape of crystals

Crystal shapes vary widely, as illustrated by the two examples in figure 4.22. What is the exact ideal shape of a crystal which emerges from nucleation and growth in some near equilibrium system? Which facets are exposed in what proportions as the crystal continues to grow? The underlying principles of equilibrium crystal shapes is very important in developing methods to form useful materials such as nanoparticles into specific desired shapes during their grow stage and even post-growth in terms of re-shaping during some annealing process.

How to approach this problem? Since we have an equilibrium situation, we look for the type of work involved to reshape. And that has to be surface work since at

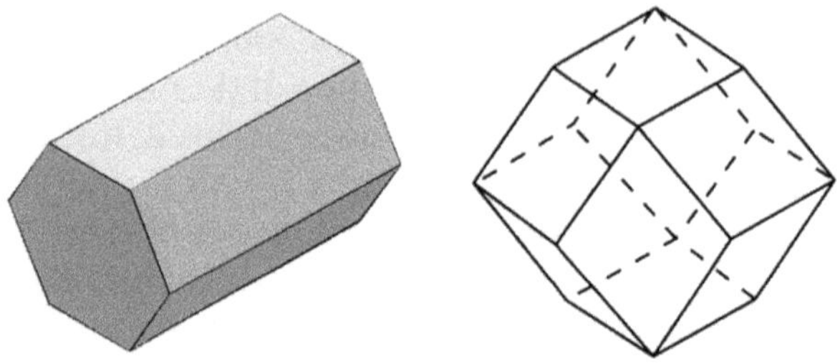

Figure 4.22. Illustrations of different crystal shapes showing distinct facets.

phase equilibrium there is no work penalty in moving molecules between phases so we can transport molecules selectively on and off different faces to change the crystal shape. Alternatively, for an isolated blob of molten material of constant mass, as the temperature cools, constituent atoms or molecules internally diffuse back and forth between the different faces that are forming, with the net flow driven by the differences in the internal surface pressures for the faces of different surface tension. For constant P, T conditions the net effect of the reversible surface work processes is changes in the total excess Gibbs free energy G^S of the crystal in which each jth facet makes a contribution according to its fractional area and G_j^S, or equivalent γ_j^S, value. In the end we should have a minimum value of the total free energy for any crystal of a fixed volume (or more properly, mass). The facets are just specific crystal planes set at specific angles to each other according to the structure of the crystal unit cell, so we need to ask the question, do different crystal planes have different values of γ? They must! Remember the missing atom model and from that you can see that different planes have different numbers of missing coordinations so the values of U^S must differ, and even neglecting the details of S^S, clearly the values of γ must therefore differ. We will learn more about how one might measure surface tension values for different crystal surface planes of a material in section 4.4, but for now let's assume we have these values in hand so we can proceed to see how to use these to predict crystal shapes.

The central requirement then is to minimize the total surface excess free energy of fixed mass as given by the integral $G^S = \int_s \gamma(\mathbf{n})\mathrm{d}s$, where $\gamma(\mathbf{n})$ = the value of the surface tension for a patch of area ds with unit vector **n** normal to the surface patch with integration over the entire surface. This equation is very general and includes surfaces with bumps and really small local features. The rigorous solution for the minimization is contained in the mathematics of well-known Wulff theorem which treats the problem in terms of a geometric surface constructed from vectors of magnitude of the local surface tensions and directions given by the [hkl] direction vector of the plane. This can be a bit tedious to explain and work with but fortunately the simple equation first proposed by Wulff for polyhedral crystal shapes (the type you are used to and ones that avoid complicated features like hills and valleys) is easy to use:

$$\boxed{\frac{\gamma_{hkl}}{r_{hkl}} = \frac{\gamma_{h'k'l'}}{r_{h'k'l'}} = \cdots = \rho,} \tag{4.21}$$

where γ_{hkl} and $\gamma_{h'k'l'}$ are the surface tension for any of the different surface planes, r_{hkl} and $r_{h'k'l'}$ are the distances from the center of the crystal to the center of each facet surface, as illustrated in the 2D projection in figure 4.23, and ρ = a constant which applies to all facets on the crystal.

As a simple example we predict the shape of a cubic crystal with $\gamma_{110} = 140$ mJ m^{-2} and $\gamma_{100} = 165$ mJ m^{-2}, which from equation (4.21) gives $\frac{r_{100}}{r_{111}} = 1.18\ \rho = 1.18$. A 2D projection is shown in figure 4.24, where the high surface tension (100) facet is seen to be considerably smaller than the low energy (110) facet. The crystal is not a perfect cube shape since the higher energy {100} family of faces can adjust their fractional contribution to the perimeter to attain equal compressive pressures with the higher length, low energy family of {110} faces. Notice that the introduction of the short {100} segments round off the crystal corners. As with bubbles which become spherically shaped to avoid sharp bumps, a crystal will attempt to round off corners by using short segments of high energy faces which are at appropriate angles to the low energy faces that dominate the facets. We see here nature's attempt to make a sphere out of every object within the constraints of having periodic surface planes that force specific alignments of the atoms or molecules according to their arrangements in the unit cell.

Why does the Wulff equation work and what is the underlying behavior of the system? The overall idea is as follows. First appreciate the fact that the surfaces of the crystal exert a compressive force on the bulk which drives up the chemical potential,

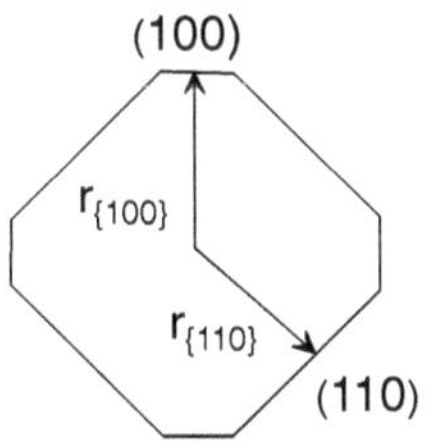

Figure 4.23. Schematic of a 2D crystal shape with (100) and (110) faces. The vectors $r_{(100)}$ and $r_{(110)}$ extend from the origin at the center of the crystal to the center of each of the faces.

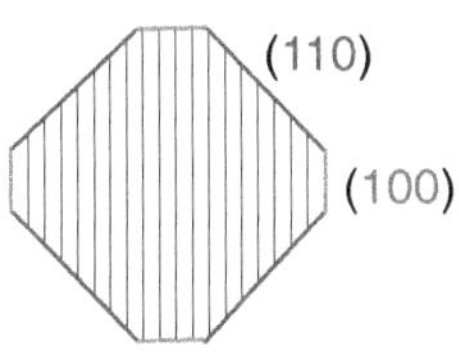

Figure 4.24. Schematic of a 2D projection of a hypothetical cubic crystal with {110} (blue lines) and {100} (red lines) facets at an equilibrium shape calculated from the Wulff equation using $\gamma_{110} = 140$ mJ m^{-2} and $\gamma_{100} = 165$ mJ m^{-2}.

just as for a simple spherical drop. This essentially means that each crystal facet is under compressive tension, trying to pump molecules back into the bulk, and all the facet surfaces compete until the pumping pressures are equal everywhere and the crystal comes to mechanical equilibrium. Since the chemical potential of a substance is directly proportional to the local pressure ($d\mu = \bar{V}dP$, equation (T3b)) you can see that this is just a way to adjust the chemical potential of the atoms or molecules to be equal everywhere. These compressive forces are the same ones that resist expanding the crystal, just like the compressive surface tension forces that wrap around a bubble and resist expansion. But now how do we describe the compressive force on a crystal given the different facets with different surface tensions?

One way to do this is just to borrow from the previous approach to the Y–L equation in section 4.3.2.4. The basis of the Y–L equation is that a patch of curved surface when expanded at constant shape to a larger area requires an overpressure proportional to the surface tension. If we replace the curved surface with a flat parallelogram of arbitrary shape and replace the radii of curvatures with the distance r_{hkl} from the crystal center to the center of the flat surface, we get equation (4.18a). This works because the proportions of the shape are maintained with increasing surface area expansion and the area, which requires γA work for expansion, is always proportional to the distance r_{hkl}.

Now some thought questions:

1. How does the Wulff relationship explain the tendency for crystals to not have sharp protrusion, edges or cusps?
2. Do you think the Wulff relationship can explain why surfaces reconstruct? Why or why not? Would there be an effect on crystal shape because of different reconstruction at different facets when the crystal comes to final equilibrium?
3. How might surfactants affect crystal shape during solution growth? What sort of surface chemistry would be needed to change shape? Does this strategy fit with the missing atom model? How?

4.4 Contacting media: interfacial tension

Up to this point we have considered only the thermodynamics of objects comprised of a single pure substance, typically in equilibrium with an adjoining vapor phase to form an interface whose properties arise from the intrinsic instability of atoms or molecules confined to a thin transition region with a characteristic thickness δz_{AA} of $\sim$1 or slightly more molecular or atomic diameters. For the case of atoms δz_{AA} can be sub-nanometer scale whereas for large molecules such as polymers, the characteristic average molecular size or correlation length can be quite large and δz_{AA} may be tens of nanometers (refer back to figure 1.3 in chapter 1).

Analogously, in the case in which two objects of different substances A and B are in intimate contact a new, intrinsically unstable A–B interface region is formed of some thickness δz_{AB} which we can expect to be at the atomic or molecular size scale. The main difference with the surface or interface of a single pure substance is that in the A–B interface region the two components must mix to some extent. In order to understand the behavior of such interfaces we need to characterize this instability not only in terms of the properties of the pure constituent substances but also their interactions on mixing. One simple way to start is to look at the work required to reversibly make a perfect break in an A–B interface by pulling it apart with an applied force. This will help us to define interfacial tension. So let's put on our work gloves and pull apart an A rod welded to a B rod with a perfect A–B interface.

4.4.1 Work of disjoining—mechanical definition of an A–B interface

Referring to the A–B rod in figure 4.25, we see a region just where the two rods join which contains the result of how the A and B atoms or molecules arrange themselves when the rods are perfectly joined. Now apply opposing forces at each end and magically pull the rod apart perfectly in a reversible way exactly at the interface region (not always easy if the weld is done by a skilled craftsman and one or both of the rods have lots of defects). Borrowing from the logic in the same experiment in figure 4.4 for a rod of a pure substance, we set the work equal to the cost of making two new A and B surfaces, which is just $\gamma_A + \gamma_B$.

But wait, is not there some work associated with the interface region itself? To get the two clean A and B surfaces we have to unscramble any mixing of A and B atoms or molecules at the A–B interface and that must contribute something. Is that a positive or negative contribution to the total work? Look at it this way:

1. Start with the knowledge that A and B are not miscible in each other, otherwise when the two rods were pushed against each other under conditions to allow mixing at the interface, the process would not have stopped and the rods would simply make an AB solid solution.
2. On the other hand, when the AB rods were joined some of the constituent A and B molecules were each pulled out of their unstable bare surface regions

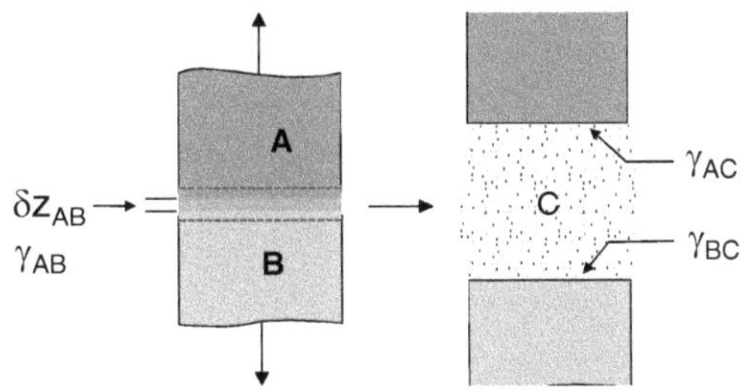

Figure 4.25. Illustration of a rod made from substances A and B which are connected at an A–B interface of thickness δz_{AB} which is described thermodynamically by the interfacial tension γ_{AB}. The rod is broken across the interface in some ideal, reversible manner by applied tensile forces to expose a pure A and pure B surface, each with surface tension γ_A and γ_B. The blue dashed lines represent hypothetical lines at which clean breaks would produce separate A and B rods with pure A and B surfaces and an isolated region which is a mixture of As and Bs.

and mixed in the interface region. This makes sense in terms of the missing atom model. Some of the As helped to make up for the missing atoms or molecules in the B surface and vice versa. But this can only go so far since obviously, in view of the immiscibility, A–A and B–B interactions are better than A–B ones and in the thin mixing layer the immiscible As and Bs are making the best of a bad situation, so to speak.

3. So now let's imagine the break occurs just a layer or so away from the interface region on each rod, as shown in figure 4.25 at the blue dashed lines, to expose two perfect A and B surfaces, which costs us $\gamma_A + \gamma_B$, and leaves some hybrid region of As and Bs including those from the interface and a few others above the break lines. What do we do with this region in order to end up at the final state of the break process with nothing but two rods with fresh surfaces? Since we already have the two rods each with a perfect A or B surface, the only reasonable thing left to do thermodynamically would be to put the interfacial region As back into the A bulk and the Bs back into the B bulk. You hopefully agree that this has to be favorable in terms of work since As and Bs are immiscible relative to their pure bulk phases so you gain in free energy putting them back in their home base, so to speak.

Writing all this out in a work per area equation we obtain

$$\boxed{\frac{\delta W}{A} = \Delta G^{S}_{\text{disjoin}} = (\gamma_A + \gamma_B) - \gamma_{AB}}, \qquad (4.22a)$$

where $\frac{\delta W}{A}$ = work per unit area at the interface, $\Delta G^{S}_{\text{disjoin}}$ = the work of disjoining the rods and γ_{AB} = the interfacial tension, which makes a positive contribution to the job of pulling the rods apart. This work is often called the work of adhesion since it is associated with the adhesive strength of the two media but this term is also used in a different way in connection with liquid drops on solid surfaces (see below) so it is

Table 4.5. Values of interfacial tensions for selected pairs of liquids.

Interface	T (°C)	γ_{AB}, (mJ m^{-2})
H_2O/cyclohexane	20–25	50.0
H_2O/hexadecane	20–25	53.3
H_2O/benzene	20–25	35.0
H_2O/ethyl acetate	20	6.8
H_2O/1-butanol	20	1.8
H_2O/heptanoic acid	20	7.0
H_2O/mercury	20	415
Ethanol/mercury	20	389

better to use the term work of disjoining. Note that since we define γ_{AB} in terms of reversible work at constant T, P then it is a free energy, ΔG, not an internal energy, ΔU and must include any ΔS contributions associated with the A–B mixing and number of ways to arrange the interactions in the interface. With that here is the formal definition of interfacial tension is the excess free energy per unit area of forming an A–B interface from pure bulk A and pure bulk B atoms or molecules.

Here is another way to view interfacial tension in terms of a mechanical model. Just as surface tension defines the compressive force that is trying to push molecules back into the bulk, interfacial tension is the compressive force or pressure trying to push A and B molecules each back to their pure phases. When you pull an interface apart everything happens at once—two new pure surfaces are formed, and interface atoms or molecules are returned to their pure phases. You cannot physically break the interface apart by itself in any process, so the interfacial tension is inextricably coupled with all the terms together in equation (4.22a) for joining or disjoining separate surfaces.

A common situation is to either form or disjoin an interface in the presence of adjoining phases of either the pure A or B substances or in the presence of a gas or liquid phase of a third substance C, in which the excess free energies of the two new surfaces formed are described in terms of interfacial tensions rather than surface tensions. For example, in the case of the presence of a substance C, equation (4.22a) becomes

$$\boxed{\frac{\delta W}{A} = \Delta G^{\mathrm{S}}_{\mathrm{disjoin}} = (\gamma_{\mathrm{AC}} + \gamma_{\mathrm{BC}}) - \gamma_{\mathrm{AB}}}. \tag{4.22b}$$

The general form of equations (4.22a) and (4.22b) is called the Dupre equation.

4.4.2 Thermodynamic basis of interfacial tension—an approximate model based on molecule–molecule pair contact energies

From the last section we learned the essential basis for describing the thermodynamics of an A–B interface but let's go a bit deeper and look at an approximate model in terms of A–B contacts or coordination and exchange interactions. Overall, we know that the interfacial tension is the reversible work or free energy per unit area to create an AB interface from pure A and B interfaces. Underlying this work is the process of transferring an A molecule surrounded by other As to be in contact with at least one B molecule so we can look at this as an exchange of coordination and treat the process as an exchange equilibrium. There are a number of possible reactant and product states, each of which depend upon how many A and B neighbors there are for each A and B species and we would need to write equilibrium constants for each of these if we were to be thorough. Instead we will make things simple and consider a limit in the exchange in which each A and B starts out in their pure phase surrounded by all As and all Bs, respectively, and then exchange places to be surrounded completely by the opposite species. On this basis we define a limiting equilibrium constant for exchange:

$$\mathrm{A(A)+B(B)} \underset{}{\overset{K_{\mathrm{xchange}}}{\rightleftarrows}} \mathrm{A(B)+B(A)},$$

which follows the pictorial description below the equation. We can define the interfacial tension in approximate terms by $K_{x\mathrm{change}}$, for which we use the simpler notation K_x:

$$\gamma_{\mathrm{AB}} = G^{\mathrm{s}}_{\mathrm{AB}} = \frac{\Delta G_x}{\bar{A}_{\mathrm{m}}} = \frac{-RT \ln K_x}{\bar{A}_{\mathrm{m}}} \sim \frac{\Delta U_x}{\bar{A}_{\mathrm{m}}},$$

where ΔG_x = free energy of exchange per mole, $\bar{A}_{\mathrm{m}}$ = area per mole of both A and B phases and we apply the further approximation that $\Delta G_x \sim \Delta U_x$. In this simple approximation all the interfacial interactions are assumed to be non-directional, attractive energy interactions, such as those that arise from van der Waals forces between nonpolar molecules. From this we obtain

$$\ln K_x \sim -\frac{\Delta U_x}{RT}, \tag{4.23}$$

which tells us that the extent of mixing at the interface, and therefore the magnitude of γ_{AB}, depend on the relative strengths of making AB interactions versus AA and BB. The proper treatment would require consideration of both energy and entropy. The latter is often messy to deal with but fortunately it turns out that for nonpolar substances on can make a reasonable approximation by neglecting entropy.

On the basis of a similar exchange equilibrium but based on an average interface environment of As and Bs, referring to figure 4.26, we will treat the interface region with following model (at constant P, V, T):

1. The interface region is localized to thickness δz_{AB} and centered at $z = 0$.

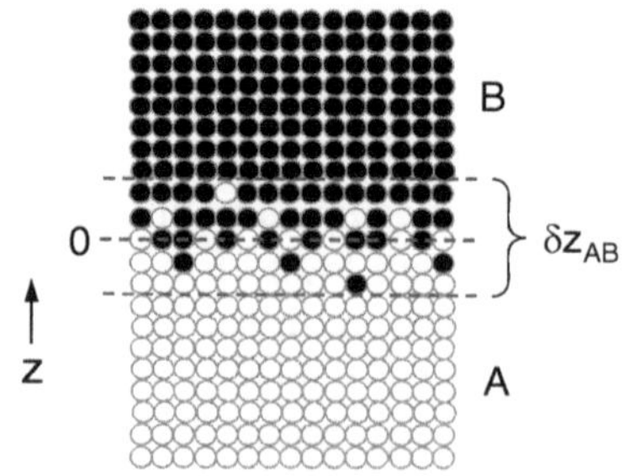

Figure 4.26. Schematic of an A–B interface showing mixing of the component particles within the interface region of thickness δz_{AB}. The z-direction is shown as increasing from the A to the B medium with $z = 0$ in the center of the region.

2. An average number of nearest neighbors are replaced on exchange with an average change in the interaction or coordination energies.
3. Coordination numbers are equal within each of the pure lattices.
4. There is no directional bonding, only uniform average A–A, B–B and A–B interactions; this is essentially the basis of a mean field approximation.
5. $\Delta S \sim 0$ for exchange; this is essentially a frozen lattice model in which we neglect any freedom of A moving around in B and vice versa.

According to equation (4.23) to avoid mixing we require $K_x < 1$ which requires $\Delta U_x > 0$, which means mixing is endothermic, consistent with less favorable A–B interactions than A–A and B–B ones. The model is based on average attraction energies so keep in mind that these are all negative since it is energetically downhill to make the bonds from isolated As and Bs. If we stated them as bond energies, implying work to pull the bonds apart, they would be positive. To adjust for that it is useful to use absolute values for comparisons when needed, rather than algebraic ones, which helps us to think about which interactions are strongest. Here are the steps:

1. Write the overall thermodynamic equation for mixing energy:

$$\Delta U_x = U_{A(B)} + U_{B(A)} - \left[U_{A(A)} + U_{B(B)}\right].$$

2. Assume all A–B bonding is the same everywhere regardless of other nearest neighbors:

$$U_{A(B)} = U_{B(A)} = U_{AB}.$$

3. Combine 1 and 2 and average the AA and BB interactions to obtain a single average energy for bonding between like atoms:

$$\Delta U_x = 2\left[U_{AB} - \left(\frac{U_{A(A)} + U_{B(B)}}{2}\right)\right] = 2\left(U_{AB} - U^{avg}_{AA,\,BB}\right),$$

where the term $(U_{AB} - U^{avg}_{AA,\,BB})$ is the difference between the average energy of A–A and B–B bonds and determines the sign of ΔU_x. If AB bonding is favorable compared to AA and BB bonding, then $(U_{AB} - U^{avg}_{AA,\,BB}) < 0$ or in absolute value terms $|U_{AB}| > |U^{avg}_{AA,\,BB}|$.

Applying this to the exchange constant we have $K_x \sim \exp\left[-\frac{(U_{AB} - U^{avg}_{AA,\,BB})}{RT}\right]$ so as AB bonding becomes stronger compared to AA and BB bonding $K_x > 1$ and the phases will tend to be miscible. What about the width δz_{AB} of the interface? It makes sense that as the exchange increases and approaches miscibility then the mixing in the interface region will increase and the interface will broaden. We can see these effects in the plot of the A–B interaction energy U_{AB} relative to the average AA/BB energies $U^{avg}_{AA,\,BB}$ versus distance as we pass through the interface in figure 4.27.

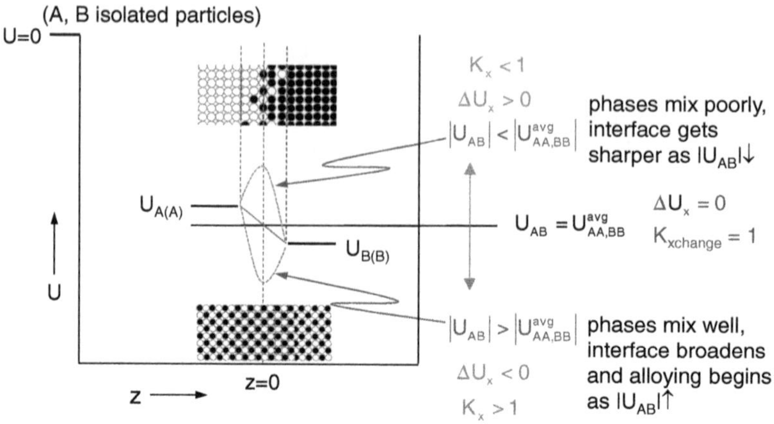

Figure 4.27. Illustration of the dependence of the A–B interface thickness on the equilibrium constant for exchange. The dependence is based on a model of the miscibility controlled by the energetics of mixing. Increasing z moves from one medium, through $z = 0$ at the middle of the interface region, and into the adjoining phase. If $|U_{AB}| < |U^{avg}_{AA,\,BB}|$ mixing is unfavorable, and the interface is sharp. Conversely, if $|U_{AB}| > |U^{avg}_{AA,\,BB}|$ mixing is more favorable, the interface becomes broad and eventually disappears as A and B become fully miscible.

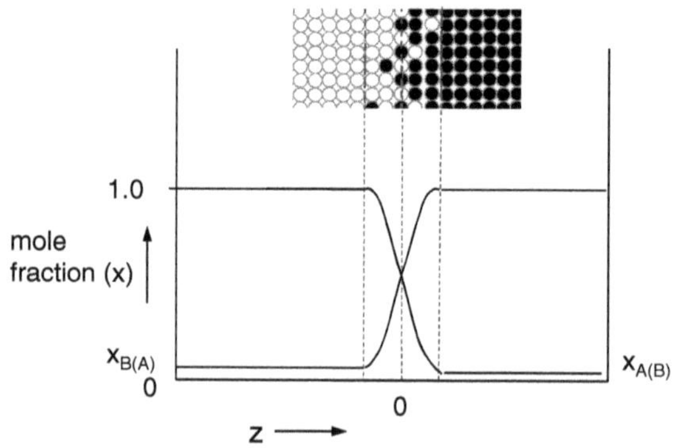

Figure 4.28. A plot of the hypothetical profile of mole fraction of each of the components through an A–B interface. In general, the profiles follow some sort of exponential shape and cross somewhere in the center of the interface region according to the strength of the A–B, A–A and B–B interactions. Notice that the mole fractions do not go exactly to zero at each side of the interface. Some material is always dissolved, even if the concentration is vanishingly low.

The main point is what you would expect. When $|U^{avg}_{AA,\,BB}| \gg |U_{AB}|$, AA and BB bonds are greatly preferred over AB, there is little mixing and the interface is sharp, as seen in the U versus z red curve which peaks uphill in energy through the interface. In contrast, when $|U_{AB}| \gg |U^{avg}_{AA,\,BB}|$ A–B bonds are relatively strong and mixing is complete everywhere, as seen in the blue curve which dips to a minimum in the interface. In between these limits the interface broadens as mixing gets better and sharpens when worse. It follows from this analysis that one can expect a plot of composition through the interface to look the hypothetical one in figure 4.28.

The actual physical thickness of the interface would depend upon the scale of the atomic or molecular sizes. Keep in mind that for different size molecules and different CNs for A in A and B in B, as well as other effects, we can expect asymmetric shaped interface profiles.

Some questions for thought:

1. Predict the relative thicknesses of these interfaces:
   a. H_2O on a pure glass slide (SiO_2).
   b. H_2O on a clean Au(111) surface.
   c. H_2O on nylon 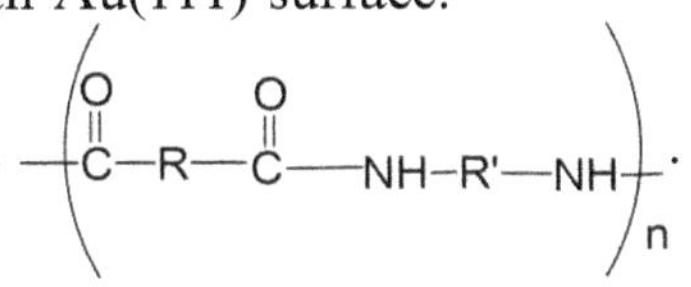
   d. Oil on water.
   e. Ether on water.
   f. Gasoline on rubber.
   g. Molten aluminum on the surface of a ceramic mold.
2. Do you think you can make a general prediction on whether the interfacial tension between two substances increases or decreases with temperature? Why or why not?
3. Would you expect that including the entropy of mixing to increase or decrease the depth of an AB interface? Why?

4.4.3 Fluid–fluid interfaces—direct measurement of interfacial tensions

Contact between two different liquid substances or liquid with vapor generates an interface which forms an equilibrium size and shape to minimize the area, in contrast to interfaces involving one or more rigid solids for which the solid resists deformation. Thus for the fluid–fluid interface the alteration of shape in response to interfacial forces provides a way to measure the interfacial tension. We have already seen how to do this for the liquid–vapor interface of a single pure substance with the example of the maximum bubble pressure method for surface tension. It follows accordingly that one can also develop various methods to measure interfacial tensions between two liquids by looking at characteristics such as the pressure to create a drop of one liquid immersed in another liquid or by measuring shapes of liquid–liquid interfaces. Using a variety of these methods over the years values of liquid–liquid interfacial tensions have been measured and tabulated and a few are listed in table 4.5. The dominant measurements have been for water with a variety of immiscible organic liquids since these represent the common systems of interest over a wide range of applications including biology, environmental science and engineering, chemical engineering and industrial manufacturing.

In the table you can notice that the size of γAB increases as the two media become more incompatible, or alternately increasingly insoluble in each other, e.g. H_2O and Hg at 415. In contrast, liquids that exhibit some significant values of solubilities are quite low, e.g. H_2O and butanol at 1.8. This fits well with the idea that as the mixing energies become increasingly favorable the interface broadens, the component molecules (or atoms) penetrate deeply into each of the two phases and ultimately the penetration is complete, and the two liquids simply dissolve in one another.

Let's look at the simplest aspects of interfacial tension in terms of the overpressure ΔP created within a spherical region surrounded by an interface, just exactly like the simple bubble or drop that we encountered earlier in sections 4.3.2.1 and 4.3.2.3. For an isolated bubble or drop of radius r the overpressure is given by equation (4.7), $\Delta P = \frac{2\gamma}{r}$. Looking at the pictures in figure 4.29 for a drop of one liquid immersed in another immiscible bulk liquid and vice versa, you can readily see that the two interfaces are identical and therefore the overpressures arising from the interfacial tension are equal, just as for the simple isolated bubble and drop systems. This gives the equation

$$\boxed{\Delta P = \frac{2\gamma_{\mathrm{AB}}}{r}}, \tag{4.24}$$

where we replace surface tension in the bubble equation with interfacial tension.

Following what was done with the Kelvin equation (equation (4.19a)), we can apply the interfacial drop equation (4.19a) to a situation in which solute molecules partition between a bulk medium A and an immersed drop of substance B of radius r, as illustrated in figure 4.30 for the case of an oil drop in water. How does the solute

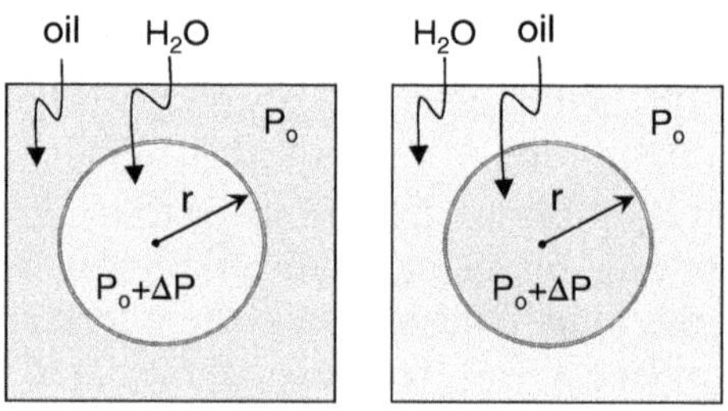

Figure 4.29. Two illustrations of the same liquid–liquid interface created in two ways. Left: A water droplet of radius r immersed in an oil. Right: A droplet of the same oil of radius r immersed in water. In both cases the internal overpressure ΔP in the drops caused by interfacial tension is identical.

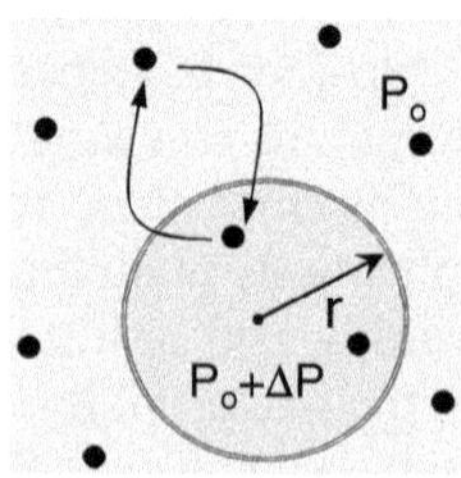

Figure 4.30. Illustration of an oil drop of radius r immersed in an aqueous medium. A solute is dissolved in the aqueous phase and partitions between the two phases to reach some equilibrium concentration in each phase with the ratio described by a partition coefficient $K_{\mathrm{o}}^{\mathrm{w}}(r)$ which is a function of the drop radius.

concentration in the drop change with the size of the drop? As you might expect, first, the smaller the drop the more unfavorable the chemical potential of any species residing in the drop, and, second, the equation we derive will apply equally well to the inverse situation of an A drop immersed in medium B. Typically, the A B combination is an organic oil type of substance in contact with water so we will use this combination with an oil drop in the example below.

This is a problem that has both surface work and mass transport work, so we need to consider the interplay between the two. To make the problem simple we can assume that the volume of the water phase is far larger than that of the oil drop such that the concentration of solute in the water phase c_{w} is constant. In contrast, the solute concentration in the oil drop, $c_{\mathrm{o}}(r)$, is a function of r.

The main relationships in the problem should be fairly clear at this point since we have encountered almost all the equations before. Here is the set-up in brief:

1. Describe the partitioning of the solute between an oil and a water phase in terms of an equilibrium constant (essentially a partition coefficient or constant):

$$K_{\mathrm{o}}^{\mathrm{w}}(r) = \frac{c_{\mathrm{w}}}{c_{\mathrm{o}}(r)}.$$

2. Write the partition coefficient in terms of the equivalent chemical potentials (assuming ideal behavior):

$$-RT \ln K_{\mathrm{o}}^{\mathrm{w}}(r) = \mu_{\mathrm{w}} - \mu_{\mathrm{o}}(r).$$

3. Within the oil drop the chemical potential of any species will vary with the local over pressure:

$$\mu_{\mathrm{o}}(r) = \mu_{\mathrm{o}}(\infty) + \int_{P_{\mathrm{o}}}^{P_{\mathrm{o}}+\Delta P} (N_{\mathrm{Av}} v_{\mathrm{s}}) dP = \mu_{\mathrm{o}}(\infty) + (N_{\mathrm{Av}} v_{\mathrm{s}})\Delta P,$$

where v_{s} = the volume occupied by a solvent molecule when in the oil phase, this typically is approximated as the molecular diameter and whatever little bit of 'rattle' space is available between the surrounding oil molecules. So as the molecule in its cage is squeezed by the pressurized surrounding molecules the chemical potential rises.

4. Substituting ΔP from equation (4.24) we obtain

$$\mu_{\mathrm{o}}(r) = \mu_{\mathrm{o}}(\infty) + \frac{2\gamma_{\mathrm{ow}}}{r}(N_{\mathrm{Av}} v_{\mathrm{s}}).$$

5. Substitute the expression from four back into that for two to obtain

$$-RT \ln K_{\mathrm{o}}^{\mathrm{w}}(r) = \mu_{\mathrm{w}} - \left[\mu_{\mathrm{o}}(\infty) + \frac{2\gamma_{\mathrm{ow}}}{r}(N_{\mathrm{Av}} v_{\mathrm{s}})\right] = [\mu_{\mathrm{w}} - \mu_{\mathrm{o}}(\infty)] + \frac{2\gamma_{\mathrm{ow}}}{r}(N_{\mathrm{Av}} v_{\mathrm{s}}).$$

6. Recognize that $[\mu_w - \mu_o(\infty)]$ is just the chemical potential difference of the solute between bulk water and an infinitely large oil drop (bulk oil) so combining with five and moving terms we have

$$RT\ln\left[\frac{K_o^w(r)}{K_o^w(\infty)}\right] = \frac{2\gamma_{ow}}{r}(N_{Av}v_s) \text{ or } K_o^w(r) = K_o^w(\infty)\exp\left[\frac{1}{RT}\frac{2\gamma_{ow}}{r}(N_{Av}v_s)\right],$$

where $K_o^w(\infty)$ = the partition coefficient between bulk oil and water phases. This equation tells us that the oil-water partition coefficient depends exponentially on the oil drop size and that partitioning of the solute from oil to water becomes exponentially better as the oil drop becomes smaller and approaches the bulk value as $r \to \infty$.

Alternately, by using the definition of the partition constants in terms of the concentrations we have

$$\frac{c_o(r)}{c_o(\infty)} = \exp\left[-\frac{1}{RT}\frac{2\gamma_{ow}}{r}(N_{Av}v_s)\right]$$

And since it does not matter which phase is the drop then to be general, we have for any solute s portioning between any spherical drop of radius r immersed in a different medium:

$$\boxed{\frac{c_o(r)}{c_o(\infty)} = \exp\left[-\frac{1}{RT}\frac{2\gamma_{AB}}{r}(N_{Av}v_s)\right]}, \tag{4.25}$$

where A and B are the contiguous phases and v_s = the volume occupied in the drop medium by the solute. Notice that this equation is very general and can be applied to any two immiscible phases. In this equation we neglect non-ideal behavior of the solute and assume that the solute partitioning applies to an equilibrium in which each medium is saturated with the other one, namely, water-saturated oil and vice versa.

An example of the partitioning versus r from equation (4.25) is given in the plot in figure 4.31 for the hypothetical case of a molecule with a molecule diameter of ~1.0 nm dissolved in a drop of oil of radius r immersed in water at 25 °C with γ_{ow} ~50 mJ m^{-2} (see table 4.5). The concentration of the solute in the aqueous phase is arbitrary. The plot shows a sharp roll off of the ratio of the solute concentration at radius r to the concentration in the bulk (infinite radius) as the oil drop radius decreases to ~50–100 nm. This illustrates an interesting point that if one were trying to extract a solute molecule out of an aqueous phase, e.g. a contaminant molecule, by shaking with a hydrocarbon or similar medium, breaking the extracting medium up into nanoscale drops is actually less efficient than larger ones on the basis of the interfacial thermodynamics. There is some advantage to smaller drops, of course, in terms of faster absorption of solute because of the larger area of interface formed with the small drop size. So, S/V ratio can work both ways.

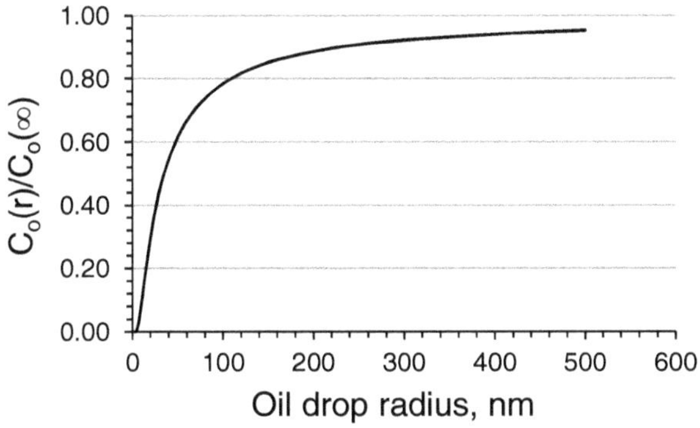

Figure 4.31. A plot of the equilibrium concentration of a solute extracted from an aqueous phase into an oil drop as a function of drop radius as calculated from equation (4.22a). The concentration is given as the ratio of the concentration for the drop radius relative to the concentration for infinite radius (or bulk oil phase). The extraction becomes highly inefficient as the drop radius decreases into the 50 nm and lower range. Details of the calculation are given in the text

4.4.4 Solid–liquid media—liquid drop contact angles

4.4.4.1 General description

In the case of liquids involving contact with solid, rigid media, only the liquid is able to deform in response to the interfacial stress. For example, a drop of water on a stiff piece of polyethylene plastic will bead up into a drop that approaches a spherical shape while the rigid plastic stays flat. While the resistance of the solid to deform limits what can be learned directly about the interfacial tension, it does allow measurement of the relationship between three of the surface thermodynamic properties of the two media. Since the experiment is very easy to perform it has become a widely used measurement. Let's see how we can use this to advantage to learn something useful. A simple illustration of a contact angle is given in figure 4.32 where a liquid drop is sitting on a solid surface in equilibrium with the vapor phase of the drop liquid. Gravity pulls on the liquid and tends to make it flatten out but this can only go so far as the liquid resists further spreading to avoid increasing its surface area with its associated cost of work of surface expansion. What holds the drop in place and what can we learn from the final shape?

4.4.4.2 In-plane balance of forces at the pinning line—the Young–Dupre equation

Of course, if we wanted a complete analysis of the equilibrium thermodynamics of the drop, we could do a full measurement of the complete drop shape and use equations such as the Y–L equation along with gravity force effects. There is a much easier way, however. If one looks at the edge of the drop, as shown in figure 4.32, around the perimeter of this line is the locus of points at which the liquid, vapor of the liquid and the solid support phase all meet. This three-phase line is pinned in place by the balance of three competing surface forces, as shown in the expanded cross section in figure 4.32 (left bottom) where at any point along the three-phase

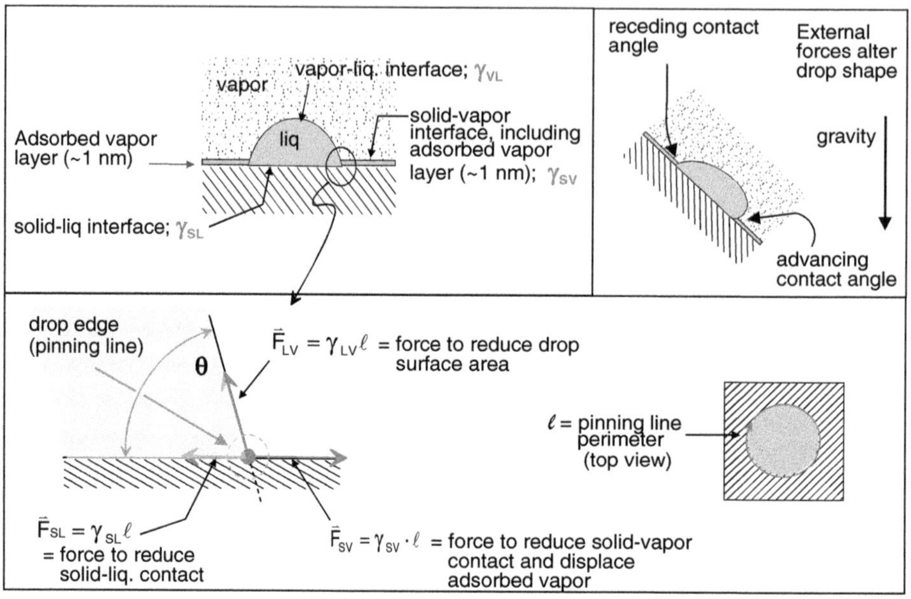

Figure 4.32. Schematics of a drop of a pure liquid on the solid surface of another pure substance. Upper left: Under the influence of gravity a liquid forms a circular drop adhering to the solid surface. At the edges the drop merges with a thin layer of vapor adsorbed on the solid surface. Each of the three interfaces has an associated characteristic interfacial tension, γ_{SV} (solid/vapor), γ_{LV} (liquid/vapor) and γ_{SL} (solid/liquid). The region circled on the right side of the drop next to the solid surface contains the line at which all three phases meet. This line is the pinning line which defines the contact perimeter of the drop. Upper right: Under the influence of gravity when the solid is tilted the drop will distort and if not sufficiently adherent to the surface will move downward. Bottom: A magnified cross section of the region containing the three-phase contact line. The red dot is the point at which the phases meet. The compressive forces of each interface to reduce the area of that interface are balanced at this point. On the right is a top view of the three-phase pinning line of perimeter length ℓ.

line each of the component interfacial tensions, γ_{LV}, γ_{SV}, γ_{SL}, where L, V, S signify liquid, vapor and surface, is competing to shrink the associated surface area and these forces come into balance. This is a line tension balance, analogous to appropriately tightening or loosening a belt until the compression just balances the resistance. Here are the three forces and the balance across their components in the z-direction of the surface plane (pay attention to the directions):

1. Given an angle θ for a tangent line defined at the drop edge (see figure), the force trying to shrink the drop surface area pulls along this direction all around the pinning line to shrink the perimeter and is given by

$$F_{LV} = -\gamma_{LV}\ell\cos\theta.$$

2. The force pulling to shrink the L/S interfacial area and shrink the perimeter is

$$F_{LS} = -\gamma_{SL}\ell.$$

3. The force pulling to shrink the S/V interface and increase the perimeter is

$$F_{SV} = \gamma_{LS}\ell.$$

When in balance we obtain for the net force

$$\cos\theta = \frac{(\gamma_{SV} - \gamma_{SL})}{\gamma_{LV}}. \tag{4.26}$$

Equation (4.26) is called the Young–Dupre (Y–D) equation, or more commonly, just Young's equation.

4.4.4.3 Practical value of contact angles and common conventions

Here is the major value of contact angles. Measurements are rather simple. You just need a telescope with cross hairs for measuring angles (it is better to use a video camera to keep from getting eye strain) and by using different liquids of known surface tension you quickly obtain qualitative information about the wetting characteristics of the surface of interest. In addition, when an experiment is set up properly with a pure liquid, well defined, clean solid and under equilibrium with vapor at a fixed temperature, you obtain a useful quantitative relationship between the three interfacial tensions. First let's look at the practical value of contact angles and after that try to dig out some of the thermodynamics of surfaces.

Let's take a quick tour to learn why contact angles are one of the most widespread surface measurements in fields ranging across biology, biomedical engineering, materials science, physics, chemistry, geology, and essentially you name it, with diverse applications such as in industrial product development and quality control. The variation of contact angles is illustrated in figure 4.33. When the substrate surface is very wax-like or hydrophobic (water hating), for example, and a drop of water is placed on top, the water would bead up with a very high contact $> 90°$, as shown on the right, and for the opposite case of say clean glass slide with an affinity for water (hydrophilic or water loving), the drop would spread to give a near zero angle, as shown on the left. These two limits are typically referred to as non-wetting and wetting. In general, with a polar liquid on what we might call a polar surface, such as clean glass, we obtain low angles whereas with polar liquids such as water on

Figure 4.33. Illustration of two limits of liquid drop contact angles. Left: the liquid interacts strongly with the support surface relative to the liquid surface tension force and vapor/solid interface force which causes the drop to spread out over and reduce the contact angle to near zero. Right: the liquid surface tensions forces dominate, and the drop beads up with a contact angle exceeding 90°.

highly nonpolar surfaces we obtain near spherical drops. In the case of some highly nonpolar surfaces constructed in special ways the contact angles with water can reach angles in the range of $\sim 140°$ or so. These are termed *super hydrophobic* and are of special interest in fields such as biomedical engineering. It is easy to see that contact angles make good qualitative tests of the types of liquids and surfaces in terms of polarity. But surface tension values by themselves are very important, not just polarity. Put a drop of mercury, which is nonpolar, on a glass slide and it makes quite a large contact angle. There is lots of information in a contact angle.

There are also some non-ideal behaviors and time-dependent effects that happen in measuring contact angles. Results do not always reflect thermodynamic equilibrium. Referring to figure 4.32 (upper right-hand panel), notice that if a uniform force is applied in one direction, such as tilting the drop support at an angle to the gravity force, you would expect that the drop would just change location, such as roll off the surface when tilted. But in most cases, especially for polar liquids such as water, the pinning line may not move until quite large tilts are used. Instead the drop shape distorts with the upper side being stretched away from the pinning line to form a lower angle, called a receding contact angle, and the lower side will distort to produce a higher angle, called an advancing angle. With increasing tilts, the drop may jump position, coming back to a distorted shape, and eventually roll off. The behavior of the pinning line to be trapped in some metastable state is due to highly local interactions at the drop edge between the liquid and asperities in the solid surface which could be local chemical or topographical fluctuations. The angle associated with no tilt is called the static angle. Drop distortion angles can also be produced by a slight forcing of liquid into the drop from a syringe or capillary tip (producing advancing angle) or by a slight withdrawal (receding angle). Advancing, receding and static angles are often measured as a set when testing the wettability of a surface with a given liquid and give qualitative information about the surface character such as roughness and chemical inhomogeneities.

Contact angles are very practical for a large number of purposes in laboratories and manufacturing facilities. Suppose you are making contact lenses, as they do at Bausch and Lomb, with the lenses made of a soft, water absorbing polymer. The exact value of the contact angle of water on the polymer can correlate with the success of the lens in its function in the eye so what could be easier than a bunch of quick contact angles with saline solution to give a quick check on new batches of polymer? Lots of other things are done as well but why not start with the easiest cheapest test. There are a huge number of surface products being manufactured daily and in a large proportion of these liquid drop contact angles are used as a test method.

4.4.4.4 Obtaining interfacial tension values from contact angles—simple approximations

So you made a contact angle measurement and now want to go on and extract the interfacial tension of the liquid on the solid? It is a simple matter to get values of the liquid (L–V) surface tension, you can probably just look that up, but unfortunately values of the other two interfacial quantities are not easy to come by and they are not separately determined from the Y–D equation. Writing the Y–D equation with the knowns on one side we have

$$\gamma_{LV}\cos\theta = (\gamma_{SV} - \gamma_{SL}) \equiv W_A,$$

where W_A is the work of adhesion. Look at the right-hand term. W_A has the units of free energy per area and consists of the difference between the two interfacial tensions involving the vapor and the liquid adhering to the surface, thus the term work of adhesion. But note that this is not the same as the work of disjoining (equation (4.19a)) so be sure not to confuse the two work terms even though they often are both called the work of adhesion. The work of disjoining in the case of the drop in equilibrium with vapor would be $W_{disjoin} = \gamma_{LV} + \gamma_{SV} - \gamma_{SL} = \gamma_{LV} + W_A$ so there is a difference. In contrast, the left-hand term of W_A is $\gamma_{LV}\cos\theta$ which involves just the measured angle and the tendency of the liquid (in the presence of its own vapor) to bead up and try and to wrap around itself. If you want to dig into the W_A term and extract the related interfacial tensions you are stuck without more information. This has been a major challenge in the field of surface wettability for well over six decades and continues to this day.

How can we pull out a value of either γ_{LS} or γ_{SV}? Fancy experiments and theories can be designed (many have come and gone) but the standard approach used is to start with estimates, which often work quite well for nonpolar materials. The simplest estimates are based on those interfaces which involve only non-directional van der Waals types of interactions which form the basis of mean field approximations such as this one:

$$\boxed{\gamma_{SL} \approx \gamma_{SV} + \gamma_{LV} - 2\sqrt{\gamma_{SV}\cdot\gamma_{LV}}}. \tag{4.27}$$

This is called a geometric mean approximation because of the square root term. You also can write the equation as $\gamma_{SL} \approx (|\sqrt{\gamma_{SV}} - \sqrt{\gamma_{LV}}|)^2$ (but notice the absolute magnitudes to avoid non-physical values) which shows that the solid–liquid adhesive interaction grows as the difference between the solid/vapor and liquid/vapor terms. So suppose you have a nonpolar liquid with a high surface tension (e.g. Hg) on a surface which has only a weak affinity for the vapor (Hg atoms in this case) then you would have a strong liquid/solid interfacial tension. But since you might expect the surface to have similar affinities for both the vapor and the liquid it seems not so easy to make accurate predictions with this approximation in many cases. None the less it does work well for a variety of simple nonpolar liquids among which CH_2I_2, bromonapthelene and hydrocarbons such as decane are often used as test liquids in carefully designed contact angle–surface composition correlations.

There has been extensive effort for years to add in polar contributions to the interfacial tension terms in a quantitative way so that more comprehensive equations than equation (4.26) can be developed to make accurate predictions of γ_{SL} for a wide variety of solid–liquid interfaces. These polar interactions can include electrical dipole–dipole, donor–acceptor or Lewis acid-base, hydrogen bonding and other sorts of interactions but generally the equations cannot be derived on a truly fundamental basis and the parameters used to define terms in the equations for γ_{SL}

are invariably based on a good extent of intuitive reasoning so only work for narrow ranges of substances.

Questions for thought:

1. The exact shape at the molecular scale of the three-phase contact line may be quite complicated, as proposed by Nobel Prize winner Pierre de Gennes. What do you think it might actually look like and why does it not seem to affect the use of measurement of the contact angle by a simple tangent line at the drop edge?
2. Related to 1, suppose you have a drop of liquid polymer (it could just be heated up to be a molten polymer or a low MW polymer which is liquid at room *T*) whose vapor pressure is essentially zero. What do you think the state of the adjacent solid surface just outside of the three-phase contact line might look like?
3. A common problem with measuring contact angles for test liquids on polymers is that the contact angle keeps drifting no matter how long you take to make the measurement. What do you think might be going on?
4. Considering the behavior of some surfaces to exhibit large advancing and receding contact angles, what do you think might be going on at the pinning line?
5. You are interested in the Wulff construct for predicting crystal shapes and need surface tension values for the different faces of metal surfaces. Clean surfaces of metals in contact with inert, nonpolar liquids should interact only by nonpolar forces, particularly van der Waals forces. How might you use this information and your skills in working with single crystal surfaces to design and experiment to measure *hkl* facet surface tensions? The experiment is not simple but is done by surface scientists from time to time.

4.4.5 Capillary effects

4.4.5.1 The Wilhelmy balance and definition of a meniscus

Now that we understand how liquids can interact with a solid surface and respond by changing shape we can tackle the phenomenon of capillary action in which liquids adhere to solid walls, often within confined spaces such as small tubes or between walls, and tend distort their shapes, often influenced by the force of gravity. The simplest case is illustrated in figure 4.34 where a clean slide is immersed vertically into a liquid, let's say water. A clean glass slide surface consists of hydrated silica (SiO_2) with very hydrophilic SiOH groups exposed that form strong hydrogen bonds with water. As a consequence, H_2O molecules have an affinity for the glass surface and the liquid starts to creep up the slide in order to increase the area of the H_2O–glass interaction, while pushing back the H_2O vapor–glass interaction which is weaker.

Meanwhile as more and more water is being lifted by this interfacial pump against the force of gravity takes over and finally the weight of the water is too much for the pump to handle and the three-phase pinning line stops at some height defined by the

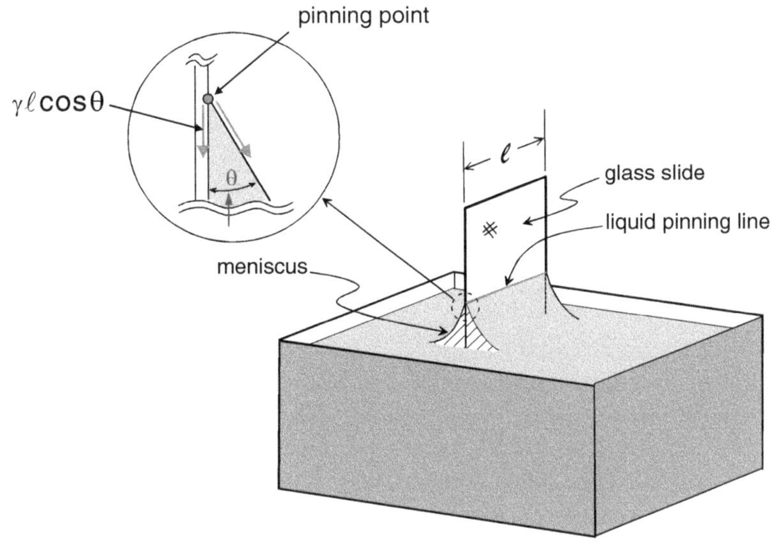

Figure 4.34. Schematic of a Wilhelmy balance experiment. A flat plate of width ℓ is immersed into a liquid which has a favorable interaction with the plate surface causing the liquid to rise against the force of gravity. At the pinning line the liquid hangs and a meniscus forms with a shape determined by the balance of the liquid surface tension forces and gravity.

balance in the z-direction between the weight of the meniscus supported and the adhesion force across the pinning line.

If you do the experiment correctly you can obtain an accurate value for the liquid surface tension (or the liquid–vapor interfacial tension if you allow the system to come into equilibrium with the vapor). The easy way to do the measurement is as follows:

1. Hang a plate with width ℓ from a hook connected to a balance.
2. Bring the plate down to almost touching the liquid and record the weight W_o.
3. Now just barely touch the liquid, let creep up to form a meniscus, record the weight W_W and measure the contact angle θ.

Now there is enough information to do the calculation as follows using the force balance:

$$\Delta W \cdot g = \gamma\left(\sum_{j=1}^{4}\ell_j\right)\cos\theta \sim 2\gamma\cos\theta,$$

where $\Delta W = W_A - W_o$ = weight of water supported in the meniscus, $\Delta W \cdot g$ = downward force of gravity pulling on the meniscus, $2\gamma\cos\theta$ = upward force of the surface tension balancing the gravity force with the assumption that the plate is sufficiently thin to neglect the perimeter contributions from the edges. From this one obtains the Wilhelmy balance equation:

$$\boxed{\gamma = \frac{\Delta W \cdot g}{2\ell \cdot \cos\theta}}. \tag{4.28}$$

This is the basis of a commercial instrument for measuring surface tension. So here is a second standard way to do the job in addition to the maximum bubble pressure method. Notice that the solid–liquid surface tension does not appear in the equation since its only role is to cause the liquid to form a meniscus on the plate. We could design take it into account of course but then the measurement and equations become much more complex. The easiest way is almost always best.

4.4.5.2 Capillary rise in tubes

A very common situation is the capillary action of drawing a liquid up a narrow diameter tube, often over quite long distances. The general phenomenon is pictured in figure 4.35. A column of liquid forms in a tube of radius r to form a meniscus, which for a cylindrical tube has a symmetrical shape approximated as a section of a spherical surface with radius R. The meniscus makes a contact angle θ at the three-phase pinning line on the wall. You should recognize immediately that we have a curved surface so there must be an over pressure ΔP pushing down on the liquid, this pressure must be caused by gravity and the meniscus stays in place by the resisting constrictive pressure from liquid–vapor interfacial (surface) tension. Again, as with the Wilhelmy plate experiment, we do not have to concern ourselves with why the liquid got up the tube, just look at the surface deformation at the pinning line. So referring to figure 4.35 for a liquid of surface tension γ and a capillary tube of radius r, let's set up the pressure balance and see how we can measure the surface tension. Keep in mind that the curvature is negative so negative signs show up of the radius of curvature since it is outside, not inside, the liquid.

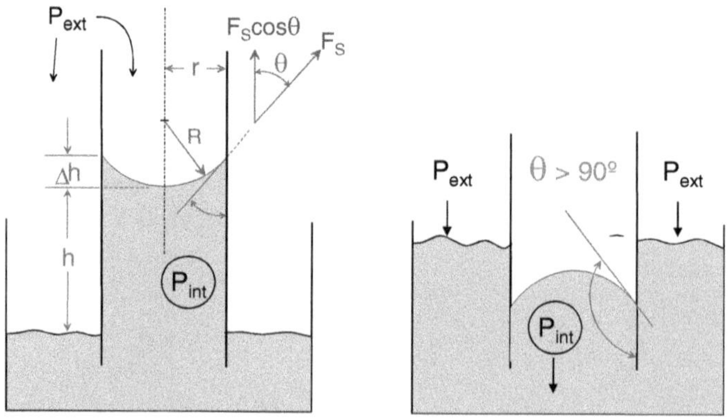

Figure 4.35. Illustrations of capillary rise up a tube of radius r. Left: The liquid–wall interfacial tension or work of adhesion is favorable for causing the liquid to rise to height h in the capillary center and form a positive meniscus which curves upward against gravity with radius R, size Δh up the tube wall and contact angle θ at the wall. Right: The liquid–wall interfacial tension or work of adhesion is unfavorable for allowing the liquid to rise and instead causes a negative meniscus of size Δh and radius $-R$ with a depression of $-h$ in the tube center and a contact angle $\theta > 90°$.

1. Assuming that the curvature is spherical then the constrictive overpressure is

$$\Delta P = -\frac{2\gamma}{R} = -2\gamma\frac{\cos\theta}{r},$$

where R = radius of curvature, which is just equal to $r/\cos\theta$, as seen in the figure.
2. The curvature is caused by gravity expressed as

$$\Delta P = -(\rho_{\text{liq}} - \rho_{\text{gas}})gh,$$

where ρ = density, g = the gravitational constant (6.67×10^{-11} m^3 kg$^{-1} \cdot$ s^{-2}) and h = the height of the liquid column from the liquid pool level to the bottom of the meniscus. The small contribution from the meniscus itself is neglected.
3. Setting the two forms of ΔP equal we obtain the capillary equation:

$$\boxed{\gamma = \frac{(\rho_{\text{liq}} - \rho_{\text{gas}})\text{rgh}}{2\cos\theta} \sim \frac{\rho_{\text{liq}}\text{rgh}}{2\cos\theta}}. \tag{4.29}$$

This provides a convenient method for measuring liquid–vapor surface tensions. In any standard measurement the liquid density is a couple of orders of magnitude greater than the vapor and gas densities so ρ_{gas} is typically ignored in the equation. It is easy to measure the liquid column height but takes more effort to get the contact angle. In cases such as aqueous solutions in clean glass capillaries where the liquid almost completely wets the surface, $\theta \rightarrow 0°$ and $\cos\theta \rightarrow 1$ so the contact angle is sufficiently small to be ignored.

But what about contact angles that are quite large, even $> 90°$? In these cases, the liquid–surface interfacial tension is so large that the liquid is essentially repelled from the surface relative to its tendency to ball up, thus giving a large θ value. This behavior causes the meniscus to depress below the liquid to minimize contact with the capillary well and form a positive radius of curvature, as shown in the right-hand side of figure 4.35. Application of equation (4.29) simply requires using $r > 0$ for positive curvature and $-\cos\theta$ when $\theta > 90°$.

The capillary equation is often written in terms of the height of the column for analysing physical systems with capillary action:

$$h \sim \frac{2\gamma\cos\theta}{\rho_{\text{liq}}rg}.$$

Notice the inverse dependence of h on r and the direct scaling with γ. As capillary diameter decreases the liquid is driven higher by interfacial forces which overcome the gravity forces, and the surface tension forces can hold the liquid meniscus together, so it does not collapse. The interfacial forces are embedded in the contact angle term. If $\theta = 0°$ then the surface is highly wetted so the interfacial forces must be strong and vice versa for large contact angles. Capillary forces are the major factor that allows fluid flow up trees to great heights via narrow microchannels in the tree. The term capillary forces is often used to describe the action of liquids in tiny spaces. This is a very general term and includes the combination of the solid–

vapor and solid–liquid interfacial tension and the liquid–vapor tension forces. In the confined geometries of nanometer-scale spaces associated with nanometer-scale objects these forces can be quite dominating in controlling the behavior of the associated systems.

> **Question**: What diameter of a highly water wettable microchannel would be required for water to be drawn up against gravity to a height of 50 feet?

4.5 Multicomponent media

4.5.1 General classes of interfaces in multicomponent media

So far, we have considered five main aspects of interface thermodynamics: (i) condensed phase objects of a pure substance in equilibrium with an adjacent phase, (ii) the effect of curvature on creating interfacial pressure differentials and their effect on shifting phase equilibria, (iii) interfaces between condensed phases of two different substances, (iv) the effect of interfacial curvature on solute phase transfer and (v) liquid/solid contact angles and capillary action. In this section we will add the general system of multicomponent solutions and focus on the composition of the components in the interface region with an adjacent phase. In the case of liquid solutions with a solute the adjacent phase is typically the vapor of the solvent while in the case of metal alloys the adjacent phase is a simple vacuum. The general classes of interface structures of interest are shown in figure 4.36. In each case there are two different media in contact and in classes I–V a solute molecule (or atom) is present. Note that in class II, the gas/solid interface, we consider the gas immersed in a vacuum as the second medium, just for purposes of consistency. Also, in class V one component can be very dilute, such as an impurity in a solid, or the concentrations comparable, as in the case of an alloy.

4.5.2 Where is the phase boundary? The Gibbs dividing line convention

In any of these interface structures where molecules or atoms are shared between media it is of interest to define where the phase boundary is in terms of accounting where the constituent species have been distributed. A standard solution example is given by the class I structure (figure 4.36), shown in more detail in figure 4.37, where a liquid solution, in equilibrium with solvent vapor, has a solute partitioning to the interface region to enrich its local concentration. Just as we saw early in the chapter for the soap film experiment (figure 4.3) the surface tension of an aqueous soap solution is much lower than water because the soap molecules preferably rise to the surface relative to the water molecules when the area is increased, thus the soap molecule composition is enriched at the surface to minimize the surface free energy. You might expect that the opposite effect for any highly polar molecules or ions dissolved in the water. We will develop equations for treating this type of behavior with the Gibbs equation for dilute solutions of the class I structure and then change conditions as needed to treat other classes of interfaces, particularly metal alloys with comparable

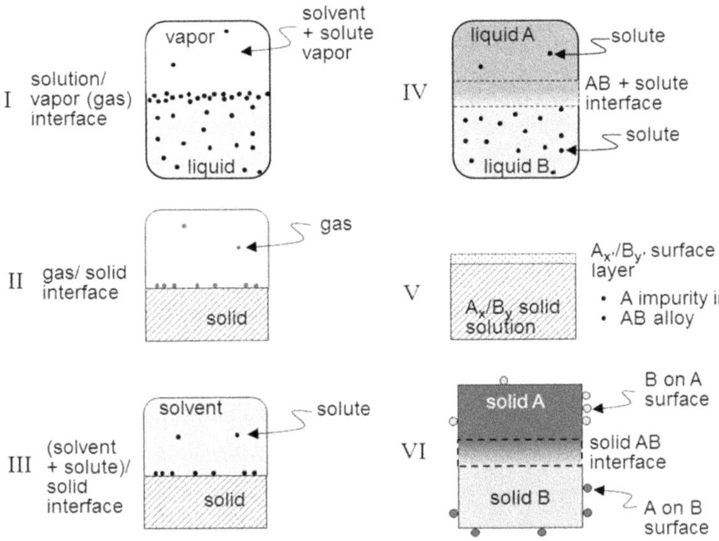

Figure 4.36. Schematic illustrations of six major classes of interfaces of multicomponent media. Each interface structure has one or two condensed phase media in contact and for classes with only one condensed phase the other medium is gas or vapor. In classes I–V a solute atom or molecule partitions between the media and the interface region. For class II the partitioning molecule is in the gas phase, effectively with vacuum playing the role of a solvent. Class V is a simple AB alloy which can have a surface region enriched in one component. Class VI is the case of a simple A–B contact in which A and B atoms (or molecules) are distributed within the interface region, which tails off into each phase, but also can be located on the free surfaces of the opposite medium. Typically, in the case of gas (or vapor)/solid and liquid/solid interface structures the molecules (or atoms) from the gas or liquid are assumed to be highly insoluble in the solid so the interface region will be quite sharp, though one can expect exceptions for soft solids such as polymers.

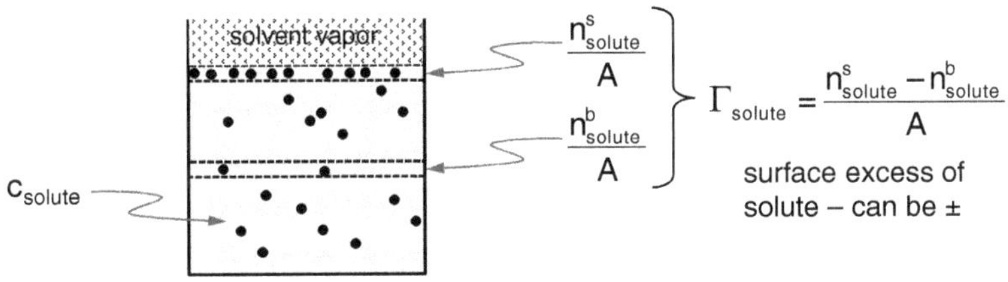

Figure 4.37. Pictorial representation of a solution containing solute molecules in equilibrium with the solvent vapor. The solute concentration is increased in the interface region relative to that in the bulk and is described in terms of the surface excess Γ which is defined as the difference between the moles per unit area in the interface and bulk regions for the same thickness layer.

concentrations of the two metals. Underlying these behaviors is a common thermodynamic treatment based on the first and seco2nd laws. The theory is quite detailed so we provide the details in appendix 1 and in these sections, we will only present a brief description so that we can go directly to the equations and apply them. The interested reader can follow the theory development in the appendix.

In order to set up the problem we first need to define the interface compositions of any species in terms of the surface excess of the moles per area of the component relative to a bulk region of the same thickness, as shown in detail in figure 4.37 for the case of a solute in a solvent where the quantity Γ_{solute} signifies the solute surface excess. To be more general, this can be easily extended to cases where there is no solvent, and the components have similar concentrations or mole fractions such as in alloys.

Next, we need to set conventions for specifying the accounting of components in terms of what part of an A–B interface is defined as belonging to the A phase and which part is the B phase. For the case of a solvent–vapor interface we accordingly choose where the solvent territory ends, and the vapor phase starts. There are different ways to do this depending on what aspect of the interface is of interest but the typical convention is that of Gibbs where the solvent concentration profile through the interface is divided exactly in half in terms of integrated amounts, as shown in figure 4.38. In the figure, moving along the z-axis starting in the liquid phase, the solvent (component 1) concentration is constant until at location z_1 it starts to drop and finally at location z_2 the solvent has completely been transformed into vapor and the vapor concentration is constant thereafter. The region between z_1 and z_2 is essentially the interface layer of thickness δz such as we discussed earlier in the interface pressure profile of a pure liquid in figure 4.7. It is not clear exactly what the solvent concentration profile $C_1(z)$ through the interface should look like in terms of some function, though one would expect some behavior similar to what is shown in the figure as a

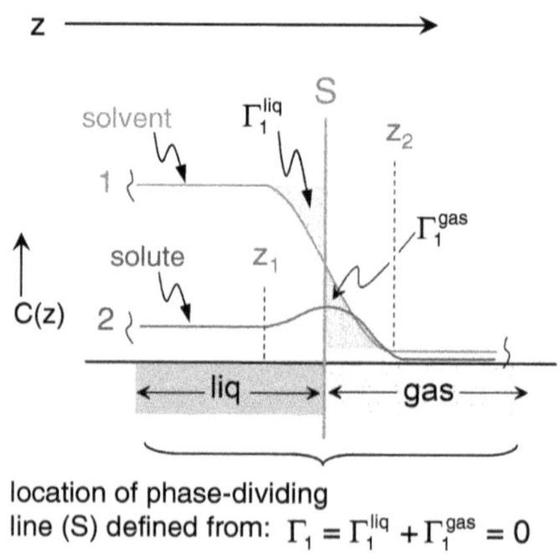

Figure 4.38. Schematic of the profiles of solvent and a solute from the solution bulk through an interface region and into the solvent vapor phase. Moving from left to right along the x-axis, the bulk solution phase ends at z_1 and at z_2 the vapor phase with constant vapor concentration starts. The interface region exists between z_1 and z_2. As the solvent concentration diminishes moving away from x_1 towards x_2, the line S is defined as that point at which the integrated amount of solvent is distributed 50:50 on either side. The solvent forms a depletion Γ_1^{liq} relative to the bulk liquid from z_1 to S and an excess relative to the gas or vapor from S to z_2.

general rule. Keep in mind for each substance δz can vary, sometimes being several molecules and the profile may be asymmetric. The solute concentration may have a maximum or minimum value at some location near S that reflects the tendency of the molecules to be in the interface region or to avoid the interface region and the peak may be way on the outer edge of the solvent (as shown by the blue line in the figure, such as one might expect for a surfactant) or towards the inner edge. If the solute were volatile (e.g. a dissolved gas or a molecule with a finite vapor pressure) the solute concentration could extend into the gas or vapor phase (see the blue line in the figure).

Now we choose a dividing line S essentially at the halfway point across δz in terms of the excess of solvent as defined by the following integral:

$$\int_{z_1}^{S} [C_1(z) - C_1^{\mathrm{o(liq)}}]\mathrm{d}z + \int_{S}^{z_2} [C_1(z) - C_1^{\mathrm{o(gas)}}]\mathrm{d}z = \Gamma_1^{\mathrm{liq}} + \Gamma_1^{\mathrm{gas}} = 0, \qquad (4.30\mathrm{a})$$

where $C_1^{\mathrm{o(liq)}}$ and $C_1^{\mathrm{o(gas)}}$ are the concentrations in the bulk liquid and continuous gas phases ($< z_1$ and $> z_2$, respectively), and Γ_1^{liq} and Γ_1^{gas} are the corresponding solvent surface excesses per area in the two halves of the interface region, each relative to the bulk and gas or vapor concentrations, respectively. You can see that S has been chosen so that $\Gamma_1^{\mathrm{liq}} = -\Gamma_1^{\mathrm{gas}}$, i.e. the solvent depletion between z_1 and S (red shaded region left of S) is compensated exactly by the solvent excess between S and z_2 (red shaded region right of S). Note that the excess value in the gas region ($S \rightarrow z_1$) is positive since the solvent concentration in this region is greater than the corresponding gas or vapor concentration until it finally levels off to the vapor concentration past z_2. So, with equation (4.30a) we have split the solvent distribution between the liquid and gas regions of the interface exactly 50:50. With this accounting system in place we can define the surface excess or depletion of any arbitrary jth component as

$$\Gamma_j = \Gamma_j^{\mathrm{liq}} + \Gamma_j^{\mathrm{gas}} = \int_{z_1}^{S} \left[C_j(z) - C_j^{\mathrm{o(liq)}}\right]\mathrm{d}z + \int_{S}^{z_2} \left[C_j(z) - C_j^{\mathrm{o(gas)}}\right]\mathrm{d}z \qquad (4.30\mathrm{b})$$

From the signs of the integral terms we see the definition of surface excess or depletion for any component becomes:

Excess: if $\Gamma_j > 0$ then the component forms a surface excess layer.

Depletion: if $\Gamma_j < 0$ then the component forms a surface depletion layer.

Please note there can be confusion about the use of the term surface excess to define Γ_j, a quantity which can have positive or negative values. When $\Gamma_j > 0$ then algebraically we call the surface excess quantity an excess; in contrast, when $\Gamma_j < 0$ then algebraically we call the surface excess quantity a depletion.

Next, we will put these conventions to work in developing the thermodynamics of solute partitioning to the interface region. Our goal will be to develop a relationship between the surface (or interface) tension of a binary solution and the concentrations or mole fractions of the components.

Some thought questions:

1. The second phase is often a solid wall (e.g. glass, as opposed to a vapor) and the solute ranges from polymers to salts.
   A. What is the meaning of depletion and excess layers in this case?
   B. How does this relate to the profiles developed for distributions through the A–B interface developed in the interfacial tension section?
2. Of the following solutions, which would you expect to form a surface excess or a surface depletion layer at the vapor interface?
   A. Aqueous surfactant solution.
   B. Aqueous KCl solution.
3. A hydrocarbon contaminant is present in the air at very low concentrations. What might the concentration profile look like next to a clay particle present in the air?

4.5.3 Bulk–surface partitioning in two-component systems

4.5.3.1 Dilute solute in solvent—the Gibbs equation

Now we consider a binary solution in equilibrium with the vapor phases of both components, as depicted for the class I structure in figure 4.36. There are two typical cases: (i) one component is in large excess and is called the solvent and (ii) both components have significant concentrations, which for solids such as metals is typically called an alloy. The basic thermodynamics that underlie the relationships between the concentrations or mole fractions and the surface or interfacial tensions is the same for both cases, but typically different simplifying assumptions are made for each. The main core of thermodynamics involves surface work, which governs the surface tension part, and chemical potentials, which govern the mass transport part to and from the interface region to the adjoining phases.

Here is the main concept. In the case of a pure substance the chemical potential of the surface region is always higher than the in the bulk (see figure 4.15, section 4.3.5.2). This reflects the basic instability of the surface, but no relaxation occurs because there is no path to decrease surface area if it already is at a minimum and constrained to shrink further by the required geometry of the object. We introduce a solute which can partition to the interface region. Now the interface region can have two components, and the solute and the original liquid molecules can replace one another anywhere in the system so an equilibrium will be established between bulk and interface region concentrations in which the final state will have the minimum overall Gibbs free energy (at constant T, P). As you might expect, the solute with the lowest surface tension in its pure state is likely to be favored at the interface (like a soap molecule), pushing the less favored molecule out of the interface (like a water molecule) back into the bulk. So effectively we have an equilibrium constant that can describe this exchange, and chemical potentials will drive the exchange balance and the size of this constant.

The famous Gibbs equation is one form of describing this balance in terms of the surface tension of a binary solution in the limit of a dilute solute in a solvent.

Since the derivation of the equation is a bit complicated, mostly because of the details of keeping track of surface excess concentration terms, we give the equation directly here and provide the details of the derivation in the appendix:

$$\boxed{\frac{1}{RT}\left(\frac{\partial \gamma}{\partial \ln(c_{\text{solute}})}\right)_{T,\,P} = -\Gamma_{\text{solute}}}, \tag{4.31}$$

where γ = the solution surface tension, and c_{solute} and Γ_{solute} = solute concentration in the bulk and the solute surface excess in moles per area (note the units on the left of moles/area). Since we are not giving the derivation here, we should look at the equation to make sure it makes physical sense. We see the RT and $d\ln(c_{\text{solute}})$ terms in the denominator and should immediately recognize the this signifies chemical potential in an ideal system (unity activity coefficients) where the usual relationship between chemical potential and concentration is $\mu - \mu_o = -RT\ \ln\left(\frac{c}{c_O}\right)$. This gives us almost everything since the minus sign also appears with the $RT\ln(c)$ term. So what the equation tells us is that as the solute concentration increases this will increase the chemical potential of the solute in the bulk, which in turn drives more solute into the interface region, forcing solvent molecules out. Is this favorable or unfavorable for the surface tension γ? Well, if you are pushing up soap molecules and pushing out water this is favorable and you expect that the surface tension of the solution will decrease with c_{soap}. So if $\left(\frac{\partial \gamma}{\partial \ln(c_{\text{solute}})}\right) < 0$ then by equation (4.31) $-\Gamma_{\text{solute}} < 0$ or $\Gamma_{\text{soap}} > 0$ and yes, just as you expected there is an excess of soap molecules per unit area in the surface region. OK, so in terms of the Gibbs dividing line diagram in figure 4.39, where are the soap molecule congregating? Just as you expected again, for $\Gamma_{\text{soap}} > 0$ the soap molecules must stick out into the gas or vapor

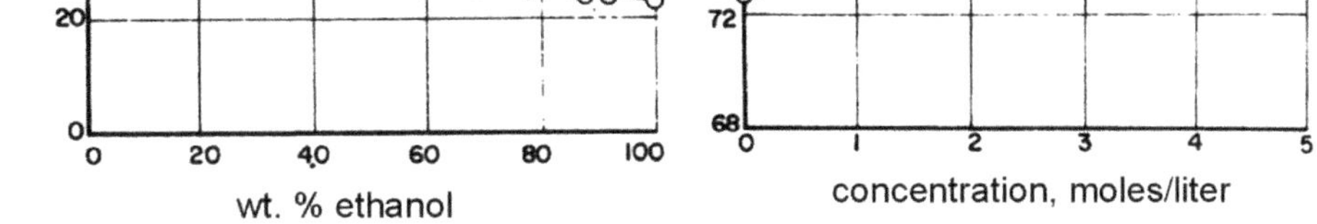

Figure 4.39. Correlations of measured surface tension versus concentration for aqueous ethanol and sodium chloride solutions. In the ethanol solution case the slope is negative, indicating $\Gamma_{\text{EtOH}} > 0$, whereas for the salt solution $\Gamma_{\text{NaCl}} < 0$.

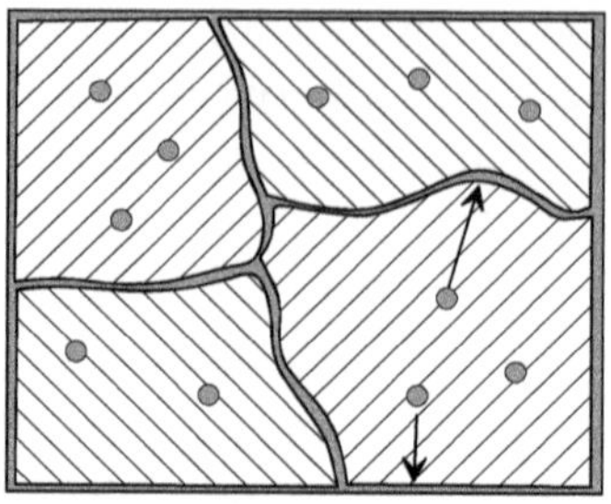

Figure 4.40. Illustration of a solid material with dissolved or trapped impurity atoms or molecules which may have a propensity to diffuse to the exterior surface and/or interior surfaces in boundary regions in the solid, such as grain boundaries or crack voids.

phase region, and according to their chemical structures the hydrocarbon tail would be protruding past the water molecules and into the vapor phase. The Gibbs equation, together with the Gibbs dividing line convention tells you all this information. Past this point you need to do some sophisticated types of surface spectroscopy or particle (photons or neutrons work well) scattering to get the structural details.

A classic type of example of surface portioning is given in figure 4.40 for two cases, aqueous solutions with ethanol and NaCl solutes, where the data is presented as plots of γ versus concentration. From the Gibbs equation you can guess the slopes ahead of time. For ethanol you expect that $\gamma\downarrow$ as $c_{EtOH}\uparrow$ and for NaCl in contrast you expect that $\gamma\uparrow$ as $c_{NaCl}\uparrow$, exactly what the plots show. The ethanol plot shows strong curvature, but this also follows other thermodynamic properties such as $\Delta H^{o}_{solution}$ and $\Delta S^{o}_{solution}$ which show strong nonlinear correlations with composition due to shifts in the solution structure. In the case of the salt solution plot the data is linear but according to equation (4.31) it should have been plotted as a log plot. If, in fact, the log plot is curved then we would conclude that changes in solvent-solute structures must change with composition. At least we know from the sign of the slope of the plot that the ions will tend to stay away from the surface where there are fewer H_2O molecules to surround the charges, thus reducing γ as ions are forced to the surface as their concentration increases.

What about the orientations of molecules with polar and nonpolar parts in the interface? A very general way to think about this question is in terms of the so-called *Langmuir principle* which states that molecules composed of different parts with different polarities will tend to place that part of the molecule away from the surface (into the ambient) which has the most unfavorable interaction with the solvent. In the case of ethanol with a nonpolar $-CH_3$ group and a polar $-OH$ group this turns the molecule with the CH_3- unit oriented towards the ambient. Experiments to determine such orientations are an active area of interface science. We will learn about some of these in the chapters on dynamics and electric properties of surfaces.

Some thought questions:

1. Can you propose more details about the molecule level structures of the ethanol and NaCl ion structures in the interfacial zone compared to the bulk? For NaCl would you expect one of the ions in the interface to be closer to the ambient phase than the other? Which one? Why? What experiments might you propose to determine these structures?
2. How might the addition of salt to a foaming aqueous liquid help to break the foam?
3. Proteins often exhibit distinct hydrophobic and hydrophilic regions on their surface in aqueous solutions (typically with a pH buffer appropriate for the particular protein. Do you think such a protein would have a preference to orient as the air/water interface? What about solid surfaces which are hydrophobic and hydrophilic, such as those with a surface self-assembled layer terminated by different functional groups?

4.5.3.2 Impurities in solids

A common practical issue in manufactured products and technological systems is the migration of dissolved or trapped impurities in solid materials to exterior surfaces and/or interior surfaces in boundary regions or crack voids, such as suggested by the illustration in figure 4.40. This type of migration can have adverse effects on the performance of the structures, e.g. the migration of impurity atoms such as S or C to transition metal catalyst surfaces where the impurity atom can act as a poison that reduces catalytic activity, or the migration of plasticizers such as oily organic molecules to the surfaces of plastics where the surface becomes sticky to the touch, compromising the usefulness of the application. You may recall examples of the latter with plastic items that you have owned that became sticky to the touch after extended use or after being exposed to sunlight (dashboards of older cars tended to do this until the plastic formulations were changed). The usual problem is that the added substance (e.g. an oil) exceeded the solubility in the material and slowly worked its way to the surface, a process sped up by heat, where it was more stable. In the case of metal catalysts with small atomic weight impurity atoms, the catalysts are often exposed to elevated temperatures in use which helps to drive the impurities to the surface.

The system of dilute impurity solutes in solid materials fits the general conditions treated by the Gibbs equation but the complexity of these systems, particularly the time scales of diffusion to and from the surface and the lack of surface tension data or measurements for solid systems, makes such treatment problematic. As a consequence, we will take a simpler approach and just give one example from the catalyst area in terms of the energetics of the location of an impurity atom in a metal catalyst.

Nickel metal is widely used as a catalyst for a number of commercial processes including hydrogenation, dehydrogenation and the Fischer Tropsch process for hydrocarbon synthesis from CO and H_2. The presence of S atom impurities on the metal surface can poison the process and it is of interest to know where the S atoms will tend to diffuse to be the most stable when present in the system.

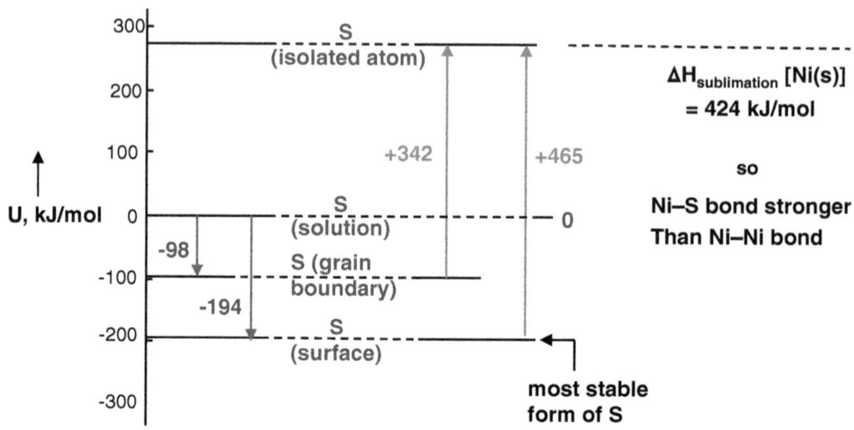

Figure 4.41. Diagram of the energy differences between different states of sulfur in a nickel metal. The least stable form is an isolated S atom. In turn, the energy of the atom lowers upon dissolution in the nickel crystal, adsorption or inclusion in a metal grain boundary, and adsorption on the metal surface, the most stable form of S. The heat of sublimation of nickel metal, shown on the right of the figure, is less than the energy required to desorb a S atom from the metal surface. Thus, for heating under high vacuum the nickel metal would eventually sublime before the S atoms desorb [4].

The thinking does involve the concepts in the Gibbs equation. For example, if the S atoms are dissolved but eventually migrate to the catalyst surface to stabilize then you can view this as driven by a tendency to lower the surface tension of the nickel surface. In fact, this is exactly what happens as shown figure 4.41 by the relative energetics of the impurity atoms located on the metal surface, in grain boundaries, dissolved in the metal lattice and as isolated atoms removed from the metal. The data show that the most energetically stable form of S is chemisorbed at the metal surface. The S–Ni bonds are extremely strong as shown by comparison with the heat of sublimation of Ni metal, which is lower in energy. Thus, upon heating under UHV (to keep the system free of other atoms) as the temperature rises the nickel metal will sublime before S atom desorption from the surface has a chance to take place. This kind of information is very important in understanding the behavior of atom impurities in catalyst systems where the impurities have the potential to act as poisons when they are stabilized at the catalyst surface. In cases such as this the S atoms can be removed by roasting in oxygen, or by other chemical means, and then reducing the oxide with hydrogen to restore the catalytically active metal surface.

4.5.3.3 The range of possible distributions of components in binary mixtures with no dominant component

The Gibbs equation deals with cases approaching the limit of dilute solute in solvent. At the other limit are mixtures with compositions with no one component dominating. For solids these are typically called alloys and there is a range of

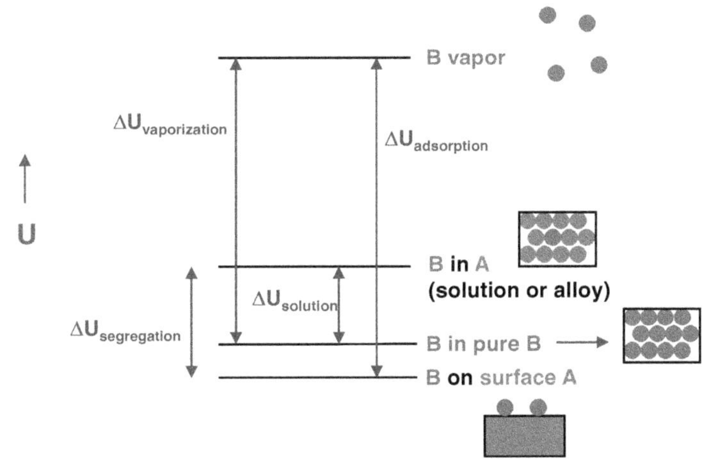

Figure 4.42. Diagram of the possible relative energies of different states of a B atom in a hypothetical AB alloy. As energy increases the states appearing are B atoms on the surface of pure A, B atoms in pure B, B atoms in A in the form of the alloy, and B vapor. In this example the most energetically stable location of B atoms is at the surface of a pure A phase.

possible phase structures with associated surface compositions. The classes of alloys can range from polymers to metals. For the case of components A and B at one limit is a uniform, single phase AB alloy or solid solution. In this case it is common for one of the components to be enriched at the surface. In the case of metal atoms if the surface stability of one component dominates, e.g. A, core–shell types of structures can arise such as nanoparticles of pure metal A with surfaces covered by a monolayer of B, similar to the case implied by the energetics in figure 4.30. At the other limit are structures of a pure phase-separated A and B with an AB interfacial region. A hypothetical energy diagram of the various possible states in terms of the disposition of B atoms in a metal alloy is given in figure 4.42.

The basic thermodynamic framework to treat alloy surface excess or depletion in alloy systems is exactly the same as used for the Gibbs dilute limit (see the appendix) but in order to end up with simple equations different assumptions are used that are suitable for the alloy composition regime. The possibility that a stable alloy will end up with an enrichment of one component at the surface is very common. Let's look at a simple model for treating an ideal case.

4.5.3.4 A simple predictive model for surface segregation and surface tensions in binary alloys and solutions of small nonpolar molecules

The model and application to prediction in an elemental alloy system

We develop a simple model which holds approximately for a number of binary alloys of similar metals and homogeneous mixtures of small molecules. The model is based on four main assumptions: (i) the surface region consists of a single layer, (ii) the mixture behaves according to the regular solution model of an ideal entropy of

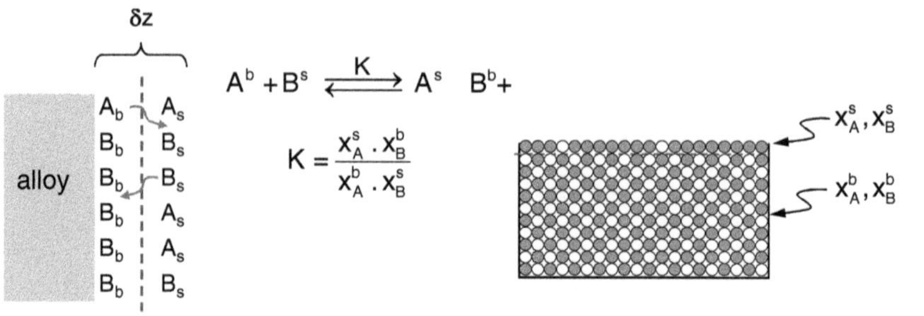

Figure 4.43. Illustrations of the surface region enrichment of one component in an ideal AB alloy. Compositions are described in terms of the mole fractions of A and B and the equilibrium composition distribution between bulk and surface by an equilibrium constant K with bulk and surface atoms interchanging. A–B interchange occurs in the surface region with one component accumulating at the surface and the other driven back into the bulk in sufficiently low dilution so as not to affect the overall bulk composition.

mixing, (iii) the component species are the same size in terms of occupying equal volumes in the bulk and equal areas in the surface region, (iv) the components exhibit equal coordination numbers. Referring to figure 4.43, in this model the surface layer reaches compositional equilibrium by simple 1–1 exchange between a bulk and a surface species (atom or molecule) and those species that go from surface to bulk become uniformly mixed. Note that we cannot base the model on the ideal solution approximation since this requires A–A, A–B and B–B interaction energies to be equal, which would mean in conjunction with ideal entropy of mixing that there would be no difference in the surface tensions of the pure phases. In spite of the approximations having a simple model is quite useful since it gives us a way to make some simple simulations of the general surface behavior of binary solutions and see the effects of changing concentrations and varying the surface tensions. Details of the thermodynamics are given in the appendix.

To start, we characterize this surface-bulk equilibrium in terms of a simple exchange equilibrium constant $K = \frac{a^{s}_{A} \cdot a^{b}_{B}}{a^{b}_{A} \cdot a^{s}_{B}} \sim \frac{x^{s}_{A} \cdot x^{b}_{B}}{x^{b}_{A} \cdot x^{s}_{B}}$, with the a and x representing the activities and corresponding mole fractions of A and B in the surface and in the bulk, respectively. This approximation involves the further assumption that activity coefficients are all equal to 1 over the entire range of mole fractions $0 < x < 1$. This assumption neglects the energy differences between an A surrounded by all As and all Bs, and vice versa, which should apply when the A–B interaction energy is not much different than A–A and B–B interactions.

The derivation follows from the thermodynamics given in the appendix and is not given here for brevity. The essence is that the surface mole fraction composition ratio of A and B is driven by the difference in surface tensions of the two pure phases as expressed by the exponential dependence:

$$\frac{x^{s}_{A}}{x^{s}_{B}} \sim \frac{x^{s}_{A}}{(1 - x^{s}_{A})} = \left(\frac{x^{b}_{A}}{x^{b}_{B}}\right) \cdot e^{\frac{\bar{A}(\gamma^{0}_{B} - \gamma^{0}_{A})}{RT}}, \tag{4.32}$$

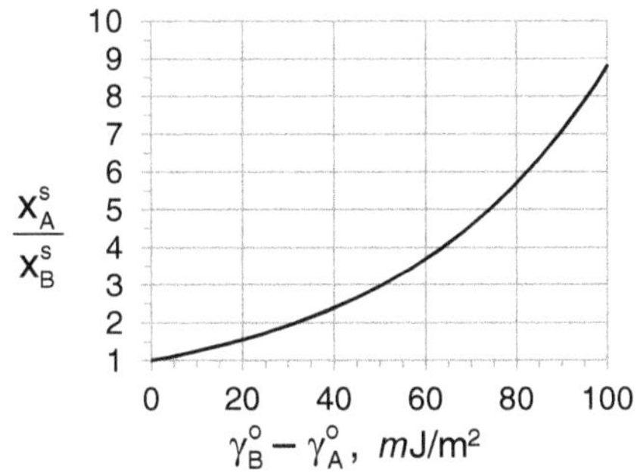

Figure 4.44. Plot of the surface enrichment of a hypothetical 1:1 AB alloy at 300 K as a function of the differences in the surface tensions of the pure components as calculated from equation (4.32). Notice the exponential surface enrichment of component A as its surface tension becomes smaller relative to that of B. The calculation is based on atoms A and B of ~0.3 nm diameter in a 1:1 solution at 300 K.

where γ_A^o and γ_B^o are the pure A and B surface tensions and $\bar{A}$ = area per mole for both A and B. As expected, with increasing values of the surface tension of a given component, the concentration of the component in the surface decreases, all other factors remaining equal. A plot of this dependence is shown in figure 4.44 for the hypothetical case of two atoms ~03 nm in diameter with $\bar{A} = 0.09$ nm^2/atom or 5.42×10^4 m^2 mol^{-1} as calculated directly from the diameter based on a simple cubic packing model.

Equation (4.32) can be rearranged to solve for the dependence of the mole fraction of one component at the surface as a function of its mole fraction in the bulk, which in terms of component A gives

$$x_A^s = \frac{f(\gamma)}{1 + f(\gamma)}, \quad \text{where } f(\gamma) = \left(\frac{x_A^b}{1 - x_A^b}\right)\exp\left[\frac{\bar{A}_m\left(\gamma_B^o - \gamma_A^o\right)}{RT}\right]. \tag{4.33}$$

Application to prediction in a two-component molecular solution

The alloy model can be extended to molecular mixtures as well but the assumptions underlying the model are quite restrictive since it is unusual to find pairs of miscible molecules with distinctly different surface tensions for which the molecules occupy equal volumes in the bulk, equal areas at the surface, have equal coordination numbers and their mixture behaves according to the regular solution model of an ideal entropy of mixing. Even chemically similar molecules of interest almost always have different sizes, shapes and solvation coordination numbers. The situation worsens when molecules can undergo H-bonding, have strong dipoles and other types of interactions that cause their solutions to be highly non-ideal. To account for this in a simple application of the alloy model we choose a pair of chemically similar, miscible, nonpolar molecules, diethyl ether (DEE) and cyclohexanone, which have equal molar volumes but distinctly different surface tensions.

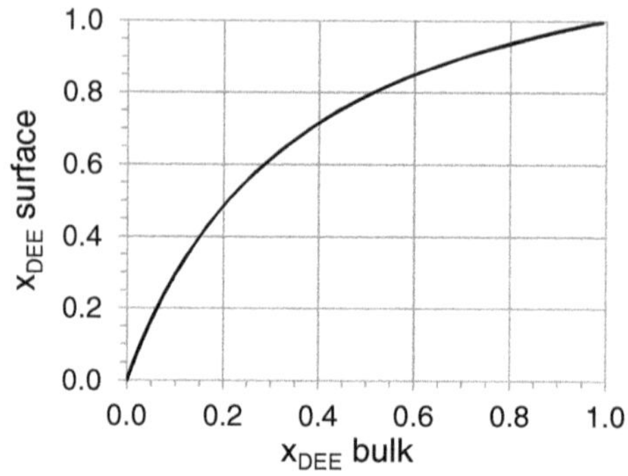

Figure 4.45. Plot of the predicted surface enrichment of diethyl ether (DEE) in a DEE and cyclohexanone solution as a function of the mole fraction of DEE in the bulk. Since x_{DEE} (surface) is directly proportional to the surface excess Γ_{DEE} the latter follows the same dependence on x_{DEE} (bulk).

A plot illustrating the use of equation (4.32) is given in figure 4.45 for a mixture of DEE and cyclohexanone, each with a surface tension, respectively, of 16.7 and 34.4 mJ m^{-2}, and the pair with equal molar volumes of 104.8 cm^3 mol^{-1}, which gives $\bar{A} \sim 1.9 \times 10^5$ m^2 mol^{-1} based on the approximation of equation (4.14b). Since the molar volumes are equal it is a good assumption that the areas occupied by each molecule in the surface region are also equal and thus the model for equation (4.32) should hold. In the figure note the predicted fast enrichment of DEE at the surface as its bulk mole fraction increases from zero, then eventually reaches the limiting bulk value as pure DEE is approached. The plot can easily be recast in terms of the surface excess Γ_{DEE} if desired since $\Gamma_{DEE}\,(\mathrm{mol} \cdot \mathrm{m}^{-2}) = \left(\frac{1}{\bar{A}(\mathrm{m}^2 \cdot \mathrm{mol}^{-1})}\right) \cdot x_{DEE}$ (surface), where the surface excess is defined in terms of the single surface layer model. No experimental data is available to test the prediction, but large deviations would not be expected since the molecules are nonpolar and should follow regular solution behavior, as required by the model for equation (4.32).

For molecular mixtures the surface tension as a function of composition is often of interest and for this purpose the same basic solution model can be readily extended. Keeping the assumptions of the model, we write the total surface tension as the sum of the excess surface free energies of each component, $\gamma = \frac{1}{\bar{A}_m}[x_A^s(G_A^s - G_A^b) + x_B^s(G_B^s - G_B^b)]$, and express each free energy term as $G = G^o + RT \ln x$. This gives the final equation

$$\gamma = \left(x_A^s \gamma_A^o - x_B^s \gamma_B^o\right) + \frac{RT}{\bar{A}_m}\left[x_A^s \ln\left(\frac{x_A^s}{x_A^b}\right) + x_B^s \ln\left(\frac{x_B^s}{x_B^b}\right)\right] = \gamma^o + \gamma_{mix}, \tag{4.34}$$

where γ^o represents the surface tension contribution with no mixing of the components, treating the surface tension as the weighted average of the surface tensions of the pure A and B phases, and γ_{mix} is the contribution of the free energy of mixing, essentially an entropy term for A and B mixing in the surface and bulk phases. Since $x_B^s = 1 - x_A^s$ and the two mole fraction ratio terms can be obtained from equation (4.32) this gives a way to calculate values of γ.

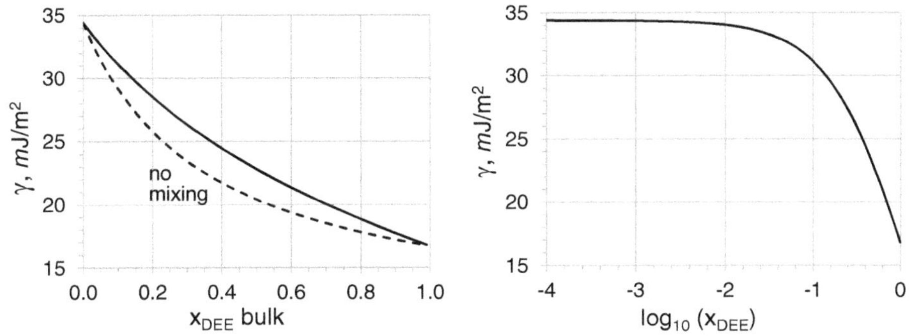

Figure 4.46. Plots of the predicted surface tension of a diethyl ether/(DEE) cyclohexanone solutions as a function of the mole fraction of DEE in the bulk. Left: Surface tension versus x_{DEE}. The dashed line represents γ°, the predicted surface tension without include the free energy of mixing in the bulk and surface regions. Right: Predicted surface tension versus $\log_{10}(x_{\mathrm{DEE}})$ in the bulk. This plot is set up according to the Gibbs equation (equation (4.31)) so the slope at any point can be used to determine Γ_{DEE}, the surface excess amount of DEE, at a given bulk BEE concentration.

Applying equation (4.34) to the DEE/cyclohexanone mixture gives the result in figure 4.46, which shows the predicted surface tension of the mixture versus x_{DEE} in both linear and $\log_{10}$ representations. The linear plot shows the contribution of the mixing free energy γ_{mix} in the calculation, as compared to just using γ^{o}, the contribution based on a simple weighted average of the pure bulk surface tension values (dashed line). Comparison of the two curves shows that γ^{o} gives a fair approximation of the actual surface tension, with the largest deviation in the middle of the concentration range, where the ideal entropy of mixing should be a maximum. The $\log_{10}$ plot is useful for determining the surface excess amount of DEE, Γ_{DEE}, directly from the slope and application of the Gibbs equation (equation (4.31)). The log plot shows that at very low DEE bulk concentrations the surface tension is essentially that of pure cyclohexanone and does not drop towards the limiting pure DEE value until $\sim 10^{-2} < x_{\mathrm{DEE}}$ (bulk) $< 10^{-1}$. As the negative value of the slope increases the value of Γ_{DEE} correspondingly grows, as expected from the associated surface mole fraction plot in figure 4.45. A quick check of the slope of the plot at x_{DEE} (bulk) $\sim 0.1(\log_{10} x_{\mathrm{DEE}} \sim -1)$ leads to a rough value of $\Gamma_{\mathrm{DEE}} \sim 1.2 \times 10^{-6}$ mol m^{-2} from the Gibbs equation, which corresponds to $x_{\mathrm{DEE}} \sim 0.24$, in rough agreement with the value taken from figure 4.45. Thus, the equations are all self-consistent.

Thought questions:

1. Would the trends shown in figure 4.44 be predictable from the energies of vaporization? What factors would you need to consider?
2. Would the ideal behavior be followed for two metals with different coordination numbers in their pure form? Why or why not?
3. Can you recall a previous example in chapter 2 of surface enrichment of an alloy? Could you have correctly predicted which component enriched at the surface?

4. As a check, estimate the slope in the $\log_{10}x_{DEE}$ plot in figure 4.46 and calculate the value of Γ_{DEE} at $x_{DEE} \sim 0.1$. Compare to the value of x_{DEE} (surface) in figure 4.45. Recall that $\log_{10}x = 2.303 \ln x$. Also, Γ is in units of moles m^{-2} and to compare with a mole fraction value x you need to apply the value of $\bar{A}$, which is in units of m^2 mol^{-1}. So this takes a couple of calculations.

The accuracy of equations (4.29)–(4.31) obviously will not hold over a wide range of molecular mixtures because of the limits of the approximations noted earlier but, in particular, it badly fails for quantitative predictions in cases of surfactant/water mixtures where the surfactant has entirely different chemical and physical properties than the water molecules and has limited solubility. Water/surfactant mixtures play such a major role in surface and interfacial chemistry that in the next section we will take a qualitative look at surface segregation and related surface film structure for a simple class of surfactant. This will give us at least an introduction to these important systems, with their numerous technological applications and critical roles in biological phenomena.

4.5.3.5 Surface segregation in surfactant/water systems—Gibbs and Langmuir films

Simple aspects of surfactant molecules and their water solutions

In this section we will look at the types of surface films that form in surfactant/water systems and some simple structure correlations. The presence of surfactants in an aqueous solution usually leads to such complex behavior that it confounds efforts to use the Gibbs and related equations for predictions. Let's see why by starting with a short detour to lay out a few of the simple aspects of typical surfactants, summarized in figure 4.47. This is a huge field and a detailed look at this subject is well outside the scope of this chapter, so the interested reader is referred to comprehensive sources (see end of chapter). In general, the major role of a surfactant is as an additive to lower surface or interfacial tension. Surfactants can have a wide variety of chemical structures and applications but for aqueous systems the molecule almost always consists of one part with an unfavorable affinity for water (water *hating* or hydrophobic), compared to an oil or air phase, covalently bonded to the other part with a comparatively strong affinity for water (water *loving* or hydrophilic). The combination of these opposite behaviors in the same molecule act as a driving force to favor partitioning to the water/air or water/oil interface with displacement of water molecules to leave each part of the surfactant molecule residing in its favored phase. The strong mismatch in chemical properties between water and each part of the surfactant molecules gives rise to highly non-ideal thermodynamic behavior for their solutions. Thus, a treatment of the thermodynamics is quite complex and difficult, so we only discuss this aspect very qualitatively.

For brevity we consider only the most common and simple surfactants formed from alkyl chains covalently bonded to a polar headgroup, as illustrated in figure 4.47. The polar group can be charged or neutral but always exhibits a strong

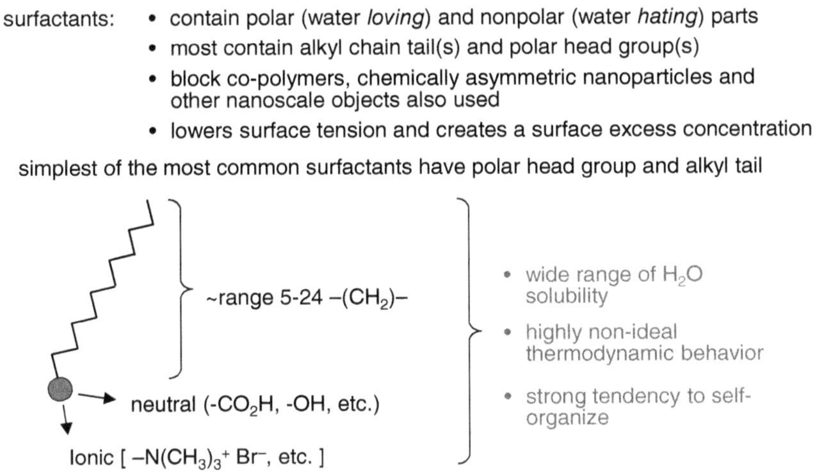

Figure 4.47. Summary of some important aspects of surfactants and a schematic of a simple example of an alkyl chain surfactant with a polar headgroup.

affinity for water. The alkyl tails have unfavorable interactions with water compared to self-interactions and thus are driven to aggregate or self-organize wherever possible; essentially, they condense to try and form a pure hydrocarbon phase as best they can, dragging their polar counterparts along with them. This can give rise to interesting structures, as illustrated in figure 4.48. At sufficiently dilute concentrations the surfactants will exist as isolated solutes but as the concentration rises eventually the alkyl self-interactions can become so strong as to induce aggregation of the molecules with self-organization into new nanostructures which disperse in solution as soluble objects and obviously make a mess of predictions from the Gibbs and related equations. For example, common structure classes include spherically shaped micelles with hydrocarbon interior cores and interacting water shells and the alternate geometry structure of bilayer sheets. Some classes of nanostructures in concentrated solutions can be quite complex with very unique geometries. Note the similarity of the chain packing in the micelles and bilayers to that of self-assembled alkyl chain types of monolayers on solid substrates. We will shortly look at this aspect and compare organization of adsorbates at liquid/air and solid/vacuum interfaces in terms of the discussion in section 3.3, chapter 3.

Formation of Gibbs and Langmuir surface layers and Langmuir–Blodgett films

Let's compare two surfactant molecules—one with a significant relative solubility in water and the other with a relative very low, near vanishing solubility, e.g. compare, a short chain ($\sim C_5$) and a highly insoluble long chain ($\sim C_{20}$) alcohol or carboxylic acid and look at their ideal behavior. As the concentration of the relatively soluble surfactant solute increases an increased partitioning to the surface will result, as suggested very qualitatively by equations (4.29) and (4.30a), and eventually, with no other factors intervening, surface coverage approaches the limit of full surface coverage or loosely, a monolayer, as illustrated in figure 4.49(top). This layer is

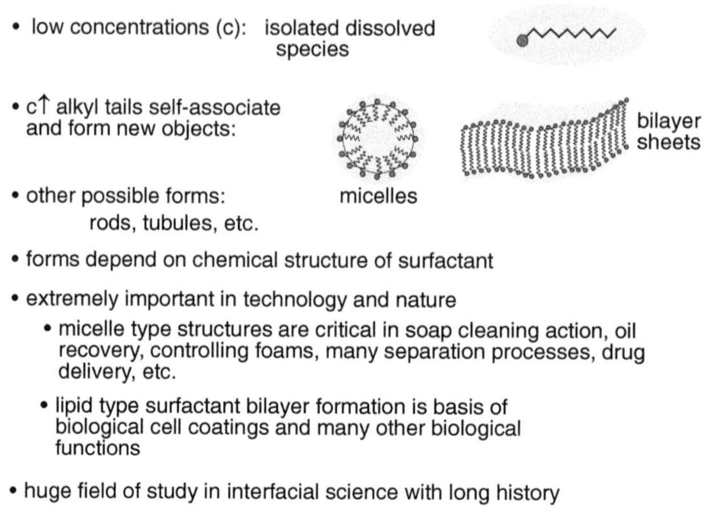

Figure 4.48. Summary of some important aspects of surfactants. The main point is that with increasing concentration in water complex, self-organized nanostructures will form and disperse in the solution.

called a *Gibbs monolayer*. If we sweep off a Gibbs surface layer one way or another (easy to do, even a paper tissue might work) then the system is out of equilibrium, the surfactant starts partitioning again and a fresh surface layer appears. This is a very general behavior and holds for any mixture with or without an aqueous phase, but the term Gibbs films are generally used for surfactant layers at water surfaces. Eventually at higher solute concentrations the solubility limit is exceeded, and bulk material will precipitate out on the surface to create scum or an oil film as a two-phase system forms.

In contrast, referring to figure 4.49 (top), as we approach the limit of an exceedingly insoluble surfactant (the solubility can never be exactly zero, of course) consider the experiment in which you take a dilute solution of surfactant in some volatile solvent and drop it on the surface, being careful to measure out only the amount of surface needed to not exceed a monolayer quantity. Once the solvent evaporates the surfactant is trapped at the surface since it is basically insoluble and cannot partition into solution, or alternatively, it is so insoluble and has such a small diffusion coefficient that the dynamics of forming a saturated solution in the reservoir subphase is exceedingly slow. Accordingly, if you sweep this layer away it will be gone for good since replenishment from the subphase does not occur, in contrast to a Gibbs layer. This surface layer is called a *Langmuir layer*.

Langmuir layers have the interesting property that in many cases you can sweep the layer by bringing up a glass plate (or similar flat surface), previously immersed in the solution below, vertically through the layer out into ambient, as illustrated in figure 4.49 (bottom). The layer usually can be easily transferred to the glass surface. The captured layer is called a *Langmuir–Blodgett* (LB) layer or film. If the surface coverage on the water is forced high enough by one means or another a highly

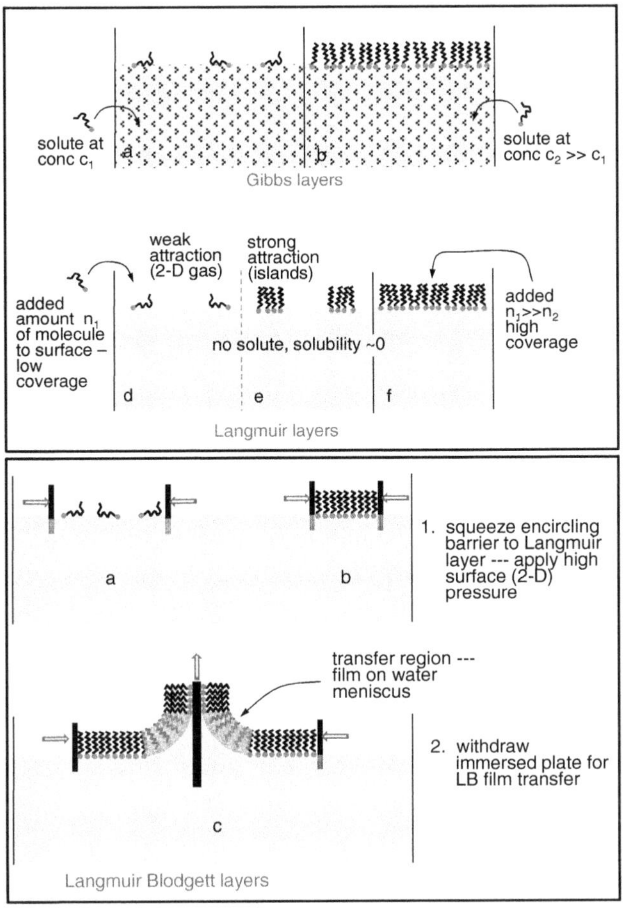

Figure 4.49. Schematic diagrams of Gibbs, Langmuir and Langmuir–Blodgett transfer films based on alkyl chain surfactant molecules with polar (red ball) and nonpolar hydrocarbon tails. Top: Parts (a) and (b) show a solution subphase with dissolved surfactant molecules, typically short chain for good solubility, at low and high relative concentrations, respectively, and their associated surface coverages (θ). At low θ the molecules are isolated in a 2D gas type of phase whereas at high θ nearing saturation the alkyl chains start to associate and the molecules self-assemble. Parts (d). (e) and (f) show Langmuir layers in which the low solubility surfactants are used and the subphase is essentially free of dissolved surfactant. In (d) and (e) surfactant added in small amounts at the surface to give low θ can result in isolated molecules but for long alkyl tails with strong self-attraction islands can form. At high θ, as with Gibbs layers, self-organization will ensue. Bottom: Parts (a) and (b) show a free Langmuir film at low θ and compressed between an encircling barrier. In (c) the barrier is adjusted to compress the film and an immersed plate, withdrawn vertically through the film, attracts the headgroups of the surfactant resulting in transfer of the monolayer.

translationally ordered Langmuir layer can form (hydrocarbon tails all crystallized nicely) and the self-organization can be maintained in the LB film if the transfer is done carefully. The transfer can be repeated over and over to prepare multilayer films. LB films have a long history of great scientific and technological interest for the past nearly 100 years since their original discovery by Irving Langmuir and

Katherine Blodgett at the General Electric research labs. Attempts have made to prepare them in single to many layers for applications such as electronic and photonic devices, as well as many other applications using surfactant molecules ranging from simple alkyl chain types to biomolecules, π-conjugated structures and various polymers.

Structures of Gibbs and Langmuir surfaces films with coverage—analogies with adsorbate structures at the gas/solid interface

Now look briefly at the surface structures of these films as given in figure 4.49, while looking back at chapter 3 and pay attention to figure 3.22 in section 4.3.2. Let's put the two pieces together. To be consistent with the latter we designate surface coverage in terms of θ, the fraction of molecules per area relative to full coverage, in this case a closest packed layer, and consider the surface molecules as an adsorbate layer, either an irreversibly trapped aliquot on a surface or as dynamically fed by a reservoir solution, analogous to a gas feeding adsorbates to a solid surface. As $\theta \rightarrow 0$, thermal energies force the molecules to be independent, like a 2D gas, with highly conformationally disordered chains and with average orientations running parallel to the surface. Now as we increase θ we need to note how this is done, by just adding aliquots to the surface (Langmuir) or by increasing the subphase concentration (Gibbs).

For discussion purposes we consider Langmuir layers first since we can ignore the subphase. As $\theta\uparrow$ but $\ll 1$, the molecules will remain as a 2D gas or perhaps will snap into islands with some degree of self-organization, depending on the relative alkyl chain–chain attraction forces. The islands will present the molecules with average chain tilt angles approaching vertical to the surface and increased conformational order compared to the 2D gas. Finally, as $\theta \rightarrow 1$ the chains will snap into a self-organized configuration with near vertical average tilts and the surface density will approach that of closest packed alkyl chains. Note that the water surface acts as a featureless substrate, in contrast to the typical solid crystalline substrate which normally exhibits definite adsorbate pinning sites in some registry with the surface lattice, e.g. three-fold hollows or atop sites. In the limit of $\theta \rightarrow 1$ with weak pinning sites, hexagonal packing should be obtained across the surface plane, like a group of aligned pencils bundled up by a rubber band.

For a Gibbs layer, we expect similar trends in surface structure versus θ. But note that since Gibbs films are typically formed from soluble surfactants, which means relatively short alkyl tails to allow sufficient solubility, the chain–chain self-interactions are weaker than for Langmuir surfactants so the onset of islands and high degrees of film organization will arise at higher coverages and the films will have less chain conformational ordering. More important perhaps is that as solute concentration (c_{solute}) increases the possibility now exists for formation of one of the various nanostructure phases such as dispersed micelles. The onset of these phases with increasing concentration (in the case of micelles, called the critical micelle concentration or CMC) can cause distinct changes in θ versus c_{solute} concentration profile as solute is consumed to form the micelles so the system behavior becomes quite complex. Generally, these effects are tracked experimentally with γ versus

c_{solute} plots, which can show strong deviations from the ideal Gibbs equation (equation (4.31)). These deviations can be analysed in terms of the new solution nanostructures. Finally, as c_{solute} approaches saturation concentration, $\theta \rightarrow 1$ but the underlying solution phases may not be simple with the presence of micelles and related structures as dictated by the length of the alkyl tails in the surfactant and other conditions (e.g. T and the exact type of surfactant).

Finally, we discuss an interesting and useful aspect of Langmuir films, the ability to transfer them to solid supports, as depicted in figure 4.49 (bottom), known as Langmuir–Blodgett (LB) transfer. This process has been very popular in many different fields over the years since it allows a way to prepare crystalline like monolayers or a variety of solid surfaces which can then be studied and used for a variety of purposes from electronic and photonic devices to biocompatibility. The key aspect of the transfer process is the application of a 2D surface pressure (π, in units of surface pressure which are the same as surface tension, millinewtons m^{-2}) to a Langmuir film by forming the film within a flexible barrier (e.g. Teflon type ribbon) and then reducing the enclosed area to laterally compress the film to increasing pressures. The pressure changes can be tracked by a simple measurement of the surface tension of the solution within the barrier, typically using a Wilhelmy balance approach with a strip of paper as the probe. The governing equation is given by $\pi = \gamma_o - \gamma_{film}$, where γ_{film} and γ_o are the surface tensions with and without the film. As the film is compressed γ_{film} decreases (less work required to make new surface as the coverage of surfactant increases). At the desired compression a previously immersed substrate is slowly drawn up through the film providing a new surface for attachment of adjacent surfactant molecules, dragging some of the water layer along with them as a meniscus. Note that for this type of transfer the substrate needs to be easily wet by water (hydrophilic) to form the transfer meniscus. Other modes of transfer are possible such as pushing a hydrophobic surface down through the film into the subphase which can result in the molecules being adsorbed upside down (alkyl tail at the substrate surface and polar group at the ambient). Multiple film stacks are also easily made by repeated cycles.

We note in closing this section that the study of surfactant behavior, phases and surface films, including LB types of transfers, is a huge, wide ranging subject of interest across many science and technology fields. The interested reader is directed to the numerous resource texts and articles available for further details.

Questions for thought:

1. Biological systems often use lipid molecules which have twin alkyl chain tails attached to a single polar headgroup, for example as a bilayer in cell walls. Using some of the basic information from this section can you think of why might this structure be favorable compared to a simpler single tail one?
2. Go back to the adsorbate structure discussion in section 3, chapter 2 (referring to figure 2.22) and explain the behavior and properties of Gibbs, Langmuir and LB films in terms of the concepts given there. Cover the aspects of structure dependence on surface coverage and the interplay between the various forces that are operating on the adsorbates.

3. Draw parallels of Gibbs and Langmuir films with self-assembled monolayers (SAMs) and discuss the differences between SAMs and the liquid surface films.
4. How do you think the plot of γ versus ln(surfactant concentration) might deviate from the ideal from the Gibbs equation in terms of two factors. (i) as the Gibbs film goes to increased θ and the molecules start to orient and (ii) for a case in which micelles form? What if the micelles form well before a full coverage Gibbs film and then in another case after? How could a γ versus ln (c) plot be used to determine the CMC?
5. Someone tries to make an LB monolayer with the headgroup attached to a clean glass slide but with a strongly hydrogen bonding group such as an alcohol at the top of the molecule and thus the film. What is the fundamental problem?
6. Langmuir films, and the related LB transfer process, can be made using a variety of liquid surfaces besides water, including molten salts and mercury. How does this work?
7. If you think about the H_2O molecule, which is bent, as a surfactant with a polar and nonpolar side how might it tend to orient at the water surface?
8. Characterization of self-assembled alkyl chain based monolayers (e.g. alkanethiolate on Au(111) surfaces) have been characterized in detail by several surface science characterization probes. Using the ones discussed in chapter 2, which might be applicable and which might not be to probe surface films such as the Langmuir and Gibbs films? Give reasons, discuss pros and cons and ways to overcome any tricky problems.
9. What do you think is the molecular-scale mechanism to explain the drop in surface tension of a Langmuir film when it is compressed to some surface pressure π?

Appendix A

Detailed derivations of the Gibbs and alloy surface segregation equations

A1 General set-up

Surface and bulk region thermodynamic balances with surface and mass transport work

We consider a solution in terms of a surface region and a bulk region of the same thickness and area, and thus equal volumes, as shown in figure A.1, at equilibrium

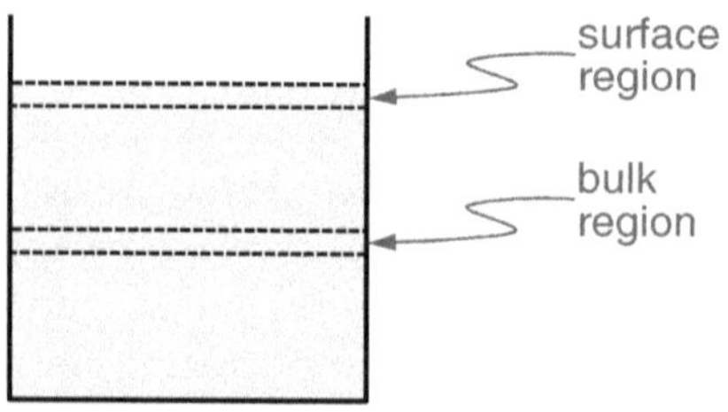

Figure A.1. Illustration of a solution with a surface and a bulk region of equal thickness and areas.

conditions of constant T and P which implies that $T^s = T^b$ and $P^s = P^b$, where we use superscripts to denote the region of interest. To each region we will apply small perturbations of the extensive quantities in the first and second laws, then do thermodynamic balances and finally take the differences between surface and bulk to get the surface excess concentrations in terms of the surface tension.

Starting with the 1 + 2 law, including mass transport and surface work as the non-PV work (μdn and γdA), applied to the surface and bulk regions, where the mass transport work sums run over all the components, we have

Surface region: $[\mathrm{d}U + P\mathrm{d}V - T\mathrm{d}S - \sum\mu_j \mathrm{d}n_j - \gamma \mathrm{d}A]^s = 0.$

Bulk region: $[\mathrm{d}U + P\mathrm{d}V - T\mathrm{d}S - \sum\mu_j \mathrm{d}n_j]^b = 0.$

It is important to note that, in general, the chemical potential and amount of each species in the surface region will vary as a function of the distance z across the region (refer to figure 4.38 and see earlier in sections 4.3.1.4 and 4.4.2). We will deal with these details shortly in the Gibbs equation.

Now scale the extensive quantities in each balance to give integral balances:

Surface region: $[\mathrm{d}U + PV - TS - \sum\mu_j n_j - \gamma \mathrm{d}A]^s + c^s = 0,$

Bulk region: $[\mathrm{d}U + PV - TS - \sum\mu_j n_j]^b + c^b = 0,$

where we add constants of integration, just to be general.

Next take the total differential of the integral balances to show all the ways the regions each can be perturbed including both extensive and intensive variable changes:

Surface region:

$[\mathrm{d}U + P\mathrm{d}V + V\mathrm{d}P - T\mathrm{d}S - S\mathrm{d}T - \sum\mu_j \mathrm{d}n_j - \sum n_j \mathrm{d}\mu_j - \gamma \mathrm{d}A - A\mathrm{d}\gamma]^s = 0.$

Bulk region: $[\mathrm{d}U + P\mathrm{d}V + V\mathrm{d}P - T\mathrm{d}S - S\mathrm{d}T - \sum\mu_j \mathrm{d}n_j - \sum n_j \mathrm{d}\mu_j]^b = 0.$

Subtracting the differential form of the 1 + 2 law from each of the expressions for the total derivatives as a real physical constraint on the changes isolates the intensive variable relationships (essentially a Legendre transformation of variables), which in this case shows how changes in μ require compensating changes in γ in the surface region. In addition we apply the conditions of constant T and P to obtain

Surface region: $A\mathrm{d}\gamma + [\sum n_j \mathrm{d}\mu_j]^s = 0$

Bulk region: $[\sum n_j \mathrm{d}\mu_j]^b = 0.$

The surface region equation is the simplest form of the dependence of surface tension on perturbations of the chemical potentials of the components in the surface region with the contribution of each potential weighted by the number of moles of that component. In contrast, the simpler bulk equation just shows that perturbations of the bulk chemical potentials are interrelated and cannot be changed independent of one another.

To show the difference between surface and bulk we subtract the bulk constraint from the surface one to get the surface excess relationship:

Surface excess equation: $A\mathrm{d}\gamma + [(\sum n_j \mathrm{d}\mu_j)^{\mathrm{s}} + (\sum n_j \mathrm{d}\mu_j)^{\mathrm{b}}] = 0$ which we rewrite as

$$A\mathrm{d}\gamma = -\sum\left[\left(n_j \mathrm{d}\mu_j\right)^{\mathrm{s}} - \left(n_j \mathrm{d}\mu_j\right)^{\mathrm{b}}\right].$$

This is a very general equation that holds for multicomponent mixtures under conditions of constant P, T. The equation states that infinitesimal, reversible perturbations of the surface tension of a homogenous mixture depend upon the numbers of each component in the surface region and in a corresponding slice of the bulk together with the corresponding perturbations of the chemical potential for each component in the surface and bulk regions.

The constraint of equilibrium partitioning of each component between surface and bulk regions

In order to make the above equations useful we need to add one more constraint. Each component j in the bulk region is at equilibrium with the corresponding component in the surface region, written as $S_j^{\mathrm{s}} \leftrightarrow S_j^{\mathrm{b}}$, where the species S_j partitions between the bulk and surface regions. To maintain equilibrium, any infinitesimal change in the chemical potential in one region must be balanced by the same change in the other region, which we express as $\mathrm{d}\mu_j^{\mathrm{s}} = \mathrm{d}\mu_j^{\mathrm{b}}$. Notice that the corresponding chemical potentials are not forced to be equal, as expected, since the surface region components suffer a loss of bonding neighbors in moving from bulk to surface and with a constant surface area there is no way to remove this effect. We can test this in the limit of going to a system of just one pure component where we know that there has to be an excess chemical potential at the surface.

Now we insert the chemical potential differential constraint into the above surface excess equation and obtain

$$A\mathrm{d}\gamma = -\sum\left(n_j^{\mathrm{s}} - n_j^{\mathrm{b}}\right)\mathrm{d}\mu_j^{\mathrm{b}}, \qquad \text{(A1.1a)}$$

which can be re-expressed based on the definition of the surface excess amount of material per area (see section 4.5.2) as

$$\mathrm{d}\gamma = -\sum \Gamma_j \mathrm{d}\mu_j^{\mathrm{b}}, \qquad \text{(A1.1b)}$$

where the region superscripts are now placed on each variable. This very useful form of the surface excess equation directly relates changes in the surface tension to small perturbations in the bulk chemical potentials of each species and their corresponding surface excess amounts. The primary way to alter the chemical potentials is to change concentrations and we now look at this relationship.

Connections between chemical potentials and concentrations

Working towards relationships that connect $\mathrm{d}\gamma$ to changes in the mixture composition the chemical potentials are expressed in the usual way in terms of the activities and standard reference states as $\mu_j - \mu_j^{\mathrm{o}} = RT\ln(a_j)$, where a_j = the thermodynamic activity of species j. So this equation tells us nothing new. It just restates the chemical potential of species j relative to a reference chemical potential μ_j^o in terms of an equivalent difference in natural logarithms of a defined quantity, the activity. It follows that $d\mu_j = RT d(\ln a_j)$. Putting the chemical potentials in terms of the activities is useful since for cases in which the species have similar interactions with each other as between themselves (some form of an ideal behavior) we can make the approximation that $a_j \sim (\frac{c_j}{c_j^o})$ where c are the concentrations, sometimes expressed as mole fractions, and c_j^o refers to the concentration of the species in some standard reference state, often selected a hypothetical state in which the species at this concentration follows some ideal behavior. If the activity–concentration approximation is not valid we can apply activity coefficients in the form $a_j \sim (\gamma_j \frac{c_j}{c_j^o})$ (be sure not to confuse activity coefficients γ with surface tension γ) to correct for deviations from the approximation. Expressing μ in terms of c we have

$$\mu_j - \mu_j^o = RT \ln\left(\gamma_j \frac{c_j}{c_j^o}\right) \quad \text{(A1.2a)}$$

$$d\mu_j = RT \ln\left(\gamma_j c_j\right). \quad \text{(A1.2b)}$$

Finally, incorporating equation (A1.2b) into equation (A1.1b) gives

$$d\gamma = -RT\sum \Gamma_j d \ln\left(\gamma_j^b c_j^b\right). \quad \text{(A1.2c)}$$

Exchange equilibria in binary component systems

So far, we have been very general in terms of systems with multiple components. Now we switch to a simple binary system since this will be the basis of the examples we will use for various conditions and applications going forward. We write the general chemical balance for moving A molecules from bulk to surface with the displacement of corresponding B molecules at the surface back into the bulk in terms of a simple exchange reaction with an associated equilibrium constant K and stoichiometric coefficients ν_A and ν_B, as shown in equations (A1.3a)and (A1.3b). Note that for non-ideal mixtures K, the coefficients and the thickness of the surface region may change with the mixture composition:

$$\nu_A A^b + \nu_B B^s \overset{K}{\rightleftharpoons} \nu_A A^s + \nu_B B^b. \quad \text{(A1.3a)}$$

The equilibrium constant K, expressed in terms of activities is

$$K = \frac{(a_A^s)^{\nu_A}(a_B^b)^{\nu_B}}{(a_A^b)^{\nu_A}(a_B^s)^{\nu_B}} \quad \text{(A1.3b)}$$

Taking the differentials of ln(*K*) with *K* as a constant gives

$$\mathrm{d}\ln K = 0 = \nu_\mathrm{A}\mathrm{d}\ln\left(\frac{a_\mathrm{A}^\mathrm{s}}{a_\mathrm{A}^\mathrm{b}}\right) + \nu_\mathrm{B}\mathrm{d}\ln\left(\frac{a_\mathrm{B}^\mathrm{b}}{a_\mathrm{B}^\mathrm{s}}\right). \tag{A1.3c}$$

The associated expression for the change in the standard chemical potentials is

$$\Delta\mu^\mathrm{o} = -RT\ln K = \nu_\mathrm{A}\left(\mu_\mathrm{A}^\mathrm{os} - \mu_\mathrm{A}^\mathrm{ob}\right) - \nu_\mathrm{B}\left(\mu_\mathrm{B}^\mathrm{os} - \mu_\mathrm{B}^\mathrm{ob}\right), \tag{A1.3d}$$

where the μ^o represent the standard chemical potentials of A and B in the surface and bulk regions and the terms have been collected in a convenient way for later use. This also can be expressed directly in equivalent terms of Gibbs free energies since the only non-PV work involved is mass transport (as long as the total surface area is kept constant—no change in shape):

$$\Delta G^\mathrm{o} = \Delta\mu^\mathrm{o} = \nu_\mathrm{A}(G_\mathrm{A}^\mathrm{os} - G_\mathrm{A}^\mathrm{ob}) - \nu_\mathrm{B}(G_\mathrm{B}^\mathrm{os} - G_\mathrm{B}^\mathrm{ob}) = \nu_\mathrm{A}G_\mathrm{A}^\mathrm{oS\prime} - \nu_\mathrm{B}G_\mathrm{B}^\mathrm{oS\prime}, \tag{A1.3e}$$

where G_A^os, G_A^ob, G_B^os, G_B^ob are the standard free energies of A and B in the surface and bulk regions and $G_\mathrm{A}^\mathrm{oS\prime}$ and $G_\mathrm{B}^\mathrm{oS\prime}$ are the surface excess free energies of A and B on a per mole basis, i.e. the work required to bring a mole of A or B from the bulk to the surface when the species are in their standard reference states, e.g. pure substances.

Starting with equation (A1.3c) as a basis we have a set of equations from which we now can develop appropriate relationships between surface and bulk concentrations once specific conditions are chosen. It is clear from the development up to this point so far that we could readily consider solutions with more than two components by just adding the appropriate chemical potential terms back in equation (A1.1a) and appropriately rewriting the equilibrium reaction equation (A1.3a) and subsequent equations.

A2 Application to the case of a dilute solute (A) in solvent (B)—the Gibbs equation

In this application we set the condition $c_\mathrm{A}^\mathrm{b} \ll c_\mathrm{B}^\mathrm{b}$ in which the solvent B is in large excess over the solute A and derive the Gibbs equation which relates the surface excess of solute A to the change of surface tension with the bulk concentration of the solute. There are two approaches depending upon how one defines the surface excess of the solute.

In the first approach, because of the large excess concentration of solvent c_B^b and c_B^s are assumed to be constant so $\mathrm{d}(\ln c_\mathrm{B}^\mathrm{b}) = \mathrm{d}(\ln c_\mathrm{B}^\mathrm{s}) = 0$. At the dilute concentrations of solute the behavior is assumed to be ideal with activities replaced by concentrations and equation (A1.2c) becomes $\mathrm{d}\gamma = -RT\Gamma_\mathrm{A}\mathrm{d}\ln c_\mathrm{A}^\mathrm{b}$ or in the more standard form:

$$\boxed{\Gamma_\mathrm{solute} = -\frac{1}{RT}\left(\frac{\partial\gamma}{\partial\ln c_\mathrm{solute}^\mathrm{b}}\right)_{P,\,T}}. \tag{A1.4}$$

This is the famous *Gibbs equation*. The equation in principle should only apply to the chosen conditions of highly dilute solute concentrations, both in the bulk and at the surface, and the surface excess Γ_{solute} is defined in terms of the average number of solute molecules per unit area in the surface region, which consists almost entirely of solvent molecules. Inspection of the units of RT and γ shows that Γ_A has units of moles m^{-2}. For $\Gamma_B > 0$, there is a net excess of the solute relative to the bulk, whereas for $\Gamma_B < 0$, there is a net depletion.

We also can choose another approach which covers a wider range of concentrations but we will have to recognize that the surface region concentrations change with location z from the end of the bulk liquid to the beginning the solvent vapor, as shown in figure 4.38, and modify our definition of surface excess. We choose the standard state of the solvent as the pure liquid and assume that both components behave ideally over the concentration range chosen so activities are replaced by concentrations and write the surface region concentrations as $c_A^s(z)$ and $c_B^s(z)$.

With these conditions and assumptions equation (A1.2c) becomes:

$$d\gamma = -RT[\Gamma_A d(\ln c_A^b) + \Gamma_B d(\ln c_B^b)].$$

Recognizing that the concentrations of A and B in the surface region have a distribution with position (z) we specify the surface excesses in terms of average or an integral over z and choose the Gibbs dividing line convention (figure 4.38), integrating as specified in equations (4.30a) and (4.30b). With this convention, since B is the solvent, $\Gamma_B = 0$ and we again obtain the Gibbs equation (equation (A1.4.4)) but now the solute surface excess is defined in terms of the Gibbs dividing line convention. Notice that the Gibbs convention strictly looks at how the solute partitions across the dividing line in the liquid/vapor (or liquid–liquid for that type of interface). So for the case of solute (such as a low surface tension oil) in a high surface tension solvent (such as water), the solute is driven to dominate surface. In the solvent convention, the solvent equally partitions across the dividing line, leaving the oil to pile up on the outer ambient side of the dividing line on top of the last bit of water at the surface. Note that as the bulk oil concentration changes the thickness of the interface layer may increase.

A3 Wide variation of component concentrations—binary metal alloy systems and simple regular solutions of nonpolar molecules

Starting with the equilibrium reaction (A1.3a), making the simplifying assumptions that the atoms A and B occupy equal volumes in the mixture and equal areas at the surface then the bulk-surface equilibrium is just a 1:1 atom (or molecule) exchange with all the stoichiometric coefficients equal to one:

$$A^b + B^s \underset{\longleftarrow}{\overset{K}{\longrightarrow}} A^s + B^b \text{ and } K = \frac{\gamma_A^s x_A^s \cdot \gamma_B^b x_B^b}{\gamma_A^b x_A^b \cdot \gamma_B^s x_B^s},$$

where, the γ represent the activity coefficients. Since we are applying the equilibrium to a wide concentration range the activity coefficients will vary with concentration to reflect any changes in the entropies of mixing in the two phases and the shifts of the solvation energies of the components as the surrounding neighbors change. Thus, for A in pure A, surrounded by A–A interactions the energy will shift going towards dilute A, surrounded only by A–B interactions. Typically, for nonpolar systems of atoms or molecules we can correct for this using mean field theory and interaction parameters such as the chi parameter for molecules. For now we put this aside since later we will assume that the activity coefficients are ~1.

Expressed in terms of the Gibbs free energy of the reaction, following equation (A1.3e), we have

$$\Delta G^{\mathrm{o}} = G_{\mathrm{A}}^{\mathrm{oS}\prime} - G_{\mathrm{B}}^{\mathrm{oS}\prime},$$

where the two terms are the surface excess Gibbs free energies on a per mole basis for A and B each going to the equilibrium surface in their pure phase. Using the connection of per mole to per area surface excess quantities we have per mole of surface,

$$\Delta G^{\mathrm{o}} = \bar{A}_{\mathrm{m(A)}} G_{\mathrm{A}}^{\mathrm{oS}} - \bar{A}_{\mathrm{m(B)}} G_{\mathrm{B}}^{\mathrm{oS}} = \bar{A}_{\mathrm{m(A)}} \gamma_{\mathrm{A}} - \bar{A}_{\mathrm{m(B)}} \gamma_{\mathrm{B}},$$

where, $\bar{A}_{\mathrm{m(A)}}$ and $\bar{A}_{\mathrm{m(B)}}$ = areas per mole of A and B atoms in their pure surfaces. But since we have assumed that A and B have equal areas at the surface this becomes $\Delta G^{\mathrm{o}} = \bar{A}_{\mathrm{m}}(\gamma_{\mathrm{A}} - \gamma_{\mathrm{B}})$. Finally, using the fundamental relationship that $\Delta G^{\mathrm{o}} = -RT \ln K$ we obtain the alloy surface segregation equation

$$K = \left(\frac{\gamma_{\mathrm{A}}^{\mathrm{s}} x_{\mathrm{A}}^{\mathrm{s}} \cdot \gamma_{\mathrm{B}}^{\mathrm{b}} x_{\mathrm{B}}^{\mathrm{b}}}{\gamma_{\mathrm{A}}^{\mathrm{b}} x_{\mathrm{A}}^{\mathrm{b}} \cdot \gamma_{\mathrm{B}}^{\mathrm{s}} x_{\mathrm{B}}^{\mathrm{s}}} \right) = \exp\left(-\frac{\Delta G^{\mathrm{o}}}{RT} \right) = \exp\left[\frac{\bar{A}_{\mathrm{m}}(\gamma_{\mathrm{B}} - \gamma_{\mathrm{A}})}{RT} \right].$$

Now we make two further simplifications: (i) the system follows Hildebrand regular solution behavior in which the entropies of mixing in both phases follow ideal behavior and (ii) the A–B, A–A and B–B interaction energies become similar, although not identical. With these two assumptions the activity coefficients approach ~1 and we write

$$\boxed{\left(\frac{x_{\mathrm{A}}^{\mathrm{s}}}{x_{\mathrm{B}}^{\mathrm{s}}} \right) = \left(\frac{x_{\mathrm{A}}^{\mathrm{b}}}{x_{\mathrm{B}}^{\mathrm{b}}} \right) \exp\left[\frac{\bar{A}_{\mathrm{m}}(\gamma_{\mathrm{B}} - \gamma_{\mathrm{A}})}{RT} \right]}. \tag{A1.6}$$

This equation shows the general exponential dependence of the surface composition on the difference between the two surface tensions of the pure components. In applying this equation, we must be mindful that the interaction energies of the components cannot be widely different, which will hold for a wide variety of nonpolar molecules with surface tension values over a limited range. Though this equation is quite approximate it is very useful as a means to estimate the trends of the dependence of surface behavior as a function of surface tensions and bulk concentrations of homogenous mixtures.

We also can rearrange this equation to solve for the dependence of the mole fraction of one component at the surface as a function of its mole fraction in the bulk. Using the relationship that $x_{\mathrm{A}}^{\mathrm{s}} = 1 - x_{\mathrm{B}}^{\mathrm{s}}$ we obtain

$$\boxed{x_A^s = \frac{f(\gamma)}{1 + f(\gamma)}} \qquad \text{(A1.7)}$$

where $f(\gamma) = (\frac{x_A^b}{1 - x_A^b})\exp[\frac{\bar{A}_m(\gamma_B - \gamma_A)}{RT}]$..

Finally, we derive a relationship between the surface tension of the solution and the bulk composition. The solution surface tension is equal to the total surface excess Gibbs free energy which, in turn, is just the sum of the excess free energies of each component at the surface, expressed as

$$\gamma = G^S = \frac{1}{\bar{A}_m}(G^s - G^b) = \frac{1}{\bar{A}_m}(\mu^s - \mu^b),$$

where the area per mole $\bar{A}_m$ is the same for both components, G^S = the surface excess Gibbs free energy of the surface region, G_s and G_b are the total free energies per mole of the surface and bulk components. Note that the free energy terms are equivalent to chemical potentials since the only work is mass transport (to and from the surface at constant surface area). Including the contributions of the excess surface free energy of both components the overall surface tension is written as

$$\gamma = \frac{1}{\bar{A}_m}[x_A^s(G_A^s - G_A^b) + x_B^s(G_B^s - G_B^b)].$$

Writing the chemical potentials in terms of the general expression $G = \mu = \mu^o + RT\ln(x)$, where we use the mole fraction x for concentration with the standard states of each component as the pure states ($x = 1$), and using μ and G interchangeably, we obtain

$$\gamma = \frac{1}{\bar{A}_m}\{x_A^s[(G_A^{os} + RT\ln x_A^s) - (G_A^{ob} + RT\ln x_A^b)] + x_B^s[(G_B^{os} + RT\ln x_B^s) - (G_B^{ob} + RT\ln x_B^b)]\}.$$

Rearranging terms gives the expression in terms of the standard excess surface free energies:

$$\gamma = \frac{1}{\bar{A}_m}\left\{x_A^s\left[(G_A^{os} - G_A^{ob}) + RT\ln\left(\frac{x_A^s}{x_A^b}\right)\right] + x_B^s\left[(G_B^{os} - G_B^{ob}) + RT\ln\left(\frac{x_B^s}{x_B^b}\right)\right]\right\}.$$

Recognize that the standard state surface excess free energy terms per mole are related to the surface tensions of the pure substances as $\bar{A}_m\gamma_A^o$, $\bar{A}_m\gamma_B^o$, where the superscript zero is added to the surface tensions to denote the pure substance values, so now we have

$$\boxed{\gamma = \left(x_A^s\gamma_A^o - x_B^s\gamma_B^o\right) + \frac{RT}{\bar{A}_m}\left[x_A^s\ln\left(\frac{x_A^s}{x_A^b}\right) + x_B^s\ln\left(\frac{x_B^s}{x_B^b}\right)\right] = \gamma^o + \gamma_{mix}}, \qquad \text{(A1.8)}$$

where γ^{o} is defined as the surface tension contribution with no mixing of the components, treating the surface tension as the weighted average of the surface tensions of the pure A and B phases, and γ_{mix} is the free energy contribution of mixing, essentially an entropy term for A and B mixing in the surface and bulk phases. Since $x_{\mathrm{B}}^{\mathrm{s}} = 1 - x_{\mathrm{A}}^{\mathrm{s}}$ and the two mole fraction ratio terms can be obtained from equation (A1.6) we now have a way to calculate values of γ as desired using equation (A1.8).

Chapter 4 problems

1. The surface tension and specific surface entropy of water are reduced upon the adsorption of n-butane (C_4H_{10}) and perfluorotributylamine [$(C_4F_9)_3N$]. Give reasons for the large reductions on $\gamma(H_2O)$ and $S^S(H_2O)$. Suggest other liquids or gases which would result in the reduction of $S^S(H_2O)$ for the same reasons.
2. For an external pressure of 1 atm, calculate the internal pressure of a water droplet of $r = 5$ nm, 5 μm, and 5 mm.
3. The contact angle of liquid aluminum on α-alumina (Al_2O_3) surfaces is $\theta = 90°$ at 1000 °C. The contact angle decreases with increasing temperature. It is $\theta = 60°$ at 1200 °C. If $\gamma_{\mathrm{lg}}(\mathrm{Al}) = 750$ ergs cm^{-2} at 1100 °C, compute γ_{sg}—γ_{sl} at 1100 °C assuming that cos θ varies linearly with temperature [5].
4. Calculate the reversible work necessary to create an additional 1 mm^2 surface area of Ag (liquid), Ag (solid), NaCl (solid) and N_2 (liquid).
5. Compute the vapor pressure of a spherical Pt particle of radius of 1 nm at 200 °C and at 800 °C using the vapor pressure data given in the literature (for example [6]). Small particles of comparable size are often used as catalysts. Could these particles be used continuously for three years assuming that the loss of material would only occur by vaporization?
6. A researcher decides to measure the surface tension of Cu(liquid). However, some of her samples are contaminated by Ag and some of them by Ni. How would these impurities affect her measurements? Could she, by comparing her results with those obtained from ure Cu samples, determine the impurity concentration, and how? If she finds that the impure liquid shows a decreasing specific surface entropy in a given temperature range while the pure molten metal does not, what conclusion could she reach about the behavior of the two different types of contaminants as a function of temperature?

Further reading

General reference texts

1. Adamson A W and Gast A P 1997 *Physical Chemistry of Surfaces* 6th edn (Wiley). A general surface chemistry text which covers surface thermodynamics and related topics such as wetting and contact angles in detail.

2. Shaw D J 1992 *Introduction to Colloid and Surface Chemistry* 4th edn (Butterworth-Heinemann). General background reading for surface thermodynamics in wet systems with an emphasis on colloids.
3. Erbil H Y 2006 *Surface Chemistry of Solid and Liquid Interfaces* 1st edn (Wiley-Blackwell). General background reading for wetting and other phenomena in surfaces.
4. Hiemenz P C and Rajagopalan R 1997 *Principles of Colloid and Surface Chemistry* 3rd edn (Marcel Dekker). A thorough text covering a broad area of topics containing a good bit of thermodynamics.

Specialized reference texts and articles

1. Penfold J 2001 The structure of the surface of pure liquids *Rep. Prog. Phys.* **64** 777–814 10.1088/0034-4885/64/7/201
2. Israelachvili J N 2011 *Intermolecular and Surface Forces* 3rd edn (Academic). A more specialized text devoted to topics which include surface thermo and related topics such as wetting and adhesion.
3. Davis H T 1996 *Statistical Mechanics of Phases, Interfaces and Thin Films* 1st edn (Wiley). A very detailed text which covers the fundamental thermodynamics and statistical mechanics theory of fluid interfaces. Very thorough but for advanced readers.
4. Evans D F and Wennerström H 1999 *The Colloidal Domain: Where Physics, Chemistry, Biology, and Technology Meet*, 2nd edn (Wiley). A very thorough presentation of theory and applications in the colloid area with some background on surface thermodynamics. Contains a number of applications to bio areas.
5. de Gennes P G, Brochard-Wyart F and Quere D 2003 *Capillarity and Wetting Phenomena: Drops, Bubbles, Pearls, Waves* 1st edn (Springer). An expert text from Nobel Prize winner Pierre de Gennes and co-authors who were key developers of continuum thermodynamic theory applied to soft material interfaces.
6. Rosen M J 2004 *Surfactants and Interfacial Phenomena* 3rd edn (Wiley-Interscience). Very extensive specialist text on surfactants and their behavior in interfacial phenomena.
7. Petty M C 1996 *Langmuir–Blodgett Films: An Introduction* (Cambridge University Press). A reference text on LB films.

References

[1] Blakely J M and Maiua P S 1967 *Surfaces and Interfaces, Chemical and Physical Characteristics* ed J J Burke, N L Reed and V Hess (Kluwer Academic)
[2] Adamson A W 1967 *Physical Chemistry of Surfaces* (Wiley)
[3] Benson G C and Yuen R S 1967 *The Solid–Gas Interface* ed E A Flood (Marcel Dekker)

[4] Somorjai G A and Li Y 2010 *Introduction to Surface Chemistry and Catalysis* (New York: Wiley) pp 293–4

[5] Brennan J J and Pask J A 1968 Effect of nature of surfaces on wetting of sapphire by liquid aluminum *J. Am. Ceram. Soc.* **51** 569–73

[6] Dreger L H and Margrave J L 1960 Vapor pressures of platinum metals. I. Palladium and platinum *J. Phys. Chem.* **64** 1323–4

IOP Publishing

Surface and Interface Science

David L Allara and Robert L Opila

Chapter 5

Surface dynamics

This chapter begins the discussion of the motion of atoms and molecules at surfaces. Here we discuss how atoms vibrate about their equilibrium sites. This chapter includes the measurement of vibrations and what happens during surface melting. The chapter goes on to discuss the simplest gas–surface interactions. These interactions involve several steps beginning with the collision of the incident particle with the surface. The incident particle experiences a potential energy as it approaches the surface. If its kinetic energy is accommodated sufficiently, it is adsorbed. One adsorbed, it may diffuse on the surface, depending on the temperature or morphology of the surface. Finally, the particle may react, or if it has sufficient kinetic energy, desorb from the surface.

5.1 Modes of motion at bare surfaces

5.1.1 Overview of internal motion, thermal energy storage, and heat transfer within the bulk and through interfaces

Condensed matter objects are undergoing constant internal motions of their constituent building blocks of atoms, or atoms formed into discrete molecules. In atomic solids the atoms are moving relative to one another in various synchronized patterns that are described as phonons with the patterns extending over large numbers of adjacent atoms (correlation lengths) in the case of periodic lattices of crystals. In molecular solids relative motions between the molecules can give rise to phonons as well; simultaneously internal atoms in each molecule are undergoing local vibrations and related motions such as hindered rotations. For atomic and molecular liquids, phonons exist but generally decay over very short distances of a few constituent neighbors since it is difficult to synchronize the motions when the constituents are not arranged in a periodic fashion. For molecular liquids each constituent molecule not only is undergoing local internal vibrations but also can undergo various types of rotations relative to its neighbors depending on the intermolecular forces.

doi:10.1088/978-0-7503-4916-1ch5

At 0 K the only motions are the lowest energy state (zero-point energy) vibrations. As the temperature is raised more and more vibrational states are activated, introducing new ways of storing the accumulating thermal energy in the object, and these motions act as the sink or source for heat transfer in or out of the object. The increase in thermal energy with temperature (T) is characterized by the heat capacity (C_V and C_P) at constant volume or pressure, and any increase in the number of ways to store energy causes a logarithmic increase in the entropy (Boltzmann's law, equations (T6A, B), chapter 4).

Now we see how the energy storage at the surface or surface regions, more properly, differs from the bulk. In chapter 4 we looked at differences in the cohesive or bonding energies between the surface and bulk regions (missing atom or molecule picture) and used these to define the surface excess energy (U^S) and related quantities such as surface excess enthalpy (H^S). Correspondingly, there will be differences between the motions of the atoms in surface and bulk regions and these differences become the basis of the corresponding thermodynamic surface excess quantities heat capacity (C_V^S, C_P^S) and entropy (S^S). Consider an infinite perfect crystal at 0 K. All the motion is confined to the lowest energy vibrations which means there are no choices for ways to store the energy; thus, the crystal entropy is zero (referred to as the third law of thermodynamics). You also can see this by considering Boltzmann's law, $S = k_B \ln W$ (see equations (T6A, B), chapter 4), where W is equal to the number of ways to arrange the system configuration. This gives the third law of thermodynamics since at 0 K every energy state is frozen in the lowest level and $W = 1$. Only upon raising T can the system have sufficient energy to allow more energetic vibrations to be accessed.

Now consider a finite crystal and imagine it shrinks in size to the nanoscale. Put a little extra energy in the system over and above what is needed at 0 K. Since in the surface regions the symmetry of the crystal is broken (missing atom model) and the surface atoms have more open space in which to move and more degrees of motional freedom than in the bulk so there are more choices for how to arrange and store the surface energy. Thus, the entropy (and heat capacity) must be larger than in the bulk. In chapter 4 we saw that $S^S > 0$ for all substances and can easily understand this from this simple model; the same argument follows for C_V^S and C_P^S. Keep in mind that these excess quantities apply to the net effect of the surface region contributions which may go as deep as several layers depending on the specific substance and surface condition.

When adsorbates are present on a surface, we have three additional, fundamental types of motion and energy transfer mechanisms to consider:

1. The adsorbates bring additional energy modes to the system; atoms can translate across the surface and undergo vibrations involving bonds to the surface while molecules themselves have further possibilities of rotations and a variety of internal vibrations. These motions contribute to $\Delta C_{P(\mathrm{ads})}$ and ΔS_{ads}.
2. The adsorbate in contact with an underlying substrate surface is constantly undergoing momentum transfer from the fluctuating amplitudes of the substrate surface phonon modes which gives rise to energy transfer channels.

In turn, the adsorbate, which is bubbling in its own way with translation, rotation and/or vibrations, can give kicks back to the surface and in the case of a neighboring liquid or gas phase the adsorbate will suffer occasional collisions by the ambient phase molecules (or atoms) and *vice versa*. Finally, at sufficient coverage adsorbates can collide with one another and transfer momentum. So the energy flow goes around in all directions, allowing the system to reach statistical equilibrium.

3. If the incoming and outgoing impulses in any of these adsorbate transfer processes are not matched in their timing, and typically they are completely random over a wide range of momenta, then this process provides a mechanism for supplying extra energy from time to time (like a double or multiple whammy) to turn on adsorbate processes such as hopping (diffusing) across the surface, chemical reactions such as are involved in catalysis and ultimately, complete desorption. Obviously, increases in T promote activating these higher energy processes and the activation energy is supplied in a random, statistical manner as the system bubbles and perks. This is the basis of the Boltzmann exponential distribution of energies at a given temperature that will be discussed shortly in an upcoming section.

We have already considered collisions and energy exchange from a contacting ambient liquid or gas phase that is at or will lead to equilibrium. But there is a variation of this that is very important in surface science and technology. Consider the non-equilibrium case of surface collisions from controlled energy atoms, ions or molecules deliberately injected into the system under vacuum and directed at the surface in the form of a beam or just generated more diffusely near the surface in some sort of discharge, e.g. a microwave generated ion plasma. These types of processes can either pump or withdraw energy into or from the surface region as well as deliver new material that can deposit and even react. There is a good deal of fundamental science to be learned about the energy accommodation processes but there are also many variations which have been developed as methods of practical advantage. Collisions with energetic particles can be used to clean impurities from a surface, remove layers of the surface region, (termed sputtering) and change the surface topography, induce chemical reactions at the surface and deposit new overlayers and thin films by delivering sputtered material from one surface to another nearby surface. Some variations also provide very useful methods for surface characterization, for example, by sputtering off nanometer-scale chunks of the surface for mass spectrometric analysis or by measuring the momentum change of ions reflected off the surface as a means to detect the specific elements at the very top surface layer. We will learn a bit about these types of surface characterization tools in a later section.

In this section we survey processes at surfaces which arise from motions of constituent atoms and groups of atoms such as molecules. In order to go deeper into the fundamentals, it is useful to start with the simplest of systems. The next step is to delve into the familiar territory of vibrations of atoms in molecules and then extend that to collective vibrations in crystalline solids and how that is related to properties such as heat capacity. The next section 5.1.2 contains details of these topics and if the reader is sufficiently familiar with some of the details the next section or simply wants to avoid the details, coming back as needed, the reader can jump to section 5.1.3.

5.1.2 Vibrations in crystalline solids

5.1.2.1 What are phonons?

Imagine a crystalline solid consisting of beads (atoms) placed on a repeating lattice with the nearest neighbors connected to one another by springs (bonds), sort of like a box spring mattress with interconnected or coupled wire spring coils. Pushing down in one place causes a distortion over a large distance because the springs are linked together in a network. Similarly if you strike a crystal at some location with a sharp blow within the elastic limit (no breakage or permanent deformation) the momentum from the impact will be felt at large distances and as the object relaxes vibrations will move all around and finally decay to leave the object at rest exactly as it was initially, with the extra heat accumulated from the energy transfer eventually released to the surroundings. This is the way that sound travels in an object and is the basis for the 'phon' part of the term phonon (from the word for sound in Greek). These types of vibrations are called acoustic. But coupled vibrations not only carry sound waves, they also synchronize in ways that do not conduct sound but rather do other things, such as interact with electromagnetic radiation depending on the electrical charge shifting during vibration. These are called optical. Overall, phonons act like particles traveling back and forth through the crystal creating local disturbances in the atom positions as they pass by. This is the basis of the 'on' part of 'phonon' which signifies a particle (such as a neutron, electron, etc). Since the particle traveling is massless, we call it a quasi-particle. Rather it is just a disturbance, similar to photons, which are just a massless local electromagnetic disturbance, traveling at the speed of light. In fact, there are deep fundamental connections between the two particles; they both obey boson statistics, which means that the different wave motions in the crystal can go on at the same time in exactly the same space and superimpose on one another. In contrast, different electrons cannot be in the exact same energy state and space, which is the basis of Fermi statistics (think about Pauli exclusion principle, for example).

Now that we have a general idea of the terms let's go look at some simple behaviors and concepts of vibrations between objects and particles. First, we look at a simple ball on a spring and then we will shrink to molecular scale where we will need quantum mechanics to drag out the correct behavior. Finally, we will consider crystalline lattices, which are just huge molecules.

5.1.2.2 A simple classical harmonic oscillator model and associated deviations

First, we look at a simple model system of a ball of mass m attached to a solid wall (in outer space to remove gravity). This will introduce us to a few simple terms and concepts. Referring to figure 5.1, stretching or compressing the spring by a displacement x from the rest position follows Hooke's force law, $F = -k_H x$, and the corresponding potential energy follows a harmonic (or parabolic) energy well law, $U = -\int_0^x F\mathrm{d}x = \frac{1}{2}k_H x^2$. Pull (or push) the ball to position $x = \pm A_o$ then

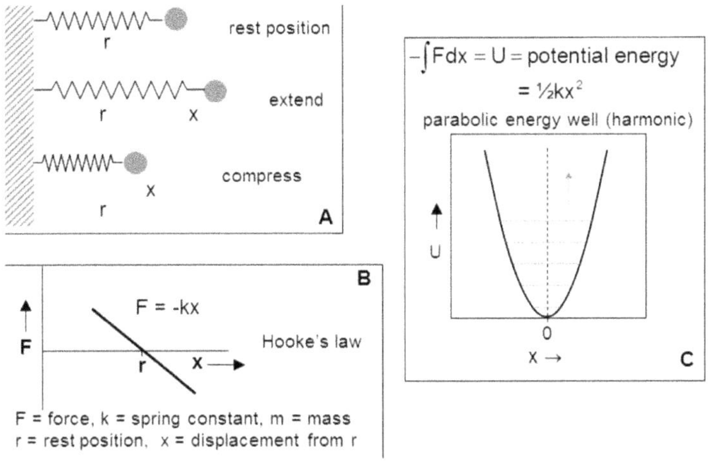

Figure 5.1. The system of a vibrating ball on a spring attached to a rigid wall. (A) Vibration system schematic. (B) Hooke's law force plot. (C) The associated harmonic potential energy well.

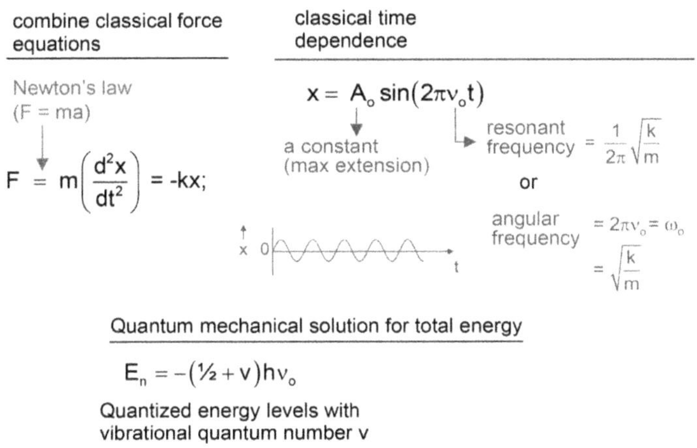

Figure 5.2. Classical and quantum mechanical (QM) solutions for the simple harmonic oscillator. Top: Classical t-dependence equation from Newton's second law and the resulting displacement versus t solution. Bottom: QM solution of the energy level spacings.

release it and start the clock at $t = 0$. The system will oscillate forever, unless interfered with, and will store the energy you put in with your initial effort to stretch (or compress) the spring, converting back and forth between kinetic and potential energy during each ¼ cycle. Applying Newton's second law of forces with the Hooke's law constraint (see figure 5.2) and the initial conditions you imposed gives the displacement time dependence as

$$x = A_o \cos\left(\sqrt{\frac{k_H}{m}} \times t\right) = A_o \cos\left(\omega_o \times t\right), \tag{5.1}$$

$$U(x) = \cancelto{0}{U_o} + \cancelto{0}{\left(\frac{dU}{dx}\right)_{x=0}} x + \underbrace{\frac{1}{2!}\left(\frac{d^2U}{dx^2}\right)_0 x^2}_{\text{harmonic term}} + \underbrace{\frac{1}{3!}\left(\frac{d^3U}{dx^3}\right)_0 x^3 + \cdots}_{\text{anharmonic terms}} \quad \text{Taylor expansion}$$

Figure 5.3. Expansion of the 1D oscillator potential in a power series to account for non-linearities in Hooke's law. Typically only a second order term is needed for most cases.

where the resonant oscillation frequency is given by $\nu_o = \frac{1}{2\pi}\sqrt{\frac{k_H}{m}}$ (units of cycles s^{-1} or Hz) or expressed in terms of angular frequency, $\omega_o = \sqrt{\frac{k_H}{m}}$ (units of radians s^{-1}). This harmonic well model is used often when discussing and describing vibrations in molecules and solids, so it provides a useful foundation to go on to more complicated problems. One thing it does well is predict the correct frequency for an oscillator in a harmonic potential well, even for molecules and in crystalline lattices.

A standard deviation from this model arises for cases in which the spring response changes with increasing displacement, especially beyond some limit. In this case k_H becomes a function of the displacement and Hooke's law becomes nonlinear. This type of deviation is called an anharmonic vibration since the shape of the potential well is distorted from parabolic and the deviations are typically described in terms of a power series, as shown in figure 5.3. These types of deviations often occur at surfaces, as will be discussed later.

5.1.2.3 Extension of the classical harmonic oscillator to molecules and crystal lattices

Starting from a chemistry background the simplest entry into the vibrations of lattices is to start with the familiar vibrations of molecules, as shown in figure 5.4 for the linear molecule CO_2. For any molecule (or object) with N atoms there are three motions (x, y, x) for each atom and three N total motions. If we lock the object in place around the center of gravity so it will not translate or rotate, this leaves $3N - 6$ or $3N - 5$ internal vibrations for nonlinear and linear molecules, respectively. The internal motions can all be active at the same time to give a very complex picture of the atoms jostling around. But these motions can be exactly decomposed into a set of independent motions, called normal modes, each of which has a pattern that does not interfere with the other normal mode patterns, as illustrated for CO_2 (figure 5.3) with $9 - 5 = 4$ normal modes for a linear molecule with three atoms. It is also common to call each of these normal modes an oscillator since the atoms involved are just undergoing periodic oscillations in some specific pattern of motion. The motions in these modes and the associated frequencies can be figured out using classical mechanics once the force constants (more properly, the potential wells around each atom) and the molecular geometry are known.

Now consider a crystalline lattice of N atoms as just a huge molecule. With large N (a semi-infinite lattice) there are $3N - 6$, or $\sim 3N$ modes which involve all the

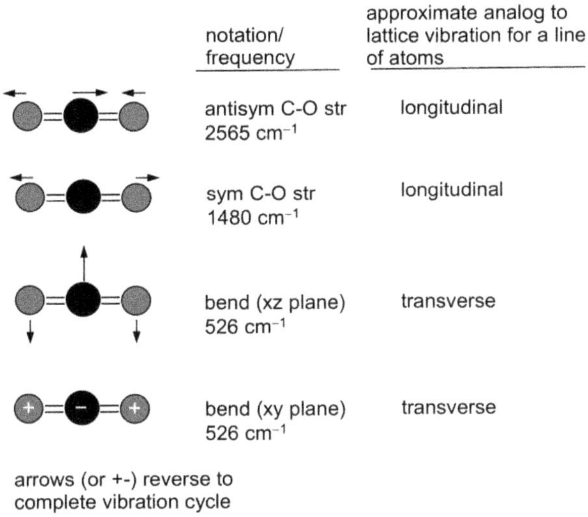

Figure 5.4. Vibrational modes of the CO_2 molecule with standard notations and analogs of the type of motion relative to the bond axis as related to crystal vibrations for a line of atoms along a crystal axis. The three atoms can oscillate around the center of gravity of the CO_2 molecule in $(3 \times 3) - 5 = 4$ independent ways. Each mode has a synchronized motion pattern that does not interfere with the motions in other modes; all the motions can occur simultaneously. In the notations sym, antisym and str stand for symmetric, antisymmetric and stretch. The directions of the motion are shown for one part of the complete oscillation cycle and reverse upon return to the initial configuration. The + − symbols show motion coming out of or going away from the plane of the page.

atoms with stretches and bends, etc, and they all can be excited independent of each other, namely, they can all superimpose their motions. For each mode the local synchronized motions of a group of neighbors can repeat across the lattice (e.g. across a stack of crystal planes) as a wave with the repeat wavelengths as some multiple of the lattice spacing between neighbor atoms. These patterns of each mode for perfect crystalline lattices can be figured out exactly from the lattice spacings and angles (essentially the lattice basis vectors). Moving ahead we will just consider a perfect crystal to simplify the discussion; but also, this will provide some standard terms and concepts used to describe surface vibrations that we will run into later. A schematic of the different types of lattice modes is shown in figure 5.5 for the example of a diatomic crystal.

The figure shows a line of alternating constituent atoms along one of the crystal axes with the atom motions directed either perpendicular to the axis (transverse modes) or in-line (longitudinal modes). The modes are labeled acoustic or optical depending on whether the different types of atoms move out of phase with each other (optical modes) or in-phase (acoustic modes). In the optical modes as the different atoms move in opposite directions any charge difference between them (local dipole across their bond or different ions such as Na^+ and Cl^-) will change which gives rise to the ability for excitation of the vibrational states by light. There are two points to note here. (i) Acoustic modes, as the name suggests, are responsible for the transmission of sound waves through the object (note the density

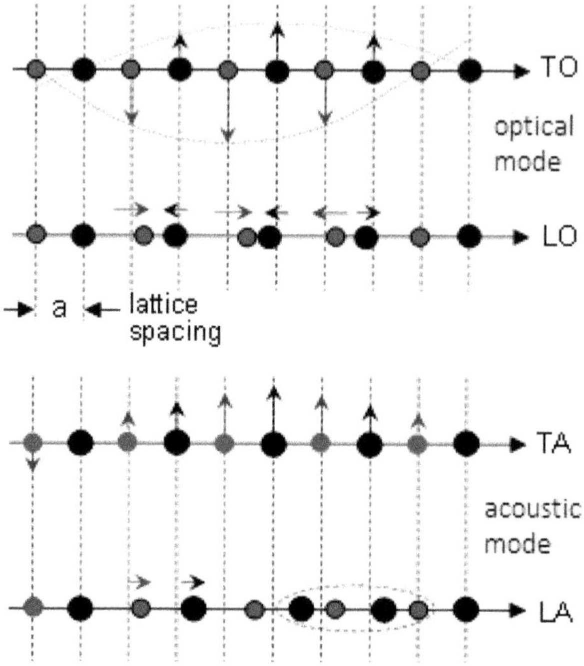

Figure 5.5. Illustrations of the four main types of vibrational modes in crystals for the example of a diatomic lattice with a heavy and a light (black, blue) atom. The light atoms will tend to move with larger amplitudes. The atoms run across a lattice axis with spacing a. Arrows indicate the directions the atoms move in one part of the vibration cycle (reverse on return). The atoms can move in-phase (acoustic; bottom) or out of phase (optical; top) and also move perpendicular to (transverse) or along the lattice axis (longitudinal). In the longitudinal acoustic (LA) mode (bottom illustration) note the bunching of the atoms (dashed red line region) as the mode is cycling, which helps to illustrate how an acoustic pulse can be carried along the axis direction. Note how the modes can be characterized by counting the number of lattice spacings for a complete wave of displacements before the next one starts along the axis. The shortest possible wave physically involves just the motions of pairs of atoms which repeats identically across the entire lattice. The light dashed lines in the top figure show a half wave for the motions of the two different atoms.

compression across the axis in the LA mode in the figure—as these compressions move, they can carry an acoustic pulse). The waves travel back and forth across the object at the speed of sound, so this parameter is important in characterizing the vibrational characteristics of material and is thus useful in analysing the thermal energy storage and heat transfer characteristics of each given material. (ii) The wavelengths λ of the wave-like motions are correlated in terms of the numbers of atoms per wave across the crystal axis and it is common to specify each particular mode in terms of the wavenumber quantity $1/\lambda = k$, which then converts the description into one in frequency or reciprocal space, which was used to describe diffraction in chapter 3 (section 3.4.2). The advantage of switching to reciprocal space is that the equations for describing the vibrations are considerably simplified, although keep in mind the phenomena can be described either way. The range of the different modes runs from one long wave across the entire crystal lattice ($\lambda_n = n \cdot a$, where n and a are the lattice row length and lattice atom spacing) to one wave over one lattice unit ($\lambda_o = a$) which for semi-infinite crystals gives $k_n \to 0$ ($\lambda_n \to \infty$) and $k_o \to \infty$ ($\lambda_o \to 1/a$). This is the basis of so-called Brillouin zone plots where mode frequency (ν_k) is plotted against the wave vector k. We leave these complexities behind and move on now to the proper, but very brief, quantum mechanical description of the vibrations and how they differ at surfaces. The interested reader can find details on phonons, reciprocal (or k) space and Brillouin zones in other sources (see resource reading at the end of the chapter). This is a very important field in physics and materials science but has complexities well beyond the scope of this chapter.

5.1.2.4 Quantum mechanical oscillators

So far, we have just discussed classical oscillators, but this is not enough to accurately describe everything needed to know about the behavior at the atomic and molecular scale. We need a deeper approach to lead us to a correct understanding of the thermal energy storage properties of a condensed matter object such as heat capacity and entropy.

Each classical oscillator can have any total energy, including zero (at rest) and take on any associated amplitude A_o since we can, in principle, stretch or compress the spring(s) as much as desired to initially inject energy initially. For this classical case, the frequency is always the same, except for problems such as reaching some region of deviation from Hooke's law to a nonlinear behavior where overtones, harmonics, etc, can arise and complicate the oscillator behavior. Now when we go to the molecular scale (for example using Br–H as a wall–ball model quite accurately here) the motion is confined within small fractions of an Å (in general, bonds only stretch and compress by a few percent of the rest length during a vibrational cycle) while the atom(s) accelerating back and forth can have sizeable momenta (moving with maximum speeds like an ideal gas molecule) sufficient to make the de Broglie wavelengths of the order of the confinement scale. So (i) we lose track of where the atoms really are and (ii) the oscillator energy is no longer continuous but is forced into discrete levels. This requires quantum mechanics to solve the motion problem (the Schrödinger equation works well) and we emerge with the familiar picture of quantized vibrational energy states with the equation

$$E_v = (\tfrac{1}{2} + v)h\nu_o = (\tfrac{1}{2} + v)\hbar\omega_o, \tag{5.2}$$

where ν = integer quantum number for each level (as shown in figure 5.2 (bottom)), h = Planck's constant, $\hbar = h/2\pi$, ν_o = the resonance frequency as before and $\omega_o = 2\pi\nu_o$; the displacement amplitudes are given as probability distributions derived from the associated wavefunction $\psi_v(x)$. This principle holds no matter how many atoms are connected together so we can conclude that each of the $\sim 3N$ normal modes of a crystal lattice will have their own set of allowed energy levels with each level described by an integral quantum number. The collective motions of the atoms give rise to phonons. The conversion from a classical oscillator to a quantum oscillator is often called first quantization. A second quantization also can be done involving all the modes in one shot as a large quantum system in the Schrödinger equation (or equivalent method); this is the proper way to frame the theory of phonons as a total set of collective modes for the crystal lattice. This treatment gives rise to a deeper concept of phonons in which the motions of the crystal at each frequency are treated as a particle, called a quasi-particle since it has no mass, moving along the lattice directions as a traveling disturbance. This is a direct analogy to photons, as mentioned earlier, which also are often treated as massless particles of a traveling electromagnetic disturbance.

Do we really need quantum mechanics for vibrational energy levels?

We are going to compare the de Broglie wavelength of a vibrating atom against the rigid wall (a big atom such as Br can serve this purpose for the H atom in HBr vibrating relative to the heavy bromine). Here are the steps:

1. The excursion or displacement (δx) of a chemical bond during one complete vibration is typically ~5% of the bond length. Given a typical bond length of ~1.5 Å estimate δx (nm and then m).
2. For a typical vibrational wavenumber of ~2000 cm^{-1} or wavelength $\lambda = 5$ μm (infrared energy region) calculate the frequency of the vibration from $\nu = \frac{c}{\lambda} = c\tilde{\nu}$, where $c =$ speed of light, 3 × 108 m s^{-1} and the time for one cycle of the vibration $\tau = \nu^{-1}$.
3. Next estimate the average speed v in m s^{-1} of the atom in going the distance of a vibration round trip in the given time τ. How does this compare to the speed of a typical gas molecule in the air of ~500 m s^{-1}?
4. Simply assign the mass of the atom as 1 g $mole^{-1}$ or 1×10^{-3} kg (H atom). Now calculate the momentum p of the single atom in the round trip from m and v. Use units of kg · m s^{-1}.
5. Finally, estimate the de Broglie wavelength from $\lambda_{dB} = \frac{h}{p}$, $h =$ Planck's constant, 6.63×10^{-34} kg · m^2 s^{-1}, and compare to the size of the total excursion δx (~5% or so of a bond length), which is the confining space of the moving atom.

Answer: λ_{dB} ~3 nm and quantum mechanics must be used. Why? Just like just like diffraction of light. If a photon with wavelength λ comes into a diffraction grating of width d and $d < \lambda$ then the photon cannot 'fit' in the spacing, so it breaks into wave properties and diffracts in specific (quantized) directions. Similarly, λ_{dB} is bigger than the confinement spacing of the vibrating atom, so the atom motion (momentum) is forced to break into wave-like properties that require discrete energy levels and we replace knowledge of the exact position of the atoms with the probabilities given as wave functions arising from solutions of the Schrödinger equation.

5.1.2.5 Vibrational frequencies in solids and the capacity to store thermal energy

5.1.2.5.1 Vibrations in molecules

At this point we have a general, qualitative, understanding of vibrations in condensed matter and lots of details about the special cases of crystalline lattices. Each nth oscillator mode has an associated set of energy levels, $\{E_n^v\}_{v=0,1,2,\ldots}$ with

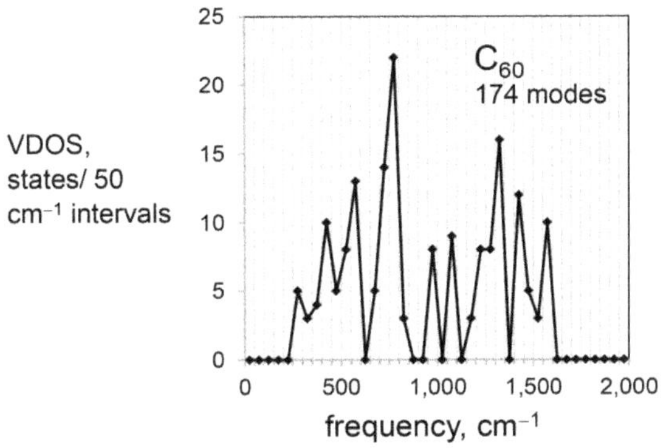

Figure 5.6. Plot of the density of vibrational states for the C_{60} molecule. There are a total of $(3 \times 60^{-6} =)$ 174 vibrational states and these are divided in the plot into the number per 50 cm^{-1} intervals with the number per increment given at the center frequency of the interval. The vibrational density of states (VDOS) spectrum is similar to an actual excitation spectrum but lacks the excitation cross sections, which are specifically determined by the type of excitation mechanism, e.g. infrared absorption, Raman scattering or inelastic neutron scattering.

associated vibrational quantum numbers ν, which can store energy. In theory there are an infinite number of energy levels but, of course, as $\nu \rightarrow \infty$ at some point the crystal melts as too much energy was absorbed. Thus, to understand the amount of thermal energy that can be stored at any given temperature T, namely, the heat capacity, we need to be able to know how many oscillator modes can be excited and how much energy they can each store in their associated energy levels at temperature, T.

An example of a distribution of vibrational mode states for a small object is given in figure 5.6 for the case of the isolated C_{60} molecule which has a total of $(3 \times 60^{-3} =)$ 174 different vibrations (all sorts of stretching, flexing, twisting, modes of the constituent C atoms). The plot shows the total number of modes in each 50 cm^{-1} interval with the number shown for the central frequency of the interval (e.g. total modes for 1400–1450 cm^{-1} range are marked at 1425). In general, the modes appear to be distributed more or less evenly across the frequency range, ending at 1575 cm^{-1}, the highest frequency mode. The C_{60} molecule is a good introduction to a semi-infinite crystal since it contains only C atoms and is a highly symmetrical nano-sphere. Notice that all the vibrations are essentially those of a surface which wraps around on itself.

5.1.2.5.2 Vibrations in semi-infinite solids

As we move to a semi-infinite solid with a huge number, N, of atoms the $\sim 3N$ vibrations are still at discrete frequencies but become so closely packed that they approach a continuous distribution. The standard way of describing this characteristic is in terms of the density of vibrational modes or states (DOS, which for vibrational states is called VDOS and for phonons, PDOS). You can see how the PDOS arises by keeping the C_{60} distribution in mind. The standard definition is given in terms of the number of oscillator states that exist per increment of energy, or equivalently, frequency (either ν or equivalently ω to hide the 2π factor, with ω

perhaps the more commonly used), as specified by $\frac{dn(\omega)}{d\omega} = g(\omega)$, where $n(\omega) =$ the number of states at a given frequency. This shows how packed together or dense the modes are at each increasing increment of mode frequency or mode energy. If the distribution is in terms of oscillator energies we just use $E = \hbar\omega_o = h\nu_o$, which is the photon energy needed to excite the oscillator from is zero (rest) state into the first excited quantum state ($\nu = 1$). This state and the higher lying ones ($v > 1$) also can be excited thermally and thus act as storage levels for the thermal energy in the solid. So, it is important to know how many of these states there are and their distribution.

Obviously just guessing from the C_{60} distribution you can imagine that each crystalline material might have fairly complex distributions (figure 5.6 above was an example of this). Since $g(\omega)$ maps out the energy storage mode distribution it can serve as the gateway into thermodynamic properties of solids through the use of statistical thermodynamic theory. Direct measurements of $g(\omega)$ over a wide frequency range are quite challenging experimentally but can be done with methods such as neutron scattering and light scattering (Brillouin scattering). Theoretical calculations can also be used and can be quite accurate using quantum mechanical density functional theory (DFT) for well-ordered crystalline solids, but calculations can be time consuming and for many complex materials are not straightforward. As such, it is useful to find some smooth, simple functional forms as approximations to facilitate estimates of thermodynamic properties. Two approximate PDOS models have been developed over the years, the simple Einstein model, in which the solid is assumed to have only one type of vibration (e.g. each atom jiggling individually against its neighbors with no synchronization) with a constant VDOS, and the Debye model, in which there is a monotonic range of different vibrations along each crystal direction up to a maximum frequency. Obviously, the Einstein model is not correct since there is always a range of frequencies in solids, but it does give surprisingly good predictions for quantities such as heat capacity. The Debye model is much closer to the real situation in solids and is commonly used to predict and estimate thermodynamic properties, in some cases with good accuracy.

5.1.2.5.3 The Debye model for the phonon density of states (PDOS)

In the Debye model, waves travel along each crystal axis, which we label as *x*, *y*, and *z* assuming a cubic symmetry with side length *L* and lattice spacing *a*, and each wavelength spans integral numbers of lattice points, with λ_{min} and λ_{max}. It is convenient to label each wave by a dimensionless integer $\kappa = \frac{L}{(½\lambda)}$ which is just the number of ½-waves (or peaks) that fit along the lattice axis (e.g. $\kappa = 1$ and $L/2a$ for λ_{max} and λ_{min}, respectively). Also note that this measure is in reciprocal space, namely, the inverse of the lattice distances. The model assumes, similar to the way we treat photons, that the frequency of each κth wave (or mode) along a given axis has a frequency $\nu_\kappa = \frac{c_S}{\lambda_\kappa} = \kappa(\frac{c_S}{2L})$, where c_s= speed of the wave, assumed to be the speed of sound (acoustic speed) and identical for all modes. The energy of the mode accordingly is given by $E_\kappa = \hbar\omega_\kappa = h\nu_\kappa = h\kappa(\frac{c_S}{2L})$. The maximum frequency is labeled the Debye frequency ν_{Debye} (or ω_{Debye}). The wave velocity is not constant at the higher frequencies as the lattice becomes more stressed during the motions,

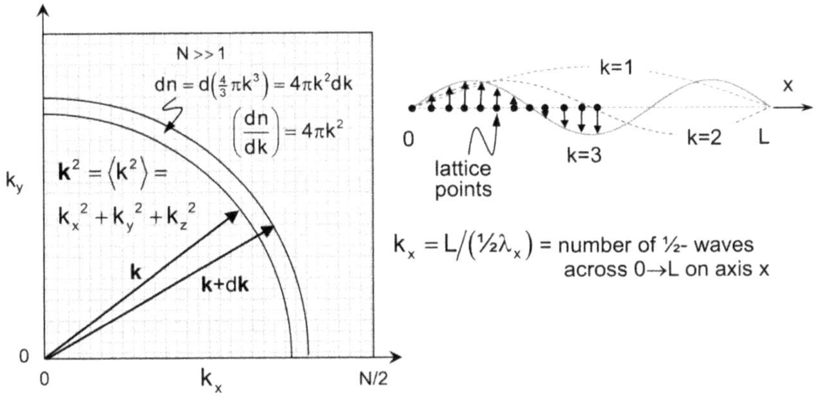

Figure 5.7. Left: 2D view of the number of amplitude peaks in lattice vibrations along the x- and y-axes in discrete values of κ_x and κ_y. Another axis exists along the z-direction but is not shown. For large N the κ values approach a continuum and the vector κ sweeps out a small spherical shell in the $\kappa > 0$ octant that contains dn intersection points (κ_x, κ_y, κ_z) representing a phonon with component waves moving along each axis. Right: Illustration for waves moving along the x-axis of length L. The κ values specify the number of half-waves that fit along $0-L$.

distorting out of harmonic behavior, so the assumption is not valid strictly speaking, but it still holds approximately.

Using the constant c_s assumption, the density of states can be determined in a simple way by treating the phonons as particles (quasi-particles) of different energies (frequencies) moving around the crystal carrying thermal energy (which they can deposit as heat), similar to treating gas molecules with different kinetic energies as quantum particles in a box with component energies (or velocities) along the x-, y-, and z-axes. The total mean square energy of a given particle is given in terms of its component energies by $\langle E^2 \rangle = E_{\kappa_x}{}^2 + E_{\kappa_y}{}^2 + E_{\kappa_z}{}^2 = \alpha(\kappa_x{}^2 + \kappa_y{}^2 + \kappa_z{}^2) = \alpha\langle \kappa^2 \rangle$, where the κ are the mode indices for the contributing waves along each axis and $\alpha = h(\frac{c_s}{2L})$. We plot the range of component modes frequencies along each of the three x-, y-, and z-directions, as shown in figure 5.7 for just the x–y-plane and locate any given state as the intersection of the κ_x, κ_y, κ_z points. As the length of the axes increase to large values, with a corresponding increase in the possible lattice points (components) we can treat the κ as continuous and the number of states $n(E)$ of average energy $E_{\text{rms}} = \sqrt{\alpha\langle \kappa^2 \rangle} = \sqrt{\alpha} \cdot \kappa_{\text{rms}} = \sqrt{\alpha} \times \kappa$ is given by $\mathrm{d}n(E) = \mathrm{d}(\frac{4}{3}\pi\kappa^3) = 4\pi\kappa^2\mathrm{d}\kappa$. This gives our PDOS as $(\frac{\mathrm{d}n(E)}{\mathrm{d}k}) \propto \kappa^2$, where we drop the proportionality constant for convenience since we are just interested, for now, in the general dependence on κ. Finally, we recast this in terms of the average frequency of a 3D phonon. Typically, the angular frequency is used so the equation becomes

$$\left(\frac{\mathrm{d}n(\omega)}{\mathrm{d}k}\right) \propto \omega^2. \tag{5.3}$$

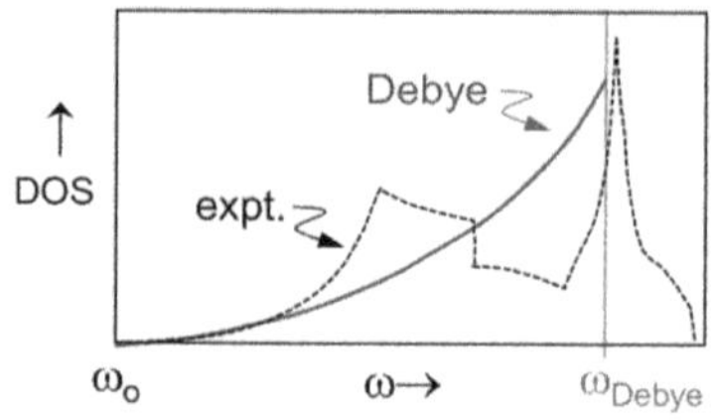

Figure 5.8. The dependence of the density of phonon states (DOS) versus their frequency (ω). (−−): Standard Debye model dependence. (●●●●●●): Representative form of a plot from experimental data. Typical data plots will be a series of data points with error fluctuations or noise so this plot is a rough envelope of the data points for a specific material. The red line indicates the cutoff frequency (ω_{Debye}) for the collection of phonon states.

Table 5.1. Debye temperatures for some common solids.

Solid	Pb	Au	Cd	H_2O (ice)	Ag	Pt	Zn	Cu	W
θ_{Debye}, **K**	96	165	186	192	225	240	300	344.5	405
Solid	Ti	Al	Ni	α-Fe	α-Mn	Cr	Si	C (diamond)	
θ_{Debye}, **K**	420	426	440	464	476	610	640	2200	

A Debye PDOS plot illustrating this ω^2 dependence is given in figure 5.8. As the average frequency of a phonon increases the number of states rises as the square and finally reaches the cutoff where the states end at ω_{Debye} (or ν_{Debye}). Also shown is the dependence for typical experimental data (dashed line) which reveals a number of regions of the frequency range with structure. This shows that the mode frequencies in real crystals look more like spectral features, which is not surprising since the data are just specific vibrational excitations of the lattice that increase and vanish as frequency increases, as one would expect to see from infrared spectra of various materials. Note that the total integral under the DOS curves is just the total number of oscillator states ($3N$) and you can see by quick inspection that the Debye model and the experimental results have roughly similar integrals. This is why the Debye model often works well for estimating thermal properties such as heat capacity.

Finally, to connect all this to quantities such as heat capacity and entropy we need to know at any given temperature T how many levels are excited and storing energy; obviously if a given level or mode at frequency ω_{k} has an energy of excitation well above the average thermal energy pool at T this mode will not participate in storing thermal energy. Each solid material has a temperature above which the thermal energy pool is sufficient to excite all $3N$ modes to a condition in which the lattice has reached the maximum number of participating modes. This temperature is called the Debye temperature (T_{Debye} or more commonly θ_{Debye}) and is straightforwardly estimated in a variety of ways such as from C_V versus T measurements, entropy, melting point, speed of sound and crystal lattice parameters, and phonon spectra. Conversely, once θ_{Debye} is known by one method it can be used to determine a variety of other properties. Table 5.1 shows θ_{Debye} for a variety of solids.

We can relate θ_{Debye} to ω_{Debye} (or ν_{Debye}), the limiting phonon frequency in a given material, by calculating the photon energy $h\nu_{Debye}$ that would be required to excite from the lowest zero-point vibrational state to the limiting Debye state and equate that to $k_B\theta_{Debye}$, the average thermal energy content of that mode at temperature θ_{Debye}. This gives the useful relationship

$$h\nu_{Debye} = k_B\theta_{Debye}, \tag{5.4}$$

where k_B = Boltzmann's constant.

5.1.2.5.4 Thermodynamic quantities from the Debye model

Now that we have a background on the Debye model, what can be done to use it. The following relationships from the Debye model are useful for estimating thermodynamic and related material properties of a crystalline substance. Note that the energy storage refers only to storage in vibrational modes. Metals also will have stored energy in electron states above the zero level (conduction bands) but these values are typically much less than the vibrational contributions.

The equations below give the associated quantities for a collection of N atoms in the solid for the high temperature condition $T > > \theta_{Debye}$ and low temperature regime $T < \theta_{Debye}$. In the cross-over region where and $T \rightarrow \sim\theta_{Debye}$ an integral involving the number of modes excited at each temperature must be evaluated (integral not shown), as given by statistical thermodynamic relationships. Note in the equations the common term of the ratio $\frac{T}{\theta_{Debye}}$, which gives a scaling of the actual temperature to the Debye temperature.

1. Internal energy (U):

$$\text{Low } T: \quad U = \frac{3\pi^4 N(k_B T)}{5}\left(\frac{T}{\theta_{Debye}}\right)^3 \tag{5.5a}$$

$$\text{High } T: \quad U \rightarrow 3Nk_B T. \tag{5.5b}$$

2. Heat capacity (C_V):

$$\text{Low } T: \quad C_V = \frac{12\pi^4 Nk_B}{5}\left(\frac{T}{\theta_{Debye}}\right)^3 \tag{5.6a}$$

$$\text{High } T: \quad C_V \rightarrow 3Nk_B. \tag{5.6b}$$

3. Entropy (S): S is determined by integrating the heat capacity, which varies as a function of temperature. The integral is given by the standard thermodynamic equation:

$$\text{Low } T: \quad S = \int_0^T C_V \frac{dT}{T} \tag{5.7a}$$

$$\text{High } T: \quad S \to \text{cons tan t} + 3Nk_B \ln\left(\frac{T}{\theta_{\text{Debye}}}\right). \tag{5.7b}$$

4. Melting point (T_m): Following Lindeman theory we have the approximate relationship where m = atomic mass, ρ = molar volume and C_L = a constant which is ~137 for metals and 200 for nonmetals:

$$T_m = \left(\frac{\theta_{\text{Debye}}}{C_L}\right)^2 m\rho^{2/3}. \tag{5.8}$$

5. Mean square lattice vibrational displacement $\langle r^2 \rangle$:

$$\langle r^2 \rangle = \frac{3h^2}{(2\pi)^2 m(k_B\theta_{\text{Debye}})}\left(\frac{T}{\theta_{\text{Debye}}}\right). \tag{5.9}$$

A number of other properties can be estimated from the Debye model but the above gives a range of the types. More accurate calculations of the properties require using accurate PDOS data. Next, we will consider surfaces and see how they differ from the bulk and what we can do to extend the Debye model to estimating surface properties.

Questions for thought:

Q. The heat capacity at low temperatures is much more accurate for the Debye model than it is for the Einstein model. Why is that?

A. The low energy (low temperature) modes in the Debye model have low frequencies and small amplitudes. These low energy modes better reflect the crystal behavior at low temperatures than the individual Einstein modes.

Q. The phonon thermal conductivity can be represented by $\kappa = \frac{1}{A}\frac{\partial Q}{\partial t}\frac{\Delta x}{\Delta T}$, where κ is the thermal conductivity, Q is the heat, t is time, Δx is an incremental distance, and ΔT is the change in temperature. This equation is analogous to Ohm's law. Can you identify the terms corresponding to the resistance, current, and voltage?

A. $\frac{\partial Q}{\partial t}$ corresponds to the current, ΔT to the voltage, and $\frac{A}{\kappa \Delta x}$ to the resistance (κ corresponds to the electrical conductivity).

5.1.3 How do surface vibrations differ from the bulk and what effects does this have on surface properties?

5.1.3.1 General principles of surface motions

The missing atom model tells the story fairly well as depicted in figure 5.9. All we need to do is use this model to interpret and estimate the changes in the vibrational properties and behavior of bulk lattices that were outlined in some detail in the

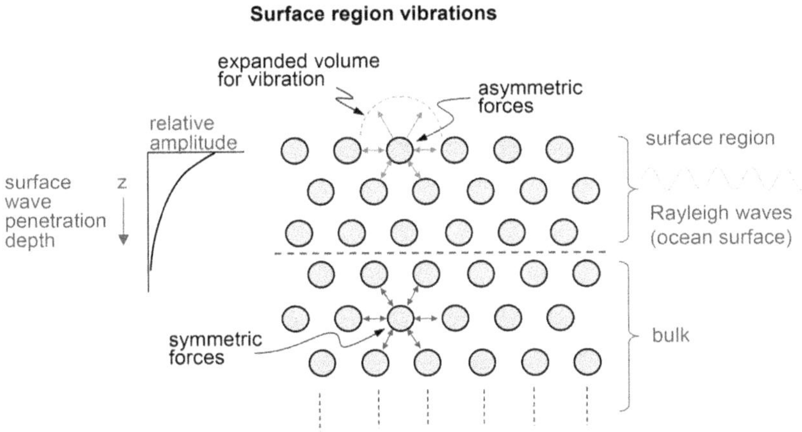

Figure 5.9. Side view schematic of a crystal lattice with the surface at the top. The forces on the atoms in the bulk are symmetrically balanced while at the surface they are asymmetric which will lead to asymmetric potential wells compared to those in the bulk. The surface atoms have more freedom to vibrate away from the surface compared to the allowed motion volume in the bulk. The perturbations in the surface decay over a short distance *z* into the bulk, typically 1–3 layers. The top layer can be viewed as a loosely connect film to the bulk and can undergo surface waves.

previous section. Here are a few important points that follow from the features in the figure.

1. The perturbations of vibrations in the surface should be limited to ~1–3 layers, the typical depth of the surface region (see chapters 2 and 3).
2. The imbalance of forces at the top layer causes the spacing to the next layer to contract in the *z*-direction, a behavior which we encountered earlier in surface reconstruction (see chapter 3).
3. The top surface atoms have more freedom of motion in the vertical (*z*) direction (no opposing compression forces from atoms above) so will have lower frequency vibrations in the *z*-direction with more closely spaced vibrational energy levels that can store more energy than for comparable bulk atoms.
4. The thin surface region layers are somewhat mechanically decoupled from the underlying bulk and can undergo lateral moving waves with amplitudes along the *z*-direction. One unique type of wave at the surface is a Rayleigh wave in which the surface layer is excited to generate waves skimming across the surface like ocean waves (transverse acoustic phonons in the *x*–*y*-plane). These waves exist in both solids and liquids and can be especially pronounced in the latter where the intermolecular forces allow greater amplitudes. The wave propagation velocity and amplitude depend on the surface tension (just like the wave character of a rope or guitar string depend on the tension across the rope or string) and measuring these waves is one way of determining surface tension values for liquids.
5. Combinations of the above points lead to the conclusion that the motions of the atoms in the thin surface region are more easily excited energetically

compared to the bulk which then means that the surface can hold more thermal energy at a given temperature and has a higher entropy than the bulk. This fits well, as it should, with the fact that the excess surface entropy and heat capacity (S^S and C_V^S) are always positive (recall surface excess thermodynamic quantities in chapter 4).

Let's explore some of these points in more detail by introducing experimental methods and a bit of theory to make relevant measurements of and calculations on surface vibrations.

5.1.3.2 Measuring surface vibrations

In chapter 3 the use of particles, ranging across photons, electrons, atoms and even neutrons, was introduced to probe surface structure, in particular by measuring their diffraction patterns to determine atom spacings on various faces of single crystal surfaces. In these measurements the particle energies (incoming = outgoing) energies were conserved, and the experiments were directed at looking at momentum changes in terms of the scattering angles relative to the incoming beam direction. These same types of experiments can also be used to measure vibrational spectra at surfaces by arranging them to measure particles exiting the surface with changed energy, as illustrated in figure 5.10. In any scattering experiment some fraction of the scattered particles exit with loss of energy due to transfer of some kinetic energy (KE) to cause

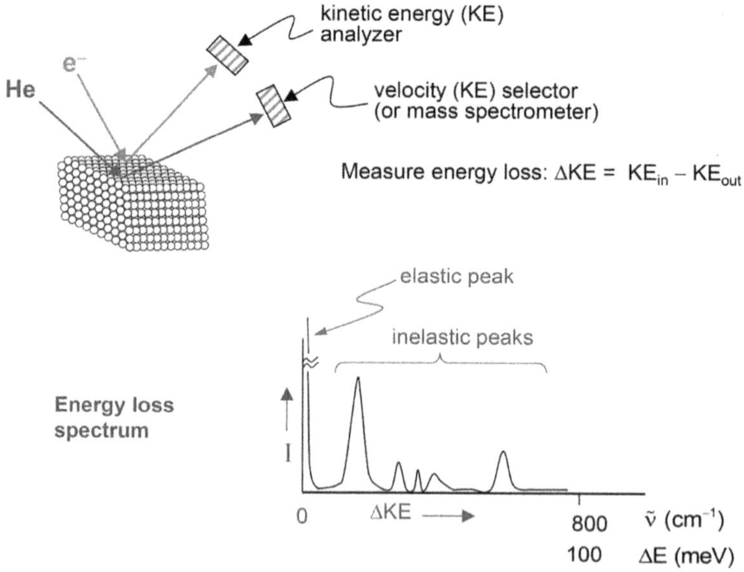

Figure 5.10. Schematic of energy loss experiments on a crystalline surface using either electron or helium atom probe beams. The relative number of the scattered particles is measured at each increment of energy loss relative to the incoming beam energy and then plotted to give a spectrum as shown at the bottom. This spectrum is then interpreted in terms of the various surface vibrational modes that were excited. The strong peak at zero loss corresponds to purely elastic scattering. The energy loss scale is given in the typical units of both wavenumbers (cm^{-1}) and meV (milli-electron volts) with the conversion 1 eV = 8066 cm^{-1}.

excitations of various surface states, particularly vibrational (phonon) states. By configuring a detector to measure the loss in energy between the incoming and outgoing beams one can obtain energy loss spectra and for single crystal surfaces these can be interpreted in terms of the various vibrational modes excited at the surfaces.

The scattered beam can be chosen to be at the specular angle (like a mirror reflection) or off specular at some other angle; each case gives useful information, but theory is generally needed to make proper interpretations in the off specular case. Typical energy loss spectra are taken using either electrons or helium atoms since unlike x-rays or other photons these particles do not penetrate into the bulk only sample the surface region; electrons perhaps penetrate only 1–3 layers before reflection and helium atoms reflect off only the top layer (refer to chapter 3). Since the phonon spectra at the surface and in the bulk for a single crystal differ only by small energy shifts and feature changes it is impossible to use photons to probe the surface since these features will be swamped by the bulk peaks. Mono-energetic electron beams are easily generated and measured because of the ability to both control the input beam and make measurements using the electrical charge properties to steer the electrons around, so these are the most common experiments. When the measurements are done with high resolution of the energy losses, they are termed HREELS for high resolution electron energy loss. Helium atom energy loss is a more challenging experiment because of the difficulty in making mono-energetic (narrow velocity) beams and the need to use a mass spectrometer or velocity selector for detection of the neutral particles in the beams. The exit beam intensities are plotted versus energy loss to generate spectra such as illustrated in figure 5.10. A typical spectrum would be obtained just using the specular beam and would show a huge peak with no energy loss followed by the appearance of peaks, each corresponding to some excitation of a surface vibrational mode (phonon for a bare crystalline solid), similar to what one would see in a simple infrared spectrum of the vibrational modes in some sample such as a thin film. We will revisit these energy loss techniques in a later section when we discuss vibrational spectroscopy of adsorbates. Meanwhile, it is useful to refer back to the related diffraction experiments (chapter 3) to see details of how electron and He atoms probes are handled in those experiments.

Examples of both He and electron energy loss spectroscopy are shown in figure 5.11 for a variety of bare, single crystal surfaces. Since the stepped surface spectra include contributions from both the terrace and step regions the effect of steps can be deduced by comparing with the spectra from the single terrace samples. Typically, the frequencies are $\tilde{\nu}_{\text{no steps}} > \tilde{\nu}_{\text{stepped}}$, which is observed for He scattering from the Ni(111), Ni(977) surfaces. The softening of the Ni(977) vibrations can be assigned to weaker out-of-plane vibrations that would arise at an exposed step edge where the coordination number drops relative to the terrace. This fits with the simple picture of the missing atom model. But note the reversal in the case of electron scattering from the Pt(111) and Pt(775), so one must be careful in interpretations and look in detail at how the particles might interact with the surface.

Inelastic Particle Scattering Examples

Incoming	KE_{in},	ΔKE, meV	Surfaces
He	thermal speed	~30	Na, Ag, Si, K / graphite
	thermal speed	14-26	Ni(111), Ni(977) } $\tilde{\nu}_{(111)} > \tilde{\nu}_{stepped}$
e^-	~2 - 10 eV	~70	ZnO
		~25	Pt(111), Pt(775) } $\tilde{\nu}_{(111)} < \tilde{\nu}_{stepped}$

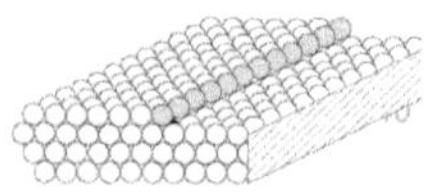

Figure 5.11. Examples of inelastic energy loss spectroscopy with He atoms and electrons on single crystal surfaces. The effect of steps can be seen by comparing the stepped and no step sample spectra. The incoming energy of the He atoms is at thermal speeds (typically a few meV) whereas the incoming electrons have ~10–100 meV energies.

Some questions for thought:

1. Explain in some detail what you think might be the problem using most photons as a surface probe? Consider a range of energies from x-rays to infrared light and discuss each range individually.
2. What kind of useful information might be in the elastic peak? What kind of experiment do you need to obtain that information?
3. What is the value of a typical vibration frequency of a solid? Give your answer in Hz (s^{-1}) and cm^{-1}. Also give the values in equivalent energy in units of eV and in wavelength in units of μm. Useful relationships:

$$c = \lambda\nu \;\; \lambda = \text{vib. wavelength, c = speed of light} = 3 \times 10^8 \text{ m s}^{-1} \text{ in vacuum}$$
$$E = h\nu \quad \nu = \text{vib. freq. , } h = \text{Planck's constant, } E \text{ is in units of J}$$
$$1 \text{ eV} = 1.602 \times 10^{-19} \text{ J} = 8066 \text{ cm}^{-1} \rightarrow 1234 \text{ nm.}$$

4. How might the frequency values of a solid depend on the lattice binding strength? If the vibration at ω_{Debye} were excited by photons what would be the photon energy? What is a very general estimate of the typical value of the thermal energy (active motions) stored in an atomic solid at room temperature? (*Hint*: recall that the thermal energy of a mole of an ideal monoatomic gas at temperature T is $E = 3/2RT$. So from this a rough estimate of thermal energy is RT per mole or $RT/N_{AV} = k_BT$ per molecule.)
5. Compare the frequencies in the spectrum in figure 5.10 with those of a typical infrared spectrum of molecules. How do these compare and does the comparison make sense in terms of the extended, collective vibrations (phonons) of many atoms in a lattice? Why? How does this relate to the thermal properties of the solid?
6. Referring to figure 5.11 and given that the He energy loss spectra for a stepped surfaces how the expected result of $\tilde{\nu}_{\text{no steps}} > \tilde{\nu}_{\text{stepped}}$ what might be going on in the reversal in the HREELS Pt(775) case? Does the latter result seem physically reasonable? What might be coming into play in the e^- scattering measurement?

5.1.3.3 Surface mean square displacements

5.1.3.3.1 Estimating mean square displacements of the bulk from θ_{Debye}

Consider the case of ambient temperatures and see how far the atoms are moving away from the surface compared to their amplitudes in the bulk. We started out early in this chapter stating that typical vibration amplitudes in molecules are only a few percent of the bond lengths so that is a number to keep in mind. Here is the approach we can use. It essentially is similar to one that we used earlier to estimate the speed of an atom during a molecular vibration in the thought question in section 5.1.2.3.

First, we give the main equation then show it comes about from fundamental relationships. We consider an average oscillator (mode) moving in an average direction over all directions (x, y, and z) with a mean squared displacement amplitude $\langle r^2\rangle$ from the equilibrium position, alternatively expressed as a root mean square $\sqrt{\langle r^2\rangle} = r_{\text{rms}}$ value. For simplicity we use the ball on a spring model and consider the oscillator as a simple atom of mass m vibrating against a fixed heavy mass. This gives the equation

$$\langle r^2\rangle \sim \frac{3h^2T}{m(2\pi\theta_{\text{Debye}})^2k_{\text{B}}}, \tag{5.10}$$

where k_{B} = Boltzmann's constant, h = Planck's constant and T = temperature.

It is straightforward to see how this equation can arise from a few fundamental relationships that we already have discussed:

A. Start with the equal distribution of kinetic and potential energies (KE and U) during each cycle of an average harmonic oscillator in x-direction (KE and U trade off during each vibration cycle): $\langle\varepsilon_x\rangle=\langle(\text{KE})_x\rangle+\langle U_x\rangle$ and $\langle\text{KE}\rangle=\langle U\rangle$,

B. Now use Hooke's law and ν_o (see equation (5.1)) to give

$$\langle U_x\rangle = 1/2\langle E_x\rangle = 1/2\langle kx^2\rangle \sim 1/2m\langle(2\pi\nu_0)^2x^2\rangle,$$

which leads to

$$\langle\varepsilon_x\rangle \sim m(2\pi)^2\left\langle\nu_0^2\cdot x^2\right\rangle.$$

C. For the condition $T \geqslant \theta_{\text{Debye}}$ (or just θ from now on) all vibrational modes are excited and contribute to the total energy with average frequency ν^θ. First, calculate the corresponding energy of the oscillator by looking at the photon energy required to excite it from the zero state (equation (5.2)): $\langle\varepsilon_x^\theta\rangle \sim h\nu^\theta$.

D. The average associated thermal energy at $T = \theta$ is given by $\sim k_{\text{B}}\theta$ so we obtain
$\nu^\theta \sim \frac{k_{\text{B}}\theta}{h}$.

E. Now set $\nu_o \sim \nu^\theta$ so the only oscillator mode is the average one which gives the equation: $\langle\varepsilon_x\rangle\sim m(2\pi)^2\left(\frac{k_{\text{B}}\theta}{h}\right)^2\langle x^2\rangle$.

Table 5.2. Calculations of mean squared displacements of vibration from the Debye temperature of a solid.

Solid	θ_{Debye}, K	AW, daltons (g mol^{-1})	m, kg	NN distance, nm	$\sim\langle r^2\rangle$, m^2	$\sim\langle r^2\rangle^{1/2}$, nm	Average % amplitude extension $= \Delta r/r$
Cu	344.5	63.5	1.05×10^{-25}	0.256	5.83×10^{-23}	0.007 63	~3.0%
Ag	225	107.9	1.79×10^{-25}	0.288			
Au	165	197	3.27×10^{-25}	0.289			
C (diamond)	2200	12	1.99×10^{-26}	0.155			

Data: $h = 6.63 \times 10^{-34}$ J · s; $k_B = 1.38 \times 10^{-23}$ J K^{-1} ; $\frac{\Delta r}{r} \sim \frac{\langle x^2\rangle^{1/2}}{\text{NN distance}} \times 100$

F. At temperature T the average thermal energy per oscillator is $k_B T$ so
$\langle \varepsilon_x\rangle \sim m(2\pi)^2\left(\frac{k_B\theta}{h}\right)^2\langle x^2\rangle = k_B T$ and $\langle x^2\rangle \sim \frac{h^2 T}{m(2\pi\theta)^2 k_B}$.

G. For average oscillations in xyz directions the mean square displacement $\langle r^2\rangle$ is given by
$\langle r^2\rangle = \langle x^2\rangle + \langle y^2\rangle + \langle z^2\rangle = 3\langle x^2\rangle$.
   a. Substituting into the previous equation gives the final result:
   $$\langle r^2\rangle \sim \frac{3h^2 T}{m(2\pi\theta)^2 k_B},$$
   b. With $\sqrt{\langle r^2\rangle} = r_{\text{rms}}$ = average radius of motion vibrating around a fixed position in the lattice in all directions.

Looking at table 5.1 pick a substance with $\theta_{\text{Debye}} \leqslant$ ambient T. Knowing θ_{Debye}, T and m we now can estimate $\langle r^2\rangle$ and r_{rms} for the substance from equation (5.3). Examples are given in table 5.2 for Cu. Go ahead and calculate these quantities for the other solids in the table.

5.1.3.3.2 Measurements from the Debye–Waller effect

A diffraction event of an impinging x-ray on a crystal takes place on a time scale of $1/\nu$ (x-ray) or $\sim 10^{-18}$ s, which is far shorter than the time of a lattice vibration ($\sim 10^{-12}$ s) so the resultant diffraction pattern is like a series of snapshots of the crystal atoms in slightly different positions, which results in a broadening or smearing of the diffraction spots. This is known as the Debye–Waller effect and applies to a variety of diffraction events. As the temperature is increased the broadening increases from the excitation of higher amplitude vibrations and the intensities of the center of the spots decreases. These data can be analysed to determine an average value (or root mean square since directions are averaged) vibrational displacement amplitude.

This method applies to surfaces and for this purpose low energy electron diffraction (LEED) is commonly used to determine vibration displacements of surface atoms, $\langle r^2\rangle^{1/2}(\text{surf}) = r_{\text{rms}}$ (surf), as illustrated in figure 5.12. Once obtained equation 5.3 can be applied to give the surface Debye temperature $\theta^{\text{surf}}_{\text{Debye}}$

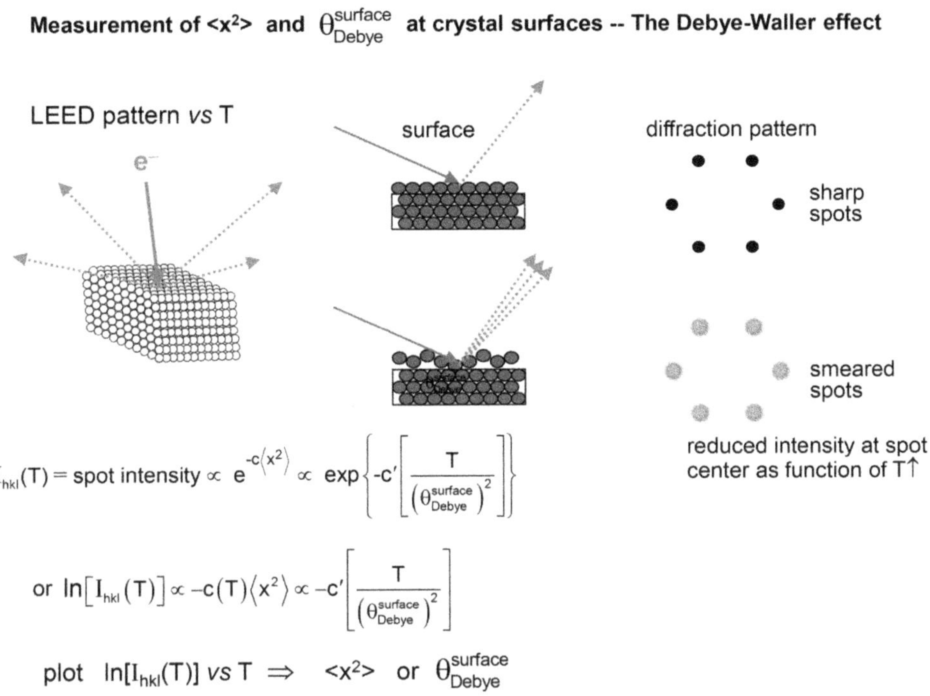

Figure 5.12. Illustration of the use of LEED to measure surface vibrational amplitudes. The diffraction angle of an electron will depend upon the exact position of the surface atoms at any given instant. Any variations caused by vibrational motion of the surface atoms will continually redirect the beam over a small range of angles (as shown in the middle pictures) which results in broadening of the diffraction spots and reduced intensity at the central angle (as shown on the right). The drop in intensity will drop with increasing temperature as the surface vibrations modes are increasingly excited which results in the equations shown on the lower left.

(or just θ_{surf}). Comparison can be made with bulk values of θ^{bulk}_{Debye} (or just θ_{bulk}) to show differences between surface and bulk vibrational properties, as illustrated in table 5.3 for a series of metals.

5.1.3.4 Surface energy, heat capacity and entropy based on surface vibrations

Whereas the measurement of the heat capacity (C_V) and entropy (S^o) of a crystalline solid is straightforward through standard thermodynamic approaches, determination of these parameters for surfaces is much more difficult, particularly to distinguish the differences between specific crystal faces. There are several approaches to this problem including:

1. Measurements of thermodynamic properties around the surface tension.
2. Statistical thermodynamic theory calculations based on the known phonon density of states (PDOS) of the surface region, obtained either from experimental measurements or directly from theory.
3. Calculations using simple models such as the Debye model with known surface θ_{Debye} and ν_{Debye} obtained by various methods.

Table 5.3. Surface rms vibrational amplitudes and θ_{Debye} from Debye–Waller LEED measurements.

Surface	$\frac{\langle r_{\perp}^{2}\rangle^{1/2}(\mathrm{surf})}{\langle r_{xyz}^{2}\rangle^{1/2}(\mathrm{bulk})} = \frac{\theta_{\mathrm{surf}}}{\theta_{\mathrm{bulk}}}$	θ_{surf}, K	θ_{bulk}, K	References
Pb(110), (111)	2.43 (1.84)	37 (49)	90	[1, 2]
Bi(001), (0112)	2.42	48	116	[2]
Pd(100), (111)	1.95	142	273	[1]
Ag(100), (110), (111)	2.16 (1.48)	104 (152)	225	[3]
Pt(100), (110), (111)	2.12	110	234	[4]
Ni(110)	1.77	220	390	[5]
Ir(100)	1.63	175	285	[6]
Cr(110)	1.80	333	600	[7]
Nb(110)	2.65	106	281	[8]
V(100)	1.52	250	380	[9]
Rh((100), (111)	1.35	260	350	[10]

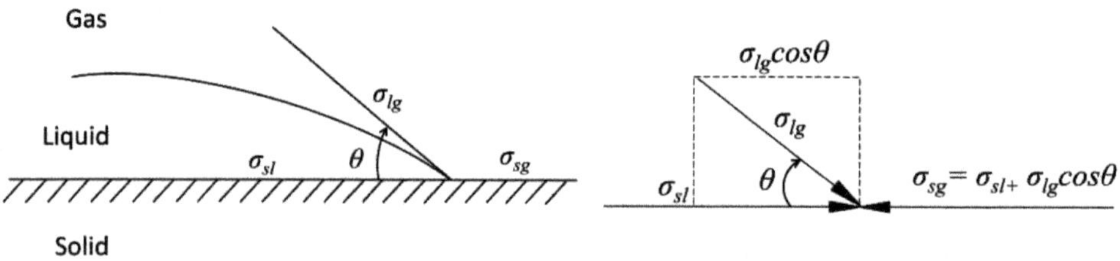

Figure 5.13. Left, drop of liquid on a solid surface. Right, derivation of Young's equation, described in text.

5.1.3.4.1 Thermodynamic properties

The surface energy, σ, is defined as the change in free energy, G, with the change in area, A, i.e. $\sigma = \left[\frac{\partial G}{\partial A}\right]_{T,\ V\dots\ \mu_i}$ at constant temperature T, volume V, and chemical potential, μ_i. The surface free energy can be determined, at least relatively, by comparing contact angles of various liquids with the solid surface.

If a small volume of a liquid is in contact with a solid surface the droplet of liquid will have a specific angle at the junction of the solid–gas, liquid–gas, and solid–liquid triple point. Experience tells us that if this contact angle is small, the liquid will tend to spread over the solid. Thermodynamically that means the liquid–gas surface energy is less than the surface energy of the solid–gas.

We can be more quantitative. In figure 5.13, at the junction of the liquid–gas, solid–gas, and solid–liquid interfaces, at equilibrium the solid–gas surface free energy, σ_{sg}, can be related to the liquid–gas, σ_{lg}, and solid–liquid, σ_{sl}, free energies by

$$\sigma_{\mathrm{lg}} \cos\theta + \sigma_{\mathrm{sl}} = \sigma_{\mathrm{sg}},$$

where θ is the angle formed between the liquid–gas interface and the liquid–solid interface. Practically, care must be taken that the drop is small, the liquid does not react with the solid, the drop is in equilibrium. In fact, at equilibrium, the drop should be neither proceeding or receding. At the end of the day, often surface energies can only be found relatively, by comparing the contact angles of two different surfaces with the same fluid.

It is found that different crystallographic orientations of surfaces of the same crystal can have very different free energies. Generally, more close-packed surfaces will have lower surface free energies than more open surfaces. As discussed above, often high index surfaces reconstruct to form faceted surfaces of low index planes.

Once the surface free energy is known, from thermodynamics, the entropy can be determined by how the surface free energy changes with temperature, i.e. $-\Delta S = [\frac{\partial\sigma}{\partial T}]_P$.The macroscopic surface entropy generally increases with temperature as the surface roughens.

There are many examples of the utility of surface thermodynamics. For example, in an alloy mixture of two metals, one metal may segregate more to the surface because it has a lower surface free energy. Liu and Wynblatt showed that, for, as the fraction of Ag on the {111} surfaces of Cu–Ag alloys at 750 K increased from 37 atom percent to 76 atom percent as the bulk Ag concentration increased from 39 atom percent to only 42 atom percent [11].

Another good example of practical applications of surface free energy is the faceting of surfaces. Some high index surfaces have large surface free energies. The overall surface free energy can be decreased by increasing the overall surface area but forming facets of low surface free energy planes.

5.1.3.4.2 Estimates of thermodynamic properties from theory

Since it is difficult to experimentally determine accurate PDOS data over a wide range of frequencies quantum theory-based calculations, particularly DFT, can be used to generate the potential energy wells around each atom and from this the complete set of the $3N$ acoustic and optical phonon frequencies can be generated, including those in the surface region. With this, statistical thermodynamic methods are used to determine C_V^{surf} and S_V^{surf} for each desired crystal face.

Work by Tewary and Fuller showed that there is a phenomenological relationship between the surface free energy and the Debye temperature for cubic solids (see figure 5.14 and [12]). They showed that the Debye temperature (in degrees kelvin) is equal to $71.9 \times \sqrt{(\Gamma/M)}$, where Γ is the surface energy (in ergs cm^{-2}) and M is that atomic weight. Here the surface energy is displayed for the solid at absolute zero. The rationale for this relationship is that the surface energy depends upon the interatomic potential by reflecting how much energy is needed to break a bond to create a surface. Similarly, the Debye temperature depends upon the solid vibrations, which also depend upon the interatomic potential.

Interestingly, in an earlier paper, Treglia and Desjonquères [13], used a relationship between the surface and bulk mean free displacements to obtain the surface Debye temperature (knowing the bulk Debye temperature), as follows $\frac{\langle u^2\rangle^{\text{S}}}{\langle u^2\rangle^{\text{B}}} = (\frac{\theta_\infty^{\text{B}}}{\theta_\infty^{\text{S}}})^2$,

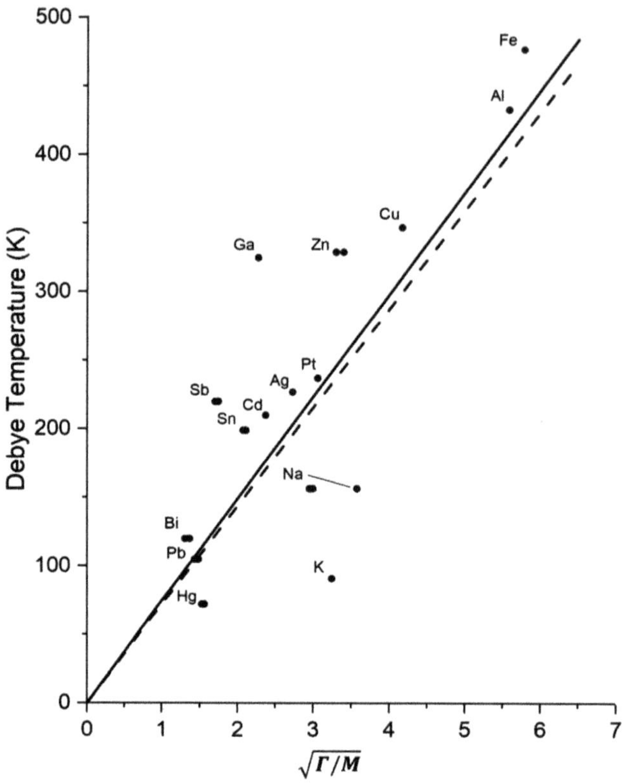

Figure 5.14. A plot of Debye temperature, θ, in kelvins as a function of $\sqrt{(\Gamma/M)}$, where Γ is the surface energy in ergs cm^{-2}, and M is the atomic weight, from [12]. The straight line is predicted in the paper and the dashed line is the least squares fit to the data. The solid points are taken from [13].

where $\langle u \rangle^2$ is the root mean square displacement for either the surface (s) or bulk (B), and θ_∞ is the Debye temperature at high temperature for the surface (s) or bulk (B). Using the relationships above, for various thermodynamic properties and the Debye temperature, the surface thermodynamic properties can be estimated.

The ratio of surface Debye temperature to bulk Debye temperature depends on knowing the ratio of the mean square displacements of the surface and the bulk. The surface mean square displacements perpendicular to the surface are different than the mean square displacements parallel to the surface. Tréglia and Desjonquères in figure 5.15 calculate that perpendicular to the surface, for face centered cubic lattices, low index (including (111), (100), and (110)) surfaces, the ratio of surface to bulk mean square displacements is about 1.4. Parallel to the surface, it depends more upon the surface orientation (packing). So for (111), (100), and (110) surfaces the ratio comes to about 1.10, 1.18, and 1.27, respectively. As a result, for vibrations parallel to the surface $\theta^{s}_{\infty,\perp} \simeq 0.7\theta^{B}_{\infty}$. The qualitative conclusions are that surfaces melt at lower temperatures than the bulk, and that loosely packed surfaces melt before closely packed surfaces.

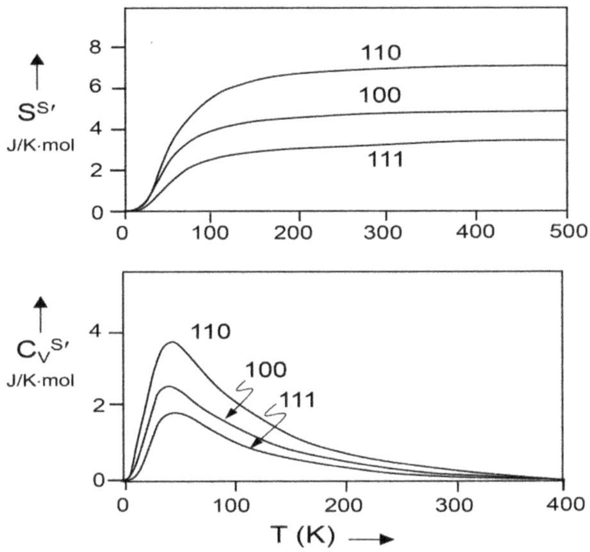

Figure 5.15. Excess entropy and heat capacity of the surface region for different faces of Pd crystals as calculated as a function of temperature by Tréglia and Desjonqueres in [14]. They look at the role of vibrations on surface vibrational and thermodynamic properties. For the (111) surface, only the top layer of atoms is considered, and for the (110) and (100) layer the top two layers are considered. (Reproduced with permission from [14]. Copyright 1985 EDP Sciences.)

Knowing the surface Debye temperature permits thermodynamic quantities such as the heat capacity at constant volume, ΔC_V, and the entropy, ΔS, for the outermost layers to be estimated. In figure 5.14 the specific heat for each surface atom is calculated as a function of temperature for the first few atomic layers of Pd. Similarly, the entropy for each surface atom in Pd is calculated for the (110), (100), and (111) surfaces. It is interesting in that both the specific heat and entropy are greatest for the more open surfaces (110) than the close-packed (111) surfaces. The more open surfaces have more degrees of freedom (more bonds are broken), which have both greater entropy and more modes available to absorb heat.

5.1.4 Surface vibrations in liquid metals

An example for thought:

Stuart A Rice and his group studied the surfaces of liquid metals, to see how much the surface of the liquid was a continuation of the ordering of the bulk liquid [15]. Using glancing angle x-ray scattering and calculations his group was able to show that in the transverse direction (parallel to the surface), the order (or disorder) of trivalent metals Al, Ga, In, and Tl was very similar to the bulk despite their differences. In the direction perpendicular to the surface, however, the metals are stratified for about three atomic layers. Thus, part of the work which is done to create a surface results in the partial ordering (at least in the longitudinal direction) of the surface. *A question for thought is to consider how this surface reconstruction is reminiscent of the ordering (reconstruction) that occurs when many single metal surfaces are created.*

The Rice group also looked at the surfaces of liquid mixtures of small amount (0.2%) of Bi in Ga at 36 °C [16]. The melting point of Bi is 271 °C. Not surprisingly, the surface is enriched in Bi. Grazing incident x-ray scattering shows that the surface contains approximately 80% Bi, and that the Bi-rich regions are separate from the Ga-rich regions. The lateral order of the Bi-rich regions is consistent with super-cooled liquid Bi. *Following the logic above, discuss how the work used to create an interface contributes to the disorder, or melting of the Bi surface layer.*

5.2 Surface diffusion

5.2.1 Types of surface diffusion

Atoms on a solid surface easily diffuse across the surface, since many of the surrounding bulk atoms are present. If there are no adsorbate atoms present, the bulk atoms (or molecules) on the surface can move around in several ways. If there are no nearest neighbors in the plane of the atom, it can hop from site to site. Eventually the diffusing species will encounter a step. It may then step up or down (depending on whether or not the step is up or down).

Once a free atom encounters a step edge, it is likely to spend a significant time at the step edge, because it now has an attractive potential to its neighbors. As a result of this stability, this edge atom will diffuse more slowly and is more likely to move along the step edge. Once it encounters a kink in the edge, it will likely stay there even longer. As discussed previously, some steps are more energetically stable than other steps. Thus, atoms from diffusing from steps will tend to remain longer on more stable step edges. As the atoms move and equilibrate the thermodynamically more stable steps will evolve. Also, because step edges are good sources and sinks of diffusing atoms, surfaces will tend to smooth with time.

One interesting observation is that when diffusing atoms encounter a step down, often the downward step acts as a repulsive barrier. The atom, in its transition to the lower step, will go through a state with only one nearest neighbor, thereby increasing its free energy in the intermediate state. On the other hand, diffusion across a terrace or along a step would have a much smaller energy barrier.

The types of diffusion on the surface are the following: diffusion on a terrace, diffusion into a step, diffusion along a step, diffusion up or down a step, as shown in figure 5.16 and enumerated in table 5.4. Almost all surface diffusion processes are one of the above or a combination. On the surface of liquids, vacancy formation (hole formation → particle passage → new hole closure) processes dominate.

5.2.2 Activated diffusion

Perfect surfaces have structure that reflects the termination of the lattice, with translational symmetries that reflect the particular crystal plane. Adsorbed atoms occupy specific sites on the surface, for example, bridge sites between two nearest neighbor or next nearest neighbor substrate atoms. Other examples might be three- or four-fold hollow sites, or even, directly on top of a surface atom. The preference

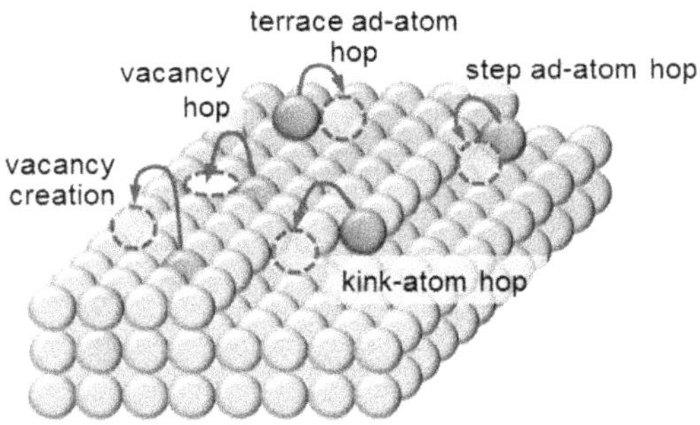

Figure 5.16. Diffusion processes around a lattice step.

Table 5.4. Types of diffusion around a terrace edge.

Process	Comments
Terrace ad-atom hop	Fast process, minimum surface coordination involved
Vacancy hop	Atom slides into adjacent vacancy defect; treated as diffusion of the vacancy which can continue to move across the surface
Step ad-atom hop	Sliding of ad-atom along the step riser
Kink atom hop	Kink atom shifts location; combination with step ad-atom hop creates step flows across surface
Vacancy creation	High energy process with large coordination loss, atom ejects from terrace site to leave vacancy defect behind

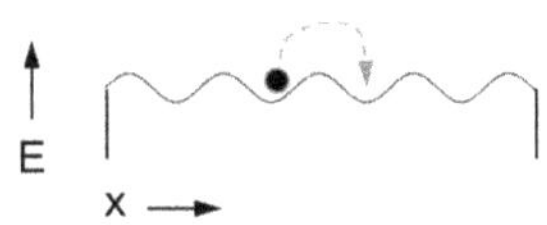

Figure 5.17. Illustrating diffusion from one low energy lattice site to another.

for each of these sites is determined by the interaction of the electrons of an adsorbed atom with the electrons of the substrate.

Since the adsorbed species is bound at a local energy minimum, to move to an adjacent location it will need to acquire some energy to hop to an adjacent site that has, presumably, equal or greater bond energy. Since the adsorbed species must acquire some energy to hop to the adjacent site, this process is an activated process (see figure 5.17). In figure 5.16 the height of the energy barrier to diffusion is ΔE^*_{act} It is reasonable to think of the adsorbed species at energy equilibrium with the surface and solid to which it is attached. Because the surface is continuously vibrating, the local (not average) kinetic energy of the adsorbed species will be continuously varying with time. The greater the temperature of the solid, the greater its vibrations, and consequently the greater the energy of the adsorbate. If the adsorbate receives

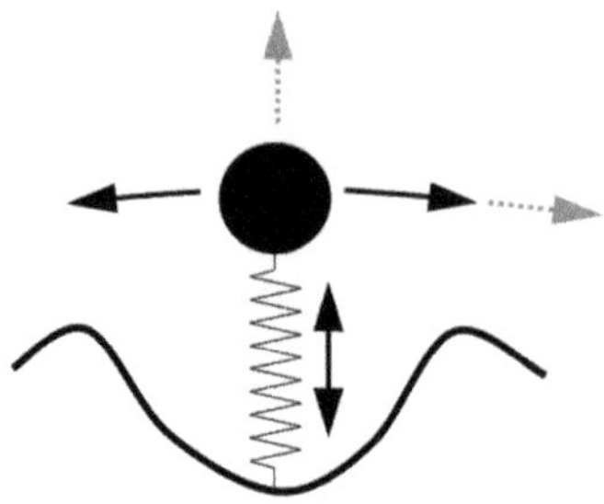

Figure 5.18. Vibrational modes of a particle bound to a low energy lattice site.

adequate energy, and momentum in a particular direction, the adsorbate will hop one lattice spacing in a particular direction. The amount of energy that an adsorbed species has at a given time will be determined by statistical factors. We usually use Boltzmann statistics to describe the energy distribution. The vibrations can be divided into those parallel to the surface and those perpendicular to the surface (see figure 5.18). If sufficient energy is present in the normal modes, the adsorbed species will desorb. If there is sufficient energy to overcome the hopping barrier, the adsorbate will hop to an adjacent site. Because diffusion is an activated process, we can express the hopping frequency, υ_{hop}, as follows:

$$\upsilon_{\text{hop}} = Z\upsilon_0 \exp\left(\frac{-\Delta E^*_{\text{hop}}}{k_{\text{B}}T}\right), \tag{5.11}$$

where Z is the number of adjacent sites, υ_0 is the oscillation (excursion) frequency, ΔE^*_{hop} is the activation energy for hopping, and k_{B} and T are Boltzmann's constant the absolute temperature, respectively. The exponential term captures the fraction of oscillations that have enough energy to hop. More generally, the diffusion expression is written as

$$D = D_0 \exp\left(\frac{-\Delta E^*_{\text{hop}}}{k_{\text{B}}T}\right),$$

where D is the diffusion coefficient (units are $\text{m}^2\ \text{s}^{-1}$) and D_0 is the pre-exponential that contains both distance per hop and excursion frequency.

It is likely that the excess energy given to overcome the barrier to hopping will be returned to the surface. The adsorbate will than remain at this site until it again obtains enough energy to hop. This activated process has several consequences.

Some sites will have greater bonding energy than other sites. Thus, adsorbates that find a step edge will likely be bonded more stably than an adsorbate on an open surface. Hence, it would remain in its position for longer than an adsorbate in an open surface. Similarly, if there were a vacancy defect that the adsorbate could fill, it would likely stay there even longer. Thus, to first order, at low temperatures, an originally rough surface would smooth as a function of time at temperature (most of the defects would fill). At high enough temperature, atoms would leave these more stable bonding conditions, the surface entropy would increase, the number of surface defects would increase, and the surface would roughen.

Another consequence of this kind of hopping diffusion, where the adsorbate re-equilibrated once it had jumped to a new spot, is the subsequent hops would be independent of each other. These kinds of processes, where each step is independent of the other are called Markovian. Processes where the probability of a step occurring depends upon the previous step are call non-Markovian. Some processes that appeared to be correlated (e.g. diffusion along a row) might still be Markovian. Jumps along the row would have lower activation energies, and thus be more probable, than jumps across the row.

5.2.3 Statistical model of random hopping

As time increases, on average the particle will be found at, or nearly at, the starting location. The particle will have taken as many left jumps as right jumps, and as many up jumps as down jumps. But as time proceeds, the process will take longer excursions from the starting location. The mean square deviation is taken as a measure of the how far, on average, the particle has gone, as shown in figure 5.19. Here the particle has moved Δx_1 after five red hops. After four more blue hops the particle has moved a distance Δr. The mean squared deviation for the starting point is defined as

$$\langle (\Delta r(t))^2 \rangle = v_{\text{hop}} \cdot t \cdot \langle d^2 \rangle, \tag{5.12}$$

where t is the observation time, and $\langle d^2 \rangle$ is the mean square hopping length (in figure 5.4 that would be the distance from the center of one square to the center of an immediately adjacent square). The diffusion coefficient can be defined, then, as

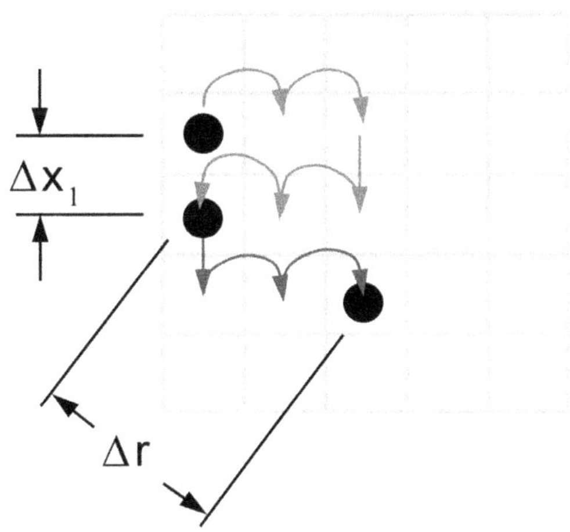

Figure 5.19. A depiction of site-to-site diffusion. Each site is separated by a distance Δx_1 and the total distance diffused is Δr.

$$D \equiv \lim_{t\to\infty} \frac{\langle(\Delta r(t))^2\rangle}{2bt} = \frac{v_{\text{hop}}\langle d^2\rangle}{2b}, \tag{5.13}$$

where $b = 1$ for a line (one-dimensional) or $b = 2$ for a plane (two-dimensional).

As time increases, on average the particle will be found at the starting location; it would take as many up jumps as down jumps, and as many left jumps as right jumps. But it the process it would take longer and longer excursions from the starting location, as measured by the mean square deviation. The probability of find the particle where it started has an increasingly large error as time proceeds.

Using the term for ν_{hop} from the activated hopping model, described above, gives and expression for the pre-exponential term, D_0, in the activated diffusion expression. Recall,

$$D = \frac{\nu_{\text{hop}}}{2b} = \frac{\langle d^2\rangle Z\nu_0}{2b} \exp\left(\frac{-\Delta E^*_{\text{hop}}}{k_{\text{B}}T}\right), \tag{5.14}$$

where $D_0 = \frac{\langle d^2\rangle Z\nu_0}{2b}$, and $\langle d^2\rangle$ is the mean square hopping distance (the distance from the center of one square to the center of a square immediately adjacent), Z is the number of possible landing sites (four in figure 5.18), ν_0 is the oscillation or vibrational frequency, and b is the dimensionality parameter (one for one dimension, two for two dimensions, and three for three dimensions).

In the presence of concentration gradient of particles on the surface, diffusion follows Fick's first law, expressed as equation (5.15), as follows,

$$\text{surface diffusion rate}\left(\frac{\text{particles}}{\text{cm}\cdot\text{s}}\right) = D\left(\frac{\text{cm}^2}{\text{s}}\right)\frac{\text{d}\sigma}{\text{d}x}\left(\frac{\text{particles}}{\text{cm}^2\text{cm}}\right), \tag{5.15}$$

where σ is the surface coverage in particles cm^{-2}. Typically, the diffusion coefficient will vary from 10^{-7} to 10^{-14} $\text{cm}^2\cdot\text{s}^{-1}$. From above, we can deduce that the room mean square excursion from a starting point,

$$\langle(\Delta r(t))^2\rangle^{1/2} = \sqrt{4Dt}\,. \tag{5.16}$$

This is a good rule of thumb. The mean distance a particle moves is equal to the square root of four times the diffusion coefficient times the elapsed time. Knowing any two of these variables allows you to calculate the third.

5.2.4 Surface diffusion of polymer chains

Consider the case of single polymer chains pinned onto a featureless surface. As the chain diffuses, it will do a self-avoiding two-dimensional random walk. We will start with a chain of N monomer units (see the figure 5.20). The attraction to the surface will weak enough to permit two-dimensional bead hopping. Each bead will be allowed to walk independently with constraints (so-called Rouse behavior). Two beads cannot occupy the same spot. The chain links will for force

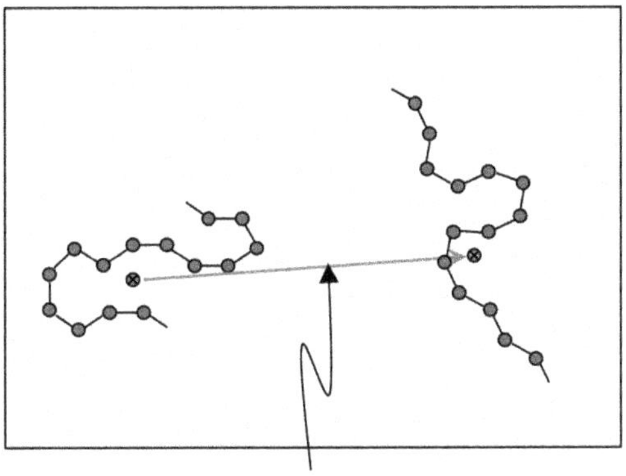

Figure 5.20. Representation of a self-avoiding 2D random walk of a chain with N monomer units. The arrow is the net diffusion of the center of mass.

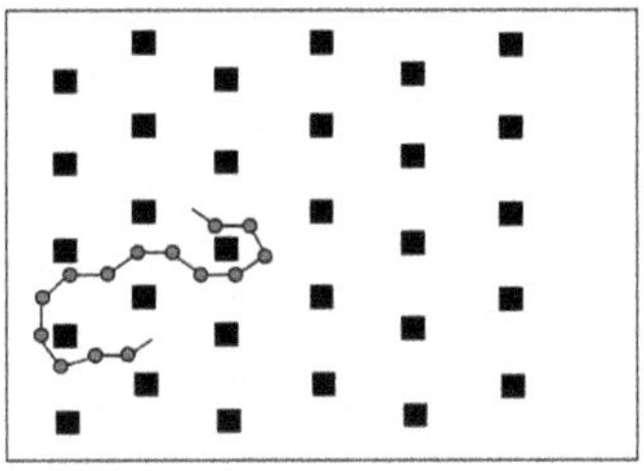

Figure 5.21. A schematic of diffusion by reptation, basically worm-like motion weaving past obstacles.

coordinated motion. The chain cannot cross-over itself. Thus, the chain is constantly changing shape, but the center of mass diffuses. It turns out that root mean square radius of gyration of the polymer is $\langle R_g^2 \rangle^{1/2} \sim N^{3/4}$, and the diffusion coefficient $D \sim \frac{1}{N}$ [17].

In the next level of sophistication, we have isolated single chains on the surface with un-climbable obstacles (see figure 5.21). At large concentrations of obstacles with respect to chain lengths, the diffusion mechanism switches from Rouse to reptation. Reptation is characterized by worm-like motion, where the chain weaves past obstacles. This motion is similar to motion through a winding pipe of tube. The diffusion coefficient for this is found to be proportional to $1/N^{3/2}$, where N is the number of monomer units.

Additional complicating factors include cross-over of chains, which slows diffusion. If there is solvent next to the surface, it allows the chains to access the vertical dimension by looping into the liquid. The loop then can come down in a new location, thereby speeding diffusion. A large surface coverage of chains creates obstacles to diffusion; these obstacles force reptation and slow the rate of diffusion.

5.2.5 Experimental measurements[1]

5.2.5.1 Tracking individual particles

Diffusion of atoms or molecules on metal surfaces attracted researcher's attention from the very beginning of surface science. E W Muller invented the field emission microscope, where a high electric field applied to a very sharp metal tip allowed one to observe single atoms on the surface of the metal tip. Gases adsorbed on the metal surface changed the local work function of the surface and either appeared darker (it their adsorption increased in the local work function), or lighter (if they decreased the work function).

Soon after hearing a seminar account of Muller's field emission microscope, R Gomer, at the University of Chicago, built one and became active in developing this instrument as a tool for surface studies. Even as early as 1959, Gomer began investigating the diffusion of small molecules and atoms on surfaces [18]. By dosing one side of the field emission tip with, for example, oxygen, he was able to observe oxygen diffusing across the metal tip and populating the initially bare tip surface. Diffusion, in this case, can be complicated, because the rate of diffusion (or the diffusion coefficient) depends upon the local concentration of the adsorbate. Since there is a sharp concentration gradient across the moving front, the diffusion coefficient is changing continuously (depending on its relationship to the concentration) as the diffusion front proceeds.

Gomer later realized that if he monitored the electron emission from a small area on the tip surface, at nominally constant adsorbate coverage, he could derive the diffusion behavior of the adsorbate as a function of coverage. Because of diffusion of adatoms into and out of the small area, there are fluctuations around the average coverage. By measuring the time constant of these fluctuations, he determined the chemical diffusion coefficient as a function of coverage in the small area. By changing the shape of the small area to a rectangle, Gomer and Tringides were able to determine diffusion coefficients in different directions depending how the probe area was oriented. In addition, they were able to relate the temperature and coverage dependence of the measured diffusion coefficients to the statistical mechanics of 2D systems of high theoretical interest, especially for understanding 2D phase transitions [19]. Years later, with Christian Uebing, many of these results were confirmed using Monte Carlo modeling of surface diffusion [20].

More recently atomic force microscopy or scanning tunneling microscopy have been used to track the same individual particle. Extremely stable instrumentation and often very low temperatures are required to make these measurements. Chemical diffusion of surface adsorbates can be followed, for example, in figure 5.22.

5.2.5.2 Macroscopic diffusion

Alternatively, the average behavior of a large number of particles could be monitored, leading to understanding of the diffusive behavior. This series of experiment is based on using Fick's second law for adsorbate coverage, σ:

[1] Some of this section overlaps with [21].

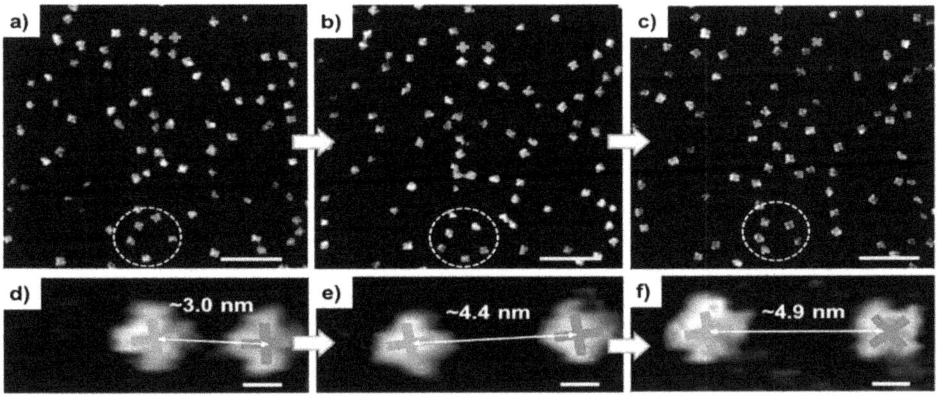

Figure 5.22. Diffusion of $Y[C_6S\text{–}Pc)]_2$, a double-decker phthalocyanine, over a Au(111) surface. Panels (a)–(c) are sequential STM images showing surface diffusion of the phthalocyanine at a 1-phenyloctane/Au(111) surface. Images (d)–(f) are expanded portion of the above images of selected molecules indicated by the green and red *x*-marks. It shows that the distance between these two molecules increase from ~3.0 to ~4.9 nm. Molecules in the circled area clearly display molecular motion. The scale bar is 5 nm for panels (a), (b), and (c). The scale bar is 1 nm for panels (e), (f), and (g). Reprinted with permission from [22]. Copyright (2022) American Chemical Society.

$$\left(\frac{\partial \sigma}{\partial t}\right) = D_{\text{avg}}\left(\frac{\partial^2 \sigma}{\partial x^2}\right). \tag{5.17}$$

Here we assume that the diffusion coefficient, D_{avg}, is constant, i.e. independent of coverage. The diffusion coefficient is taken to be activated,

$$D_{\text{avg}} = D_0 \mathrm{e}^{\left(\frac{-\Delta E_0^*}{RT}\right)}, \tag{5.18}$$

where ΔE_0^* is the activation energy and D_0 is the pre-exponential. In one example, a uniformly covered surface is pulsed with a laser and the adsorbate atoms from the region are desorbed. The observed back flow in that spot is observed and fitted with a diffusion equation. In another example, a surface is partially placed in an aqueous solution. The solute attaches to the surface and diffuses up the surface.

Somorjai did a series of experiments investigating the self-diffusion of a radioactive metal on that metal's surface and investigated how well it fit equation (5.17) and equation (5.18) [23]. In this case, the diffusion coefficient would have no concentration dependence, because the concentration of the metal would be constant. Nevertheless, he found large deviations ideal behavior. For the case of Ni diffusion on Ni(111), the close-packed surface of Ni, they found

$$c(x, t) \approx \frac{c_0 \mathrm{e}^{-\alpha x}}{x^{1/2}}, \tag{5.19}$$

$$\text{where } \alpha^2 = \frac{2}{D\delta}\left[\frac{D_V}{(\pi D_V t)^{1/2}} + \frac{\lambda_1 \nu_0 \delta}{2}\mathrm{e}^{-(5/6)\Delta H_s RT}\right]. \tag{5.20}$$

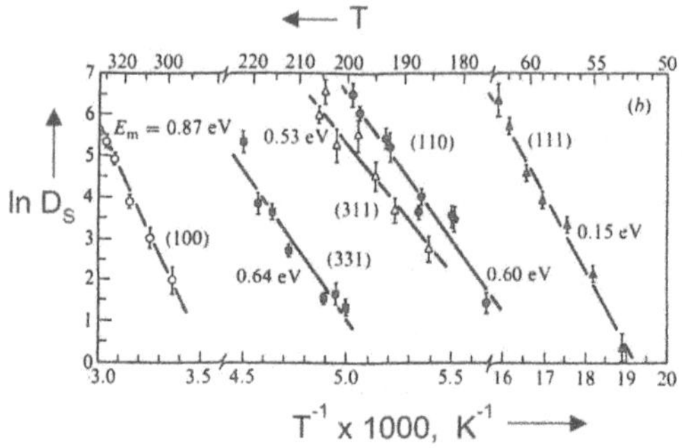

Figure 5.23. Diffusion of rhodium adatoms on single crystal planes of rhodium. The natural logarithm of the rate constant is plotted again $1000/T$ (K^{-1}). The absolute temperature is plotted across the top.

Here, D and D_V are the surface and bulk diffusion coefficients; λ_1 is the average number of interatomic distances between ledges on the surface; δ is the surface layer thickness (here, one atomic layer) and ΔH_s is the heat of sublimation. This equation captures the idea that diffusion on a surface plane is very rapid, and the diffusing particles spend a great deal of time captured at the step edges. They looked at the temperature dependence of the diffusion, and found that the activation energy for surface diffusion, $\Delta H_D^* = 38 \pm 4$ k cal mole^{-1}. The Arrhenius equation for bulk diffusion was used to derive this expression. They also found that the step widths, λ_1, were five atomic spacings and the thickness of the surface layer, δ, was 0.25 nm. In summary, the free Ni atoms diffuse quickly across the terraces, and the rate limiting step is diffusion over the step edges.

The diffusion coefficient depends upon the crystal face. In figure 5.23, we show the diffusion coefficient for rhodium (face centered cubic) as a function of inverse temperature for a series of crystal faces. Interesting, diffusion is most rapid on the smooth (111) crystal face, and slowest on the rougher (100) crystal face. The activation energies are approximately equal, and the pre-exponential varies considerably.

At high temperatures (approximately 75% of the melting temperature) face centered cubic temperatures begin to roughen (see figure 5.24), where the logarithm of the surface diffusion coefficient is plotted against the reciprocal temperature reduced with respect to the melting temperature, T_m. The metals are gold, silver, copper, and nickel. Notice that at approximately $T_M/T \approx 1.3$. When the surface begins to roughen the step edges become more mobile, and as a result, atoms formerly trapped at step edges begin to migrate more easily.

Gomer and Chen used the field emission method described above to investigate the diffusion of oxygen on the (110) plane of W [24]. This is most close-packed plane of W, which is bcc. They measured diffusion at two coverages, O/W = 0.5 and O/W = 0.2. At O/W = 0.5 the energy of activation for diffusion was roughly twice (22 kcal mole^{-1}) versus O/W coverages of 14 kcal mole^{-1}. The high activation energy for diffusion follows from the observation that O forms an ordered overlayer on W at that coverage. This ordered overlayer suggests that there is no place for O atoms to diffuse to once the ordered layer is formed.

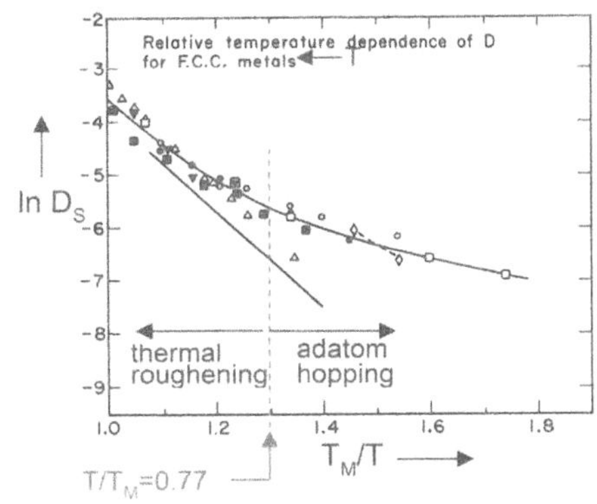

Figure 5.24. Logarithm of the surface self-diffusion coefficients of several face centered cubic metal as a function of the reduced reciprocal temperature. Gold □, Δ; copper, ■,▼,○; nickel, ●; silver, ◊.

Thought questions:

1. Why might ad-atom diffusion rate depend on the crystal face?
2. Can you explain this dependence on the hopping barrier dependent on the surface unit cell structure?
3. Think of a reason why thermal roughening is associate mostly with lower T_m metals.

White and co-workers performed a series of experiments studying the diffusion of H and D on Ni (100). Here, the surface was uniformly coated with hydrogen or deuterium. After coating the surface, a laser pulse heats a small area of the surface leading to desorption of the H from that region. The depopulated region is allowed to partially refill, and desorption from a subsequent laser pulse [25] (see figure 5.25 below). They were able to solve Fick's second law for this configuration, and able to extract an effective diffusion coefficient (taken to not vary as a function of concentration). Perhaps as expected, H diffuses faster than D. the activation energy of H was found to be slightly lower than that of D. The pre-exponential, reflective of the vibrational frequency with the surface, is found to be slightly lower for D than H as expected.

5.3 Dynamic phenomena in adsorption–desorption processes

5.3.1 Overview

A gaseous species incident on a surface can have a variety of interactions with the surface, depending on the nature of the gas and surface, the energy of the gas, and the temperature of the surface. Most simply, the gas could just scatter from the surface. This scattering might involve transfer of energy with the surface (inelastic scattering) or no transfer of energy (scattering). Alternatively, the gas might be adsorbed on the surface. It might adsorb dissociatively or non-dissociatively. Finally, if the incoming species is energetic enough it might sputter species from the surface.

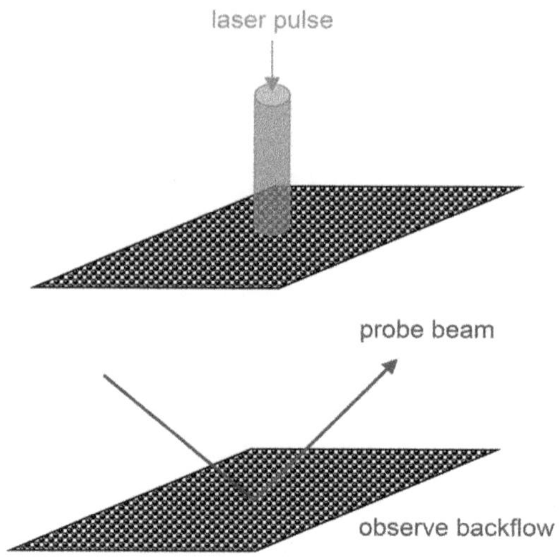

Figure 5.25. Measurement of bulk surface diffusion by using a laser pulse to clean an area of the surface. A probe of the initially clean area measures the diffusion of adsorbates into that region.

5.3.2 Surface collision rates

The flux of gas phase molecules striking the surface can be calculated. A flux, *F*, is

$$F = \frac{N_{\mathrm{Av}} P}{\sqrt{2\pi MRT}}, \tag{5.21}$$

where N_{Av} is Avogadro's number, *P* is the pressure, *M* is the mass of the molecule per mole, *R* is the gas constant, and *T* is the temperature. In common units

$$F\left(\frac{\text{molecules}}{\text{cm}^2\ \text{s}^{-1}}\right) = 3.51 \times 10^{21} \cdot P\ (\text{Torr}) \cdot [M\ (\text{g} \cdot \text{mol}^{-1})\ T\ (\text{K})]^{-1/2}.$$

Using these units we can define an approximate pressure (in torr) which leads to $\sim 1 \times 10^{15}$ gas molecules striking 1 cm^2 of surface during 1 s exposure. This dose is call a 'Langmuir' after Irving Langmuir. The formal definition of a Langmuir (L) is the gas dose at 300 K with delivers $\sim 10^{15}$ molecules cm^{-2} to a surface, equivalent to $\sim$ 1 monolayer coverage if all the incident molecules are adsorbed. Since N_2, O_2, and H_2O are typical background gases, a definition for an ultra-high vacuum system would have pressures less than 10^{-9} Torr; at 300 K, the exposed surface would stay below monolayer coverage for 1000 s.

5.3.3 Rate of thermally induced desorption of adsorbed species

The energetics of adsorption are shown in figure 5.26. Here the adsorbate is bound to the surface with a binding energy, ΔH_{ads}. The energy required to remove, or desorb the species, is an activation energy, E_{a}, or $\Delta H^{\ddagger}$.

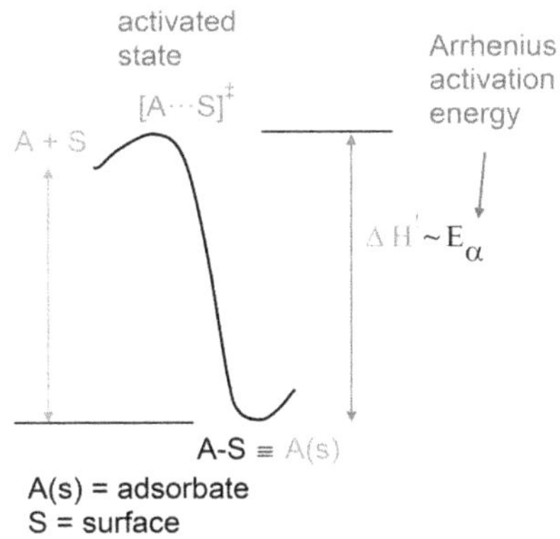

Figure 5.26. The energetics of adsorption onto a surface are plotted versus reaction coordinate. The adsorbing species approaches the surface from the left.

The rate of leaving the surface (assuming no reaction or interaction) is

$$\frac{d[A(s)]}{dt} = k_{des}[A(s)], \qquad (5.22)$$

where $[A(s)]$ is the surface concentration of adsorbates, and the rate factor $k_{des} = A_a \exp(E_a/RT)$. Here it is usually assumed that $\Delta H_{ads} \sim \Delta H^{\ddagger} \sim E_a$. It is usually assumed values of A_a are $\sim 10^{12} - 10^{12}\ s^{-1}$, close to the vibrational frequency. Often, however, much lower values of the pre-exponential are found.

The average time that a species, in thermal equilibrium with the surface, resides on the surface is $\sim 1/k_{des} \equiv \tau$, where $\tau = \tau_0 \exp(\Delta H_{ads}/RT)$. Typical values of τ_0 between 10^{-12} and 10^{-13} s; recall that τ_0 is ${A_a}^{-1}$ above.

Thought question:
Consider that $\tau_0 \sim 10^{-12} - 10^{-13}$ s. Let's see if that is related to the approximate time for a vibration on the surface. If a surface vibrational energy is about 200 cm^{-1}, we can use the Heisenberg uncertainty principle to estimate the time: $\delta\tau \cdot \delta E \sim h$. Well, $\delta E \sim 200\ cm^{-1}\left(\frac{1.60 \times 10^{-19}\ J}{8066\ cm^{-1}}\right) \sim 4 \times 10^{-21}$ J. Taking Planck's constant $h = 6.63 \times 10^{-34}\ J \cdot s^{-1}$, what is $\delta\tau$ equal to? Is it close to τ_0? What does this mean?

To help ascertain the strength of bonding between an adsorbate and a substrate temperature program desorption is often used. The sample is slowly heated, and the gaseous species evolved as a function of time and temperature are monitored with a mass spectrometer. More strongly bound species are evolved at higher temperatures. The mass spectrometer helps determine if adsorbate has reacted on the surface. The experiment is displayed in the figure below.

In the example displayed in figure 5.27 the species AB is adsorbed on the surface. At the lowest temperatures the species AB is desorbed. As the temperature is increased the species A is evolved and the finally the species B is evolved. An example of temperature programmed desorption is the adsorption of CO onto W.

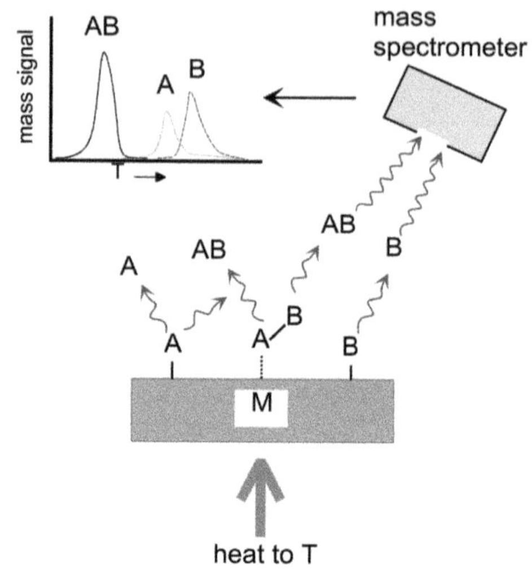

Figure 5.27. A sample substrate with adsorbate AB is heated to increasingly higher temperature in UHV. As the temperature increases, the adsorbates desorb and are detected in a mass spectrometer. The mass signals are plotted as a function of time.

From the plots ΔH_{ads} can be estimated. If a temperature heating rate $T = T_0 + \beta \cdot t$, where T_0 is the initial temperature, t is the time of heating, and β is the heating rate. The rate of desorption at any time is given by

$$\frac{\mathrm{d}[A(s)]}{\mathrm{d}t} = k_{\text{des}}[A(s)],$$

where

$$k_{\text{des}} = A_{\text{a}} \exp\left(E_{\text{a}}/R(T_0 + \beta t)\right). \tag{5.23}$$

For a first order reaction, that is, there is no reaction before desorption, and the also if the desorption rate constant and activation energies do not depend on the adsorbate coverage,

$$\frac{E_{\text{a}}}{RT_{\text{p}}^2} = \frac{k_{\text{des}}}{\beta} \exp\left(-\frac{E_{\text{a}}}{RT_{\text{p}}}\right), \tag{5.24}$$

where T_{p} is the peak temperature in the thermal desorption curve. Changing the heating rate will change the position of the peak. Of course, if the activation energy and pre-exponential are coverage dependent the analysis will be very complicated. Similarly, if the desorption is not first order, all of the analyses must be changed.

Schmidt looked at the TPD spectra of H on W(100), W(110), Mo(100) and M (110) [26]. They were able to resolve at least two temperature programmed peaks for each of these surfaces. Rather than reflecting two different binding states, it was concluded that as the hydrogen desorbed, and the coverage decreased, the hydrogen was able bind more strongly to the surface.

Temperature program desorption is a kinetic experiment, not thermodynamic. The ΔH_{ads} estimated from the activation energies are only estimates. As seen for the case of hydrogen on single crystal W and M surfaces the TPD spectra can be very complex. This technique is good for 'fingerprints' of the chemistry occurring on the surface.

5.3.4 Limiting cases of adsorption–desorption rate processes

5.3.4.1 Approach to equilibrium with adsorption rates = desorption rates
Consider the case when the adsorption rate equals the desorption rate. As the temperature is changed the adsorbate coverage is changed. Similarly, as the gas pressure is changed at a given temperature the coverage also changes.

Let's assume the following basic assumptions:

1. The gas molecules strike the surface at random position.
2. All surface sites have the same adsorption energy (E_{ads}) and the adsorption energy is not a function of coverage.
3. If an incoming molecule hits a filled site, it does not stick.
4. If an incoming molecule lands at an open site, it sticks until it desorbs at an average time τ.
5. The desorption probability is independent of coverage; that is, the desorption kinetics are first order.

If θ represents the fraction of total surface sites that are filled. The number of sites per unit area filled is σ, and the total number of sites (per unit area) is σ_0. Then the number open sites per unit area is just $\sigma_0 - \sigma$. The coverage and site occupation can be related, as follows

$$\theta = \frac{\sigma}{\sigma_0}, \quad \text{and} \quad (1-\theta) = \left(\frac{\sigma_0 - \sigma}{\sigma_0}\right). \tag{5.25}$$

If every atom that strikes an open site adsorbs, the adsorption rate will be just

$$(1-\theta)F = (1-\theta)\frac{N_{Av}P}{\sqrt{2\pi MRT}} = k_{ads}(1-\theta)P = k_{ads}\left(\frac{\sigma_0 - \sigma}{\sigma_0}\right)P. \tag{5.26}$$

The desorption rate will just be the number of adsorbed molecules per unit area divided by the average adsorbate lifetime, or

$$\frac{\sigma}{\tau} = A_a e^{-\Delta H_{ads}/RT} \cdot \sigma = k_{des}\sigma. \tag{5.27}$$

If we set the forward and reverse rates equal

$$\left(\frac{\sigma_0 - \sigma}{\sigma_0}\right)k_{ads}P = k_{des}\sigma. \tag{5.28}$$

Solving for σ we obtain the Langmuir isotherm:

$$\sigma = \left(\frac{\sigma_0 - \sigma}{\sigma_0}\right)\left(\frac{k_{\text{ads}}}{k_{\text{des}}}\right)P = \left(\frac{\sigma_0 - \sigma}{\sigma_0}\right)K_\theta' P, \tag{5.29}$$

where

$$K_\theta' = \frac{N_{\text{Av}}/\sqrt{2\pi MRT}}{Ae^{-\Delta H_{\text{ads}}/RT}}.$$

Solving for σ,

$$\sigma = \frac{\sigma_0 K_\theta' P}{\sigma_0 + K_\theta' P}, \quad \text{or} \quad \frac{P}{\sigma} = \left(\frac{1}{K_\theta'}\right) + \left(\frac{P}{\sigma_0}\right). \tag{5.30}$$

Solving for θ,

$$\theta = \frac{P}{\left(\frac{\sigma_0}{K_\theta'}\right) + P} = \frac{P}{c' + P}. \tag{5.31}$$

These last two expression are standard forms used for Langmuir isotherm plots.

K'_θ is the equilibrium constant associated with the adsorption–desorption equilibrium. Following the equations above

$$K_\theta' = \frac{\sigma}{\left(\frac{\sigma_0 - \sigma}{\sigma_0}\right)P} = \frac{\theta\sigma_0}{(1 - \theta)P}, \tag{5.32}$$

P is the gas pressure, θ is the adsorbate surface concentration (in molecules per unit surface area) and $(1 - \theta)$ is the fraction of open sites on the surface. K_θ' carries the units of σ_0/P.

K_θ is associated with the equilibrium formulation and is based on the reactant activities or mole fractions, so it is a proper equilibrium constant with no units:

$$K_\theta = \frac{\theta}{(1 - \theta)(P/P_{\text{m}})} \quad \text{and} \quad K_\theta' = \left(\frac{\sigma_0}{P_{\text{m}}}\right)K_\theta, \tag{5.33}$$

where P_{m} is the pressure at which one monolayer, exactly, is adsorbed.

The temperature dependence of the Langmuir isotherm is, as follows,

$$K_\theta' = \left(\frac{N_{\text{Av}}}{\sqrt{2\pi MRT}}\right)A_{\text{a}}e^{(+\Delta H_{\text{ads}}/RT)}. \tag{5.34}$$

The exponential is the dominant temperature dependence or K_θ'. As adsorbate bonding strengthens, ΔH_{ads} increases, causing an increase in K_θ' which shifts the equilibrium to increasing the surface concentration. The temperature in the radical has only a minor effect in the value of K_θ'.

5.3.4.2 Random sequential adsorption

In random sequential adsorption (RSA), particles are added to a system randomly, and if they do not overlap with any previously adsorbed particle, they adsorb and remain fixed for the duration of the process [27]. The process can be studied experimentally, by mathematical analysis, or computationally. Consider the case of circular particles on a two-dimensional surface. The first particles added have a very small probability of overlapping with previously adsorbed particles. As the coverage increases, the probability of adding subsequent particle decreases dramatically. For uniform circular particles on a surface, the saturation coverage is about 0.55. Two-dimensional, uniform, aligned squares have a saturation coverage of approximately 0.56.

The simplicity of the model comes with a cost. It is assumed that particles are monodisperse, external forces are negligible, particles do not interact, all particles adsorb directly on the surface. However, extensions of the RSA model to include bilayer and multilayer adsorption, surface heterogeneity, interparticle interactions, and eternal fields.

It has been shown that the packing fraction $\theta(\tau)$ grows with τ following the power law

$$\theta(\tau) = \theta(\infty) - A\tau^{-\frac{1}{\alpha}}, \tag{5.35}$$

where $\theta(\infty)$ is the packing fraction at saturation, A is a proportionality constant, and α equals 2 for disks. While this equation is obeyed for a large number of cases, it is not obeyed by, for example, particles placed on a lattice.

The kinetics of adsorption depend on the sticking coefficient (probability of successful deposition of a particle). This also depends upon the blocking function or available surface function. The transport of particles to the surface from the bulk can by modeled by Fick's second law,

$$\frac{\partial c(z, t)}{\partial t} = D\frac{\partial^2 c(z, t)}{\partial z^2} \tag{5.36}$$

with D as the diffusion coefficient, and $c(z,t)$ is the concentration of particles a distance z from the surface at a time t. Assuming a constant diffusion coefficient and an initial homogeneous concentration profile gives reasonable results, in agreement with experiment.

5.4 Free particle-solid surface collisions

5.4.1 Energy transfer during scattering

When a molecule strikes a surface, it can lose or gain energy. Looking at the figure 5.28, the molecule must lose energy if it is going to be captured in the potential well. The incoming molecule will have translational, rotational, electronic, and vibrational. When it scatters from the surface energy in any of these modes may be transferred to the surface, and the surface may also contribute energy into any one of these modes. Thus, the incoming molecule will have

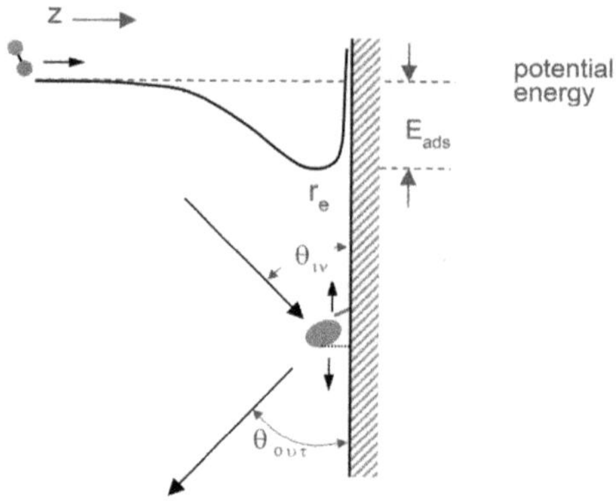

Figure 5.28. Top: The potential energy is plotted as a function of distance from the surface as a diatomic molecule approaches the surface. Bottom: the incoming angle differs from the outgoing angle as the particle's translational, rotational, vibrational angle are transferred to and from the surface.

$$E^{\text{initial}} = [\varepsilon_{\text{trans}} + \varepsilon_{\text{rot}} + \varepsilon_{\text{vib}} + \varepsilon_{\text{el}}]^{\text{initial}}. \tag{5.37}$$

Upon desorption, the molecule will have the following energy:

$$E^{\text{final}} = [\varepsilon_{\text{trans}} + \varepsilon_{\text{rot}} + \varepsilon_{\text{vib}} + \varepsilon_{\text{el}}]^{\text{final}}. \tag{5.38}$$

The translational, rotational, vibrational, and electronic energies need not be the same before and after scattering from the surface. In fact, energy from any one of these modes could be transferred to or from the surface. Overall, of course, energy must be conserved,

$$\left(E_{\text{gas}}^{\text{final}} + E_{\text{solid}}^{\text{final}}\right) - \left(E_{\text{gas}}^{\text{initial}} + E_{\text{solid}}^{\text{initial}}\right) = 0. \tag{5.39}$$

Molecular beam experiments where the translational, rotational, vibrational and electronic energies of the incoming molecule are controlled, the surface temperature is known, and the translational, rotational, vibrational and electronic energy of the scattered molecule are measured. For example, the Cardillo group scattered hyperthermal Xe atoms as a function of energy from single crystalline GaAs(110), Ag (100) and Ge (100) [28]. The angle distributions of the reflected Xe resulted from the topography of the surface. The large energy losses reflected coupling to local modes of motion.

The scattering of inert gases from surfaces is a good example. Because the incoming species are monoatomic, there is no rotational or vibrational energy. Each of the atoms is in its electronic ground state. As can be seen in figure 5.29, for inert gases at 300 K scattered from graphite at temperatures up to 2000 K, the incoming inert atoms acquire some energy, but do not desorb (scatter) at temperatures equal to the substrate temperature.

There is a term for the relative energy accommodation between the incoming species and the surface. The energy accommodation coefficient, α_E, is defined as

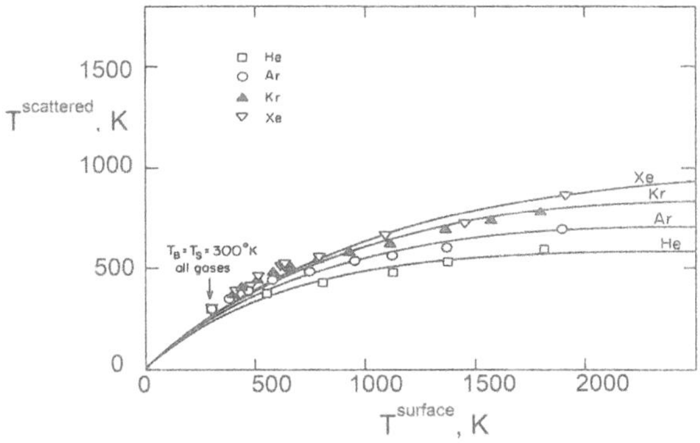

Figure 5.29. The temperature of inert gas atoms scattered from graphite as a function of temperature of the graphite. Reprinted from [29], Copyright (1972), with permission from Elsevier.

$$\alpha_E = \frac{E^{\text{init}} - E^{\text{scattered}}}{E^{\text{init}} - E^{\text{surface}}}. \tag{5.40}$$

The limits for the accommodation coefficient follow. If $E^{\text{init}} = E^{\text{scat}}$, $\alpha = 0$, and there is no accommodation. if $E^{\text{scat}} = E^{\text{surf}}$, $\alpha = 1$, and the particle has entirely accommodated with the surface. In the plot above, as the temperature increases beyond 300 K, the atoms accommodate less and less well with the surface.

Thought question:
The smallest atoms show the poorest energy accommodation as temperature increases. What is a possible explanation?

Typically, adsorption results if $\varepsilon_{\text{trans}}^{\text{init}} < \varepsilon_{\text{vib}}^{\text{surf}}$. If the incoming translational energy is only slightly greater than the surface vibrational energy, i.e. $\varepsilon_{\text{trans}}^{\text{init}} \geqslant \varepsilon_{\text{vib}}^{\text{surf}}$, often the incoming species will convert the excess surface vibrational energy to diffusion and will eventually desorb. If $\varepsilon_{\text{trans}}^{\text{init}} \gg \varepsilon_{\text{vib}}^{\text{surf}}$, the molecule or atom will undoubtedly desorb.

Each energy mode has its own accommodation efficiency. That is,

$$\alpha_M = \frac{E_M^{\text{init}} - E_M^{\text{scat}}}{E_M^{\text{init}} - E^{\text{surf}}}, \tag{5.41}$$

where M is translational, rotational, vibrational, electronic, or any other energy mode. For example, the accommodation coefficients for the translational, vibrational, and rotational modes as a function of surface temperature of scattering NO from Pt(111) are approximated in figure 5.30.

Thought question:
Translational and vibrational energy are better accommodated the rotational energy at a surface with increased temperature. What is a possible explanation?

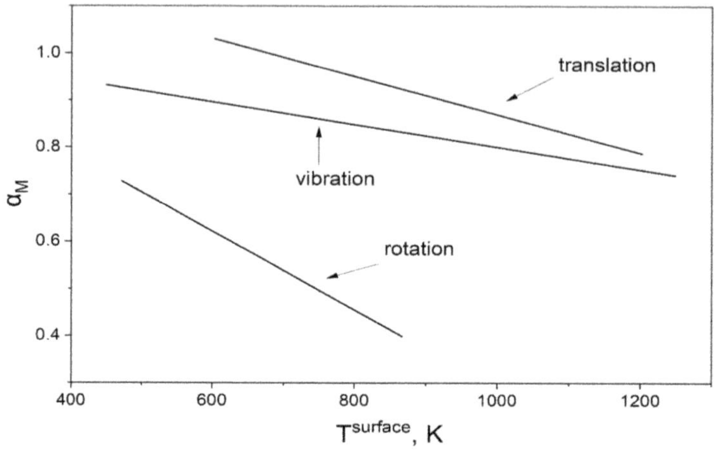

Figure 5.30. Translational, vibrational, and rotational energy accommodation coefficients as a function of energy for NO scattered from Pt(111) as a function of crystal temperature [30].

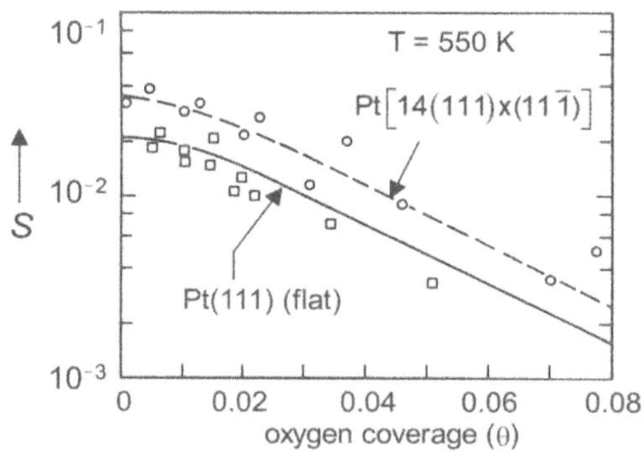

Figure 5.31. The sticking coefficient of O_2 on flat Pt(111) and stepped Pt[14(111)x(111)].

5.4.2 Sticking probability

The sticking probability, *S*, is defined at the adsorption rate divided by the collision rate. Generally, the sticking probability goes up as the enthalpy of adsorption increases. An incoming molecule sticks when there is sufficient attraction energy to overcome recoil energy of the incident particle. If a simple chemical bond forms between the surface and the molecule, and the molecule remains intact, this is referred to as molecular chemisorption. On the other end, in the case of formation of strong chemical bonds, the molecule dissociates into surface-bonded fragments. This is known as dissociative chemisorption. Many examples are a mixture of simple molecular and dissociative chemisorption.

In one example the sticking probability depends upon the current coverage of molecules on the surface. In figure 5.31, O_2 reacts with a stepped and flat (111) surface of Pt at 550 K. As the coverage of oxygen increases the sticking probability decreases. Interestingly, the sticking coefficient appears to increase with the presence of step edges.

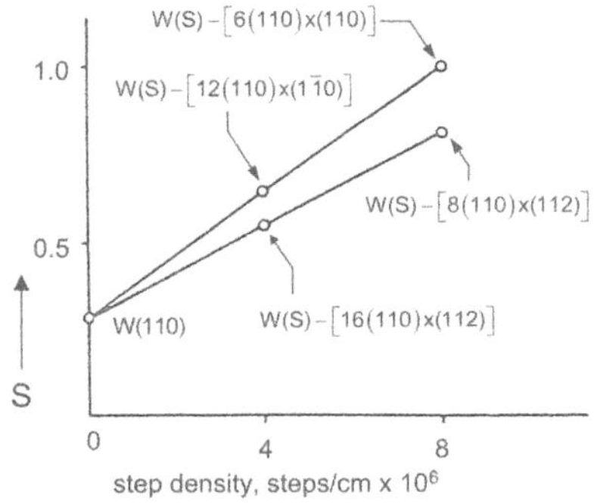

Figure 5.32. The sticking coefficient of N_2 on W as a function of step density [31].

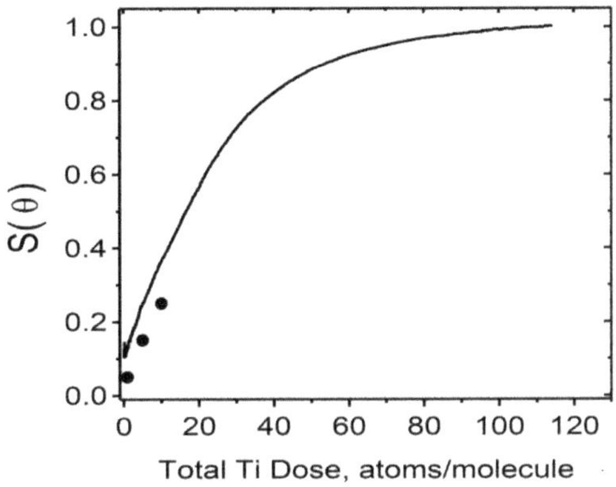

Figure 5.33. The sticking coefficient of Ti as a function of total Ti dose. The solid line is from quartz crystal microbalance and the points are derived from XPS data. Reprinted with permission from [32]. Copyright (2005) American Chemical Society.

Thought question:
Why is the sticking coefficient often a strong function of the number of molecules already absorbed?

The effect of step edges is shown in figure 5.32 for N_2 sticking on W(110) as function of density of step edges. Here the sticking probability increases with density of step edges. The sticking probability is lowest on a flat terrace and increases as the step density increases.

Another interesting case is presented by the scattering of Ti atoms from self-assembled monolayer of hydrocarbon ($CH_3(CH_2)_{15}S$–) on Au shown in figure 5.33 [32]. As exposure of the hydrocarbon surface to Ti begins (no adsorbed Ti) the sticking probability is approximately 0.1. As the total exposure of Ti atoms approaches 100 atoms per hydrocarbon molecule the sticking coefficient approaches unity (see figure below).

Thought questions:
Why does the sticking probability increase as the density of steps increase?

What are possible reasons that the initial incident Ti atoms recoil and do not react?

Once atoms start sticking, why do following incident atoms have a great probability of sticking?

For dissociative chemisorption, the sticking probability is equal to the dissociation probability. In other words, every molecule that sticks dissociates. A good example is the exchange of hydrogen and deuterium on Pd catalyst surfaces. These catalysts are typically high surface area particles on inert substrates.

Somorjai studied H_2 and D_2 incident on a stepped Pt surface (Pt[6(111)x(100)] at low pressures [33]. In this work the H_2 and D_2 are supplied to the surface separately in molecular beams. HD leaves the surface in large quantities, and the results suggest that H_2 and D_2 are adsorbed dissociatively. They then recombine, often forming HD, following the so-called Langmuir–Hinshelwood mechanism. The mechanism is inconsistent with a molecular species reacting with a dissociated species at the surface (the Eley–Rideal mechanism). This will be discussed in more detail in chapter 8 (Catalysis).

Because the H_2–D_2 exchange mechanism depends upon the dissociative adsorption of H_2 and D_2, factors that affect the dissociation of H_2 and D_2 will affect the exchange probability. As shown above, steps affect the probability of molecular adsorption. Similarly, as seen in the table below, step edges affect the reaction probability of H_2 and D_2.

Surface	Reaction probability
Stepped Pt(332)	0.9
Flat Pt(111)	$\sim 10^{-1}$
Defect-free Pt(111)	$\leqslant 10^{-3}$

5.4.3 Ion scattering spectroscopy

Ion scattering spectroscopy, abbreviated as ISS, is a surface characterization probe. Typically, an inert gas ion, is incident on a surface. The ion usually has unity charge, and energy of about 1 keV, and comes in at an angle of 45°. The output angle is variable. The energy of the scattered ion is measured. The energy loss ΔE is equal to $E_{in} - E_{out}$.

The collisions can be considered classically, basically as billiard ball type transitions. A low mass incident ion, such as He^+ or Ne^+, is used to avoid sputtering atoms from the surface. He^+ is usually used because they rebound classically the best, analogous to a BB scattering from a ball bearing. Only the top layer of atoms is involved. The energy of the scattered ion depends upon the mass of the surface atom that scattered the incident ion.

Knowing ΔE, the conservation of energy and momentum gives the mass of target atom. The ratios of the fraction of scattered atoms at each ΔE, or alternatively E_{out}/E_{in} gives the relative surface atom composition.

Campbell *et al.* studied the adsorption of Ca onto poly(methyl methacrylate) (PMMA) using a variety of techniques including ISS [34]. On pristine PMMA the first half of monolayer absorbed is invisible to ISS. This is because the Ca is binding to ester groups under the CH_3/CH_2 terminated PMMA surface. Above 0.5 layer the Ca starts to grow as three-dimensional clusters on top of the surface. This stage of adsorbate growth is measurable using ISS. If the PMMA surface is exposed to electrons before Ca exposure, the Ca immediately begins to grow on the surface. This new mode of growth is attributed to electron damage to the PMMA surface that now acts as nucleation centers.

5.4.4 Destructive collisions—sputtering

In the sputtering process an incoming projectile strikes the surface and knocks out secondary particles. The incoming projectile might be an ion or an atom. Ions are easy to produce and accelerate by electric field. To accelerate atoms to sufficient energy to produce secondary particles, often ions are accelerated in electric fields and then neutralized. Typical energies of incident particles vary from 10s of eV to 100 keV. Typical energies are 10–20 keV. Ions range from atomic to molecular. Their typical charge is +1. Atomic species include Ar^+, Xe^+, In^+, Cs^+ and Ga^{+3}. Atomic clusters in include $(Au)_n^+$ Common molecular species used for sputtering are O_2^+ and $C_{60}^{+\cdot}$

The secondary particles emitted from the surface include positive and negative ions and neutrals. Often molecular species are emitted; they might be fragments or intact molecules. The relative intensity of each of these species depend on the type and energy of the incident projectile. This will be discussed in more detail when the analytical technique secondary ion mass spectrometry (SIMS) is described below.

After the material is ejected from the surface an impact crater remains after the substrate material was ejected. The width and depth depend upon the projectile energy and type. Continued bombardment etches away the sample. Often the incident beam is rastered across the surface to get a more uniform etching. What remains on the surface is often analysed as function of sputtering time, or, if calibrated, depth. A sputter depth profile results. Similarly, if the ejected secondary particles are analysed, usually with a mass spectrometer, a sputter depth profile remains. Because the incident beam is energetic, usually the layer is not removed uniformly. Some mixing with the underlying layer occurs, and the amount of mixing depends upon the energy of incident projectile.

For these depth profile measurements, the ideal would be to etch away the surface region layer by layer. Of course, this does not happen precisely. Etch rates depend upon the material; inorganic compounds etch slowly; organics can be much faster. The etch conditions are set so as to remove nanometers in a second to several minutes. The underlying surface composition can be monitored with techniques such as x-ray photoelectron spectroscopy and Auger electron spectroscopy. The removed species can be monitored with secondary ion mass spectrometry (described in more detail below). Of course, the ion beam mixes the remaining 2–10 nm convoluting the profile data. Large molecule projectiles (i.e. C_{60}) minimize the churning.

Often one would like to analyse the pristine surface. However, upon exposure to the ambient the surface becomes contaminated with a variety of hydrocarbon and the surface oxidizes. Thus, the analyst often sputters the surface briefly to remove the surface contaminants with minimal damage to the original surface. Given the complications from sputtering care must be taken into the analysis to be aware of potential artifacts created by the sputter cleaning.

Another use of sputtering is deposition. In this application a target is sputtered, and the sputtered atoms and ions are deposited on a substrate. Often the sputtering ion is Ar^+. Usually, the target is very pure metal. Alloys can be used, but because some species are preferentially sputtered over others, the composition of the deposition might be significantly different than the composition of the target. With careful calibration, the composition of the target that gives desired deposition can be found. Occasionally organic targets (e.g. polymers) are used; new types of organics often result from fragment recombination. The rates of deposition from different targets can vary widely. Inclusion of gases in the sputter gas mixture can yield new oxides, nitrides, and carbides. Because sputter deposition can be a very robust process it is often used in commercially for thin film deposition of metals, alloys, oxides, and other materials.

A powerful surface analytical technique that analyses the ejected secondary ions in secondary ion mass spectrometry (see figure 5.34). The ejected ions are analysed with a mass spectrometer and yield chemical information about the sample surface.

In static SIMS it is the intent to acquire information about the composition of the topmost monolayer without disturbing its structure and composition. This is done by using a very low current of primary ion, so that it is improbable to impact the same spot twice. In the other extreme, in dynamic SIMS high primary current densities are used so that a monolayer may exist for only 10^{-3} s. The fast erosion supplies information about the chemical composition of the specimen as a function of depth (actually, sputtering time).

Many types of projectiles can be used. Ar^+ can be used since high sputter rates and well-focused beams are obtained. O_2^+, O^-, and N_2^+ primary beams increase the sputter ion yield. Cs^+ primaries give higher yields of negative ions. These ion

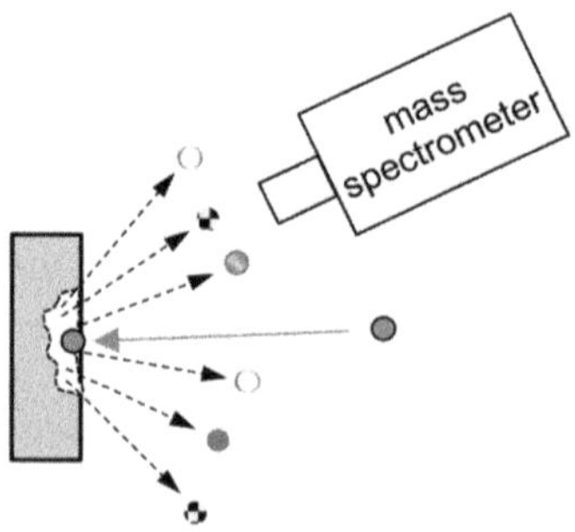

Figure 5.34. A schematic of secondary ion mass spectrometry. An energetic incident ion (red) strikes the surface, and a population of neutral and ionic species are sputtered from the surface. The ionic species are analysed in a mass spectrometer.

Table 5.5. Advantages and disadvantages of surface composition via ion projectile techniques FOR ISS and SIMS.

Technique	Advantages	Disadvantages
Ion surface scattering	Very surface selective	UHV only
	Quantitative composition	Sample damage is possible
Secondary ion mass spectrometry	High surface sensitivity	UHV only
	High surface selectivity	Conducting samples are best
	Good for organics and bio	Requires high resolution mass detection
	Image to better than 100 nm	Difficult to quantify

sources are often available in the same instrument so that positive and negative analysis can be done in the same instrument. C_{60}^{+} is occasionally used as an ion source because it produces to least damage to organic samples (both organic adsorbates and polymer and biological surfaces). In ideal conditions whole molecules or large polymer fragments can be sputtered. For the case of polymers, the fingerprint mass spectra of the surface allow qualitative determination of the top layer chemical composition.

One method to attain high mass resolution is the use of time-of-flight (ToF-SIMS). In this technique a short pulse of ions strikes the surface. All the emitted ions are accelerated through an electric field. By the conservation of energy, the small ions pick up more energy, and thus more velocity, than the heavier ions. By knowing how long it took for the ions to reach the detector, the mass can of the ion can be determined to better than 1 part in 10^4.

In all SIMS techniques it is difficult to quantify the chemical composition because secondary ion yields, which differ by orders of magnitude from one ionic species to another, can vary dramatically with slight changes in the matrix structure or composition (table 5.5).

Chapter 5 Problems

1. What is the physical meaning of the Debye–Waller factor?? How would it differ form low energy electron diffraction, helium atom diffraction, and x-ray diffraction from the same solid.
2. With the use of the Debye model, compute the mean square displacement of Ni at 300 K and at its melting point. What is the fractional displacement of the metal atoms relative to the interatomic distance at the melting temperature?
3. The melting of surfaces as the temperature of the crystal is increased has been theoretically treated. One such experiment is described in [35]. How may such events be detected?

4. A ^{187}W isotope was deposited on one end of a 5 cm long W slab. Estimate the time necessary to detect the radioisotopes at the other end of the slab at 1000 °C.
5. The surface self-diffusion of Cu was measured by Collins [36] and Bradshaw [37]. Review these articles and suggest possible reasons for the discrepancy of their data.
6. Adsorption probability or ticking probability of diatomic molecules have been measured for many transition metal surfaces. These experiments are carried out as a function of temperature and coverage and also as a function of the energy content and the angle of incident molecules. Describe the results of the experiment described in [11], in particular, how does the sticking probability depend upon the surface structure and the temperature of the transition metal.
7. Calculate the distance traveled by a Cu surface atom on a Cu surface in 1 h at 300 K.
8. In a TPD experiment, the maximum desorption rate of CO from Pt (111) occurs at 480 K. Assuming first order desorption kinetics and a heating rate of 30 K s^{-1}, calculate the activation energy for desorption.
9. The translation energy accommodation and transfer between a monotonic gas and a metal surface depend on both the translational energy (E_T) of the incident atom and the temperature of the solid, $E_{surface}$. Describe the nature of energy transfer in the two extremes, when (a) $E_T \gg E_{surface}$ and (b) $E_T \ll E_{surface}$.

Further reading

General reference texts

1. Somorjai G A and Li Y 2010 *Introduction to Surface Chemistry and Catalysis* 2nd edn (Wiley). This text provides general advanced undergraduate and beginning graduate level material and chapter 2 is focused primarily on surface structure, including the cases of adsorbates.
2. Oura K, Katayama M, Zotov A V, Lifshits V G and Saranin A A 2003 *Surface Science: An Introduction* 1st edn (Springer). A general advanced undergraduate/beginning graduate level text. Throughout the text the focus is primarily on surface structure.
3. Desjonqueres M-C and Spanjaard D 2002 *Concepts in Surface Physics* (Springer). An advanced text covering the physics of surfaces including phonons.
4. Ibach H 2006 *Physics of Surfaces and Interfaces* (Springer). An advanced text covering the physics of surfaces including phonons and detailed sections on HREELS.
5. Zangwill A 1988 *Physics at Surfaces* (Cambridge University Press). Intermediate level text covering a variety of topics in surface physics.
6. Bechstedt F 2003 *Principles of Surface Physics* (Springer). An advanced text covering the physics of surfaces including phonons.

7. Somorjai G A 1972 *Principles of Surface Chemistry* 1st edn (Prentice-Hall). This text provides general advanced undergraduate and beginning graduate level material and chapter 3 is focused primarily on kinetics and dynamics of surfaces.

Specialized reference texts and articles

1. Yates J T and Madey T E 1987 *Vibrational Spectroscopy of Molecules on Surfaces* (Springer). A specialized text, a bit out of date, but with many useful examples of the use of vibrational spectroscopy for characterization of molecules in surface science applications.
2. Alford T L, Feldman L C and Mayer J W 2007 *Fundamentals of Nanoscale Film Analysis* 1st edn (Springer). Graduate level text on common surface analysis techniques including ion scattering.
3. Vickerman J C and Gilmore I S 2009 *Surface Analysis: The Principal Techniques* 2nd edn (Wiley). A thorough reference text covering the principal surface analysis techniques including ion scattering and secondary ion mass spectrometry.
4. Woodruff D P and Delcharm T A 1994 *Modern Techniques of Surface Science* 2nd edn (Cambridge University Press). A thorough reference text covering common surface analysis techniques.
5. Gomer R 1990 Diffusion of adsorbates on metal surfaces *Rep. Prog. Phys.* **53** 917 10.1088/0034-4885/53/7/002. A review article of diffusion on metal surfaces.

References

[1] Goodman R M, Farrell H H and Somorjai G A 1968 Mean displacement of surface atoms in palladium and lead single crystals *J. Chem. Phys.* **48** 1046–51

[2] Goodman R M and Somorjai G A 1970 Low-energy electron diffraction studies of surface melting and freezing of lead, bismuth, and tin single-crystal surfaces *J. Chem. Phys.* **52** 6325–31

[3] Morabito J M, Steiger R F and Somorjai G A 1969 Studies of the mean displacement of surface atoms in the (100) and (110) faces of silver single crystals at low temperatures *Phys. Rev.* **179** 638–44

[4] Lyon H B and Somorjai G A 1966 Surface Debye temperatures of the (100), (111), and (110) faces of platinum *J. Chem. Phys.* **44** 3707–11

[5] Macrae A U 1964 Surface atom vibrations *Surf. Sci.* **2** 522–6

[6] Goodman R M 1969 Low Energy Electron Diffraction Studies of Phase Transformations at Metal Surfaces *PhD dissertation* University of California, Berkeley

[7] Kaplan R and Somorjai G A 1971 Observation of zero-point atomic motion near the Cr (110) surface by low-temperature diffraction of slow electrons *Solid State Commun.* **9** 505–9

[8] Tabor D and Wilson J 1970 The amplitude of surface atomic vibrations in the (100) plane of niobium *Surf. Sci.* **20** 203–8

[9] Cheng D J, Wallis R F, Megerle C and Somorjai G A 1975 Mean-square displacements of (100) surface atoms in vanadium determined by low-energy electron diffraction *Phys. Rev.* B **12** 5999–607

[10] Castner D G 1979 Chemisorption and Reactivity Studies of Small Molecules on Rhodium Surfaces *PhD dissertation* University of California, Berkeley

[11] Liu Y and Wynblatt P 1994 Nucleation of two-dimensional phases on the (111) surface of Cu–Ag alloys *Surf. Sci.* **310** 27–33

[12] Tewary V K and Fuller E R 1990 A relation between the surface energy and the Debye temperature for cubic solids *J. Mater. Res.* **5** 1118–22

[13] 1974 *The Handbook of Materials Science* ed C T Lynch (CRC Press)

[14] Tréglia G and Desjonqueres M-C 1985 Bulk and surface vibrational and thermodynamical properties of FCC transition and noble metals: a systematic study by the continued fraction technique *J. Phys.* **46** 987–1000

[15] Zhao M, Chekmarev D and Rice S A 1998 Comparison of the structures of the liquid–vapor interfaces of Al, Ga, In, and Tl *J. Chem. Phys.* **109** 1959–65

[16] Flom E B, Li M, Acero A, Maskil N and Rice S A 1993 In-plane structure of the liquid–vapor interface of an alloy: a grazing incidence x-ray diffraction study of bismuth:gallium *Science* **260** 332–5

[17] Desai T, Keblinski P, Kumar S and Granick S 2006 Molecular-dynamics simulations of the transport properties of a single polymer chain in two dimensions *J. Chem. Phys.* **124** 084908

[18] Gomer R 1959 Surface structure and diffusion *Faraday Discuss.* **28** 23–7

[19] Tringides M and Gomer R 1985 Anisotropy in surface diffusion: oxygen, hydrogen, and deuterium on the (110) plane of tungsten *Surf. Sci.* **155** 254–78

[20] Uebing C and Gomer R 1991 A Monte Carlo study of surface diffusion coefficients in the presence of adsorbate–adsorbate interactions. I. Repulsive interactions *J. Chem. Phys.* **95** 7626–35

[21] Opila R L 2024 Robert Gomer: A Bibliographical Memoir *National Academcy of Sciences* http://biographicalmemoirs.org/pdfs/gomer-robert.pdf

[22] Rana S, Johnson K N, Gurdumov K, Mazur U and Hipps K W 2022 Scanning tunneling microscopy reveals surface diffusion of single double-decker phthalocyanine molecules at the solution/solid interface *J. Phys. Chem.* C **126** 4140–9

[23] Somorjai G A 1972 *Principles of Surface Chemistry* (Prentice Hall) ch 3

[24] Chen J R and Gomer R 1979 Mobility of oxygen on the (110) plane of tungsten *Surf. Sci.* **79** 413–44

[25] Mullins D R, Roop B, Costello S A and White J M 1987 Isotope effects in surface diffusion: hydrogen and deuterium on Ni(100) *Surf. Sci.* **186** 67

[26] Schmidt L D 1972 *Proc. 2nd Int. Conf. Adsorption–Desorption Phenomena* vol 2 *(Florence, April 1971)*F Ricca (London: Academic) p 341

[27] Kubala P, Batys P, Barbasz J, Weronski P and Michal C 2022 Random sequential adsorption: an efficient tool for investigating the deposition of macromolecules and colloidal particle *Adv. Colloid Interface Sci.* **306** 102692

[28] Amirav A, Cardillo M J, Trevor P L, Lim C and Tully J C 1987 Atom–surface scattering dynamics at hyperthermal energies *J. Chem. Phys.* **87** 1796–807

[29] Siekhaus W J, Schwarz J A and Olander D R 1972 A modulated molecular beam study of the energy of simple gases scattered from pyrolytic graphite *Surf. Sci.* **33** 445–60

[30] Asscher M, Guthrie W L, Lin T-H and Somorjai G A 1983 Energy redistribution among internal states of nitric oxide molecules upon scattering from Pt(111) crystal surface *J. Chem. Phys.* **78** 6992–7004

[31] Raval R, Harrison M A and King D A 1990 Nitrogen adsorption on metals ed D D King and D P Woodruff *Chemisorption Systems, Part B. The Chemical Physics of Solid Surfaces and Heterogeneous Catalysis* **vol 3** (New York: Elsevier)

[32] Timothy B T, Thomas A D, Zhu Z, Uppili S, Winograd N and David L A 2005 Evolution of the interface and metal film morphology in the vapor deposition of Ti on hexadecanethiolate hydrocarbon monolayers on Au *J. Phys. Chem.* B **109** 21006–14

[33] Lin T H and Somorjai G A 1984 Angular and velocity distributions of HD molecules produced by the H_2–D_2 exchange reaction on the stepped Pt(557) surface *J. Chem. Phys.* **81** 704–9

[34] Zhu J, Goetsch P, Ruzycki N and Campbell C T 2007 Adsorption energy, growth mode, and sticking probability of Ca on poly(methyl-methacrylate) surfaces with and without electron damage, *J. Am. Chem. Soc.* **129** 6432–41

[35] Stoltze P, Nørskov J K and Landman U 1988 Disordering and melting of aluminum surfaces *Phys. Rev. Lett.* **61** 440–3

[36] Collins H E and Shewmon P G 1966 The effect of absorbed sulfur on the surface self-diffusion of copper *Trans. Metall. Soc. AIME* **236** 1354

[37] Bradshaw F J, Brandon R H and Wheeler C 1964 The surface self-diffusion of copper as affected by environment *Acta Met.* **12** 1057–63

IOP Publishing

Surface and Interface Science

David L Allara and Robert L Opila

Chapter 6

Electrical properties and interactions at surfaces

In this chapter we discuss the distribution and transport of charges, both electrons and ions, at the surface and interfaces of metals, semiconductors and insulators. The surface region electrons are distributed as clouds described by wave functions undulating over the core ions formed by the nuclei. The electron density extends just above the physical termination of the core nuclei and is important in electron emission, charge injection and adsorption phenomena. Under applied electrical potential the electrons in the solid move and any ions in an adjacent liquid phase to the surface also move. Finally, we discuss several ways of probing the electrical and chemical surface aspects with incoming electrons and photons, and outgoing electrons and photons. This discussion takes place using classical physics, and quantum mechanics, while very important, is used only implicitly.

Can we develop a simple set of rules or a qualitative guide to understand and predict the electrical behavior of surfaces, like we did for atomic/molecular properties with the missing atom model? Things become a bit more complicated for electrons, but the main principle is easy: according to electrostatics, opposite charges attract, like charges repel. We also need to add some of the simplest quantum mechanics necessary to describe particles in confined spaces (electrons tethered to nuclei pushed into close proximity). Let's see where this goes.

1. Of course, electrons are always on the outside of atoms, so a cloud of electrons is the last layer encountered when leaving the surface of an uncharged object to move into the vacuum. This creates a surface dipole layer, negative side pointing outward, unless other factors intervene:
   - The work of ejecting electrons from the object into a vacuum includes overcoming the electrostatic barrier of this dipole.
   - Changing the surface dipole by adsorbates or by changes in surface structure (e.g. crystal face with different atom and thus electron densities) changes the ejection energy according to simple electrostatic effects.

- The strength of adsorbate chemical bonding is determined by quantum chemical interactions at the surface electron layer, e.g. covalent and donor–acceptor interactions.

2. Atoms in increasingly close proximity force increasing divergence (splitting) of their initially, identically matched electronic energy levels, leading to near-continuum band structures.
3. The electron in the highest energy level controls the chemical potential of the object with regard to reversible electron transfer (Fermi level).

6.1 Review of basic electrostatic quantities and relationships

The fundamental basic relationship between electrical charge and its related force are given by Coulomb's law. Coulomb's law in vector form is $\vec{F}_{12} = \frac{1}{4\pi\varepsilon_0}\frac{q_1 q_2}{|\vec{r}_2 - \vec{r}_1|^2}\vec{r}_{12}$, with the charges and their positions given in figure 6.1. F is the force between particle 1 and 2 with charges q_1 and q_2, respectively, with $k_1 = 8.99 \times 10^9\ \mathrm{J \cdot m \cdot C^{-2}}$, or in vector notation $\vec{\mathbf{F}} = k_1 \frac{q_1}{q_2}\vec{e}_r$, where $\vec{e}_r$ is the unit vector in the direction of the electric field, as shown in figure 6.1.

These fundamental relationships will underlie most of the electrical charge interactions needed in this chapter. In general, we will not be using vector equations to cover subsequent topics, but it is useful to see these fundamental relationships to understand the electrical properties of simple materials including insulators, semiconductors and conductors.

First a note on energies. Typically for electrons and electrical quantities we use electronvolts (eV). Alternatively, it is common to use the standard units of joules, where $1\ \mathrm{eV} = 1.602\,176\,634 \times 10^{-19}\ \mathrm{J}$. To relate to standard thermodynamic

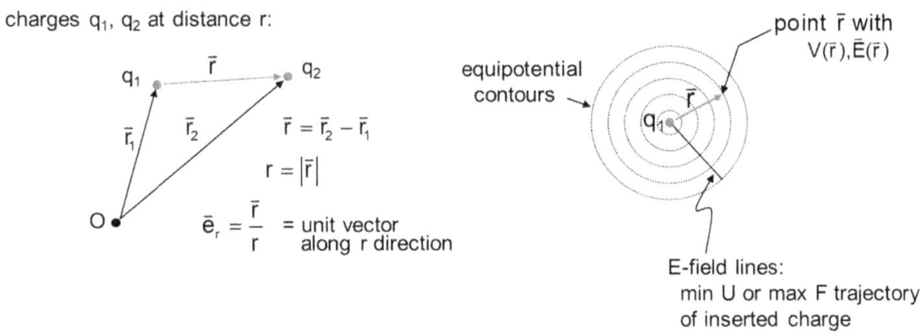

Figure 6.1. Diagram of the basic elements of electrical forces between charges 1 and 2 positioned relative to origin O. This interaction also can be described in terms of the electric field $\vec{E}_r$ generated around q_1, described as $\vec{E}(\vec{r}) = k_1 \frac{q_1}{r^2}\vec{e}_r$ with $k_1 = \frac{1}{4\pi\varepsilon_0} = 8.99 \times 10^9\ \mathrm{J \cdot m \cdot C^{-2}}$ and $\vec{e}_r =$ the unit vector along the r direction. In terms of the $\vec{E}$ field, the force on charge q_2 at distance r is given by $\vec{F}(\vec{r}) \equiv \vec{E}(\vec{r}) \cdot q_2$. The equal electrical potential and force lines around charge q_1 are also shown with the maximum force on charge q_2 increasing from the outer contours to the center. This line indicates the maximum trajectory of a charge approaching the contour fields and shows the increasing potential energy (U) as q_2 approaches q_1.

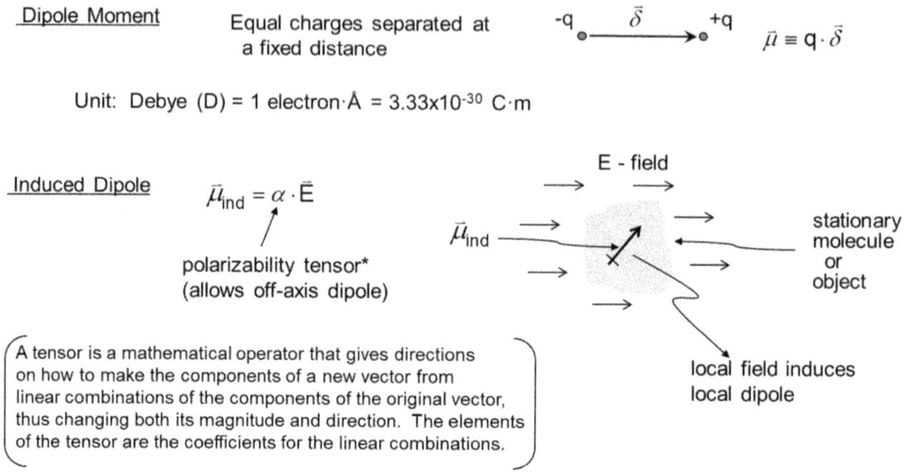

Figure 6.2. Description and definition of an electrical dipole produced when a polarizable object is placed in an electrical field. The response of electrons in the object is to shift positions according to the force field. In a non-conducting solid the electrons shift away from the positive core ions and create a dipole.

energies per mol of a species or particle, 1 eV is equivalent to 96.486 kJ mol^{-1} (per particle)

The definition of an electrical dipole is given in figure 6.2. Electrical dipoles play a prominent role in the surface properties of materials and play critical roles in controlling phenomena such as electrochemistry, catalysis, gas adsorption and semiconductor electronic devices. Dipoles in solids typically arise when an electron is separated at some distance away from its stable position next to a positively charged core ion. The separation can occur by application of an electric field or by a chemical process which shifts position of charges by a chemical reaction.

At this point it is useful to clearly define the terms electrical potential (or the equivalent term, electrostatic potential) and potential energy, arising from some separation of charges from their normal, stable positions, to avoid confusion of these terms. This is given in figure 6.3.

6.2 Charge at solid interfaces

6.2.1 The solid/vacuum interface

A general picture of a solid material with a surface exposed to some adjoining medium is given in figure 6.4 for the example of a crystalline solid such as a metal. The lattice consists of localized positively charged core ions (charge just shown as + for simplicity), each surrounded by tightly bound core electrons and more loosely bound valence electrons (shown collectively as just simple black dots). Also shown for the case of conductors are some 'loose' electrons (carriers) that move around in the solid when an electrical potential field is applied. In the case of a vacuum interface there is no adjoining medium, only vacuum.

• Potential Energy

U = potential energy = -work ($r \to \infty$) against resisting force [$F_{apply} = -F$]

$$-\text{work}(r \to \infty) = -\int_r^{\infty} (-\vec{F})\cdot d\vec{r} = \int_r^{\infty} \vec{F}\cdot d\vec{r} = \int_r^{\infty} \frac{q_1 q_2}{r^2} dr = \frac{q_1 q_2}{r} = U \qquad U < 0 \text{ attractive}$$

$$\text{inverse:} \quad \vec{F} = -\nabla U \qquad \left(\text{recall:} \quad \nabla U \equiv \left(\frac{\partial U}{\partial x}\right)\vec{e}_x + \left(\frac{\partial U}{\partial y}\right)\vec{e}_y + \left(\frac{\partial U}{\partial z}\right)\vec{e}_z \right)$$

$$\text{for } q_1, q_1: \quad \vec{F} = -\frac{dU}{dr}\vec{e}_r$$

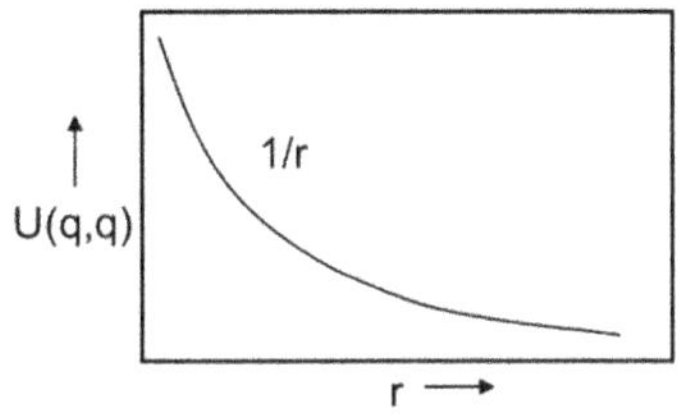

• Electrical Potential

around q_1 define a scalar potential Φ

such that when q_2 is placed at $\vec{r}$: $\quad U \equiv \Phi q_2 \quad$ and from above: $\quad \Phi = \frac{q_1}{r}$

V also is often used as the symbol to denote electrical potentials

Figure 6.3. Description and definition of electrical potential and electrical potential energy. When a charge is placed in an electrical field a force arises which can cause the charge to move. The potential energy of the shifted charge is defined as U in terms of an energy, and the electrical potential, typically given as Φ, or commonly V, is the energy per unit of charge. In an isolated environment the potential energy decreases as $1/r$. At infinite distance the potential vanishes. This configuration used as the reference energy for the potential.

From a quantum chemical point of view, the electrons in the different media are properly described in terms of their wave functions, which extend away as a probability cloud from their core nuclei. This cloud is a map of the underlying electron density. At the surface of a solid this can be pictured in terms of an 'electron spill out cloud', as shown in figure 6.5. Inside of the solid the electron wave functions show an oscillating amplitude tracking the oscillating positions of the core ions. Near the surface, where the core ions end, the wave functions still extend from the surface a short distance before the amplitude decays. In the case of a vacuum interface, where no interactions with an adjacent medium are possible, the tail into the vacuum region gives rise to a surface dipole and extends ~2–3 Å into the vacuum, a very important aspect that controls many phenomena associated with the surface. For example, molecular or atomic species in an adjoining material medium can experience some interaction of their wave functions with the surface electron cloud, which can lead to the formation of chemical bonds or even a charge transfer to the adjoining species.

Looking at this aspect, note that the exact definition of a surface cannot be made precisely since the effects of the surface decay with distance are not sharp, rather than exhibited as a mathematically sharp boundary.

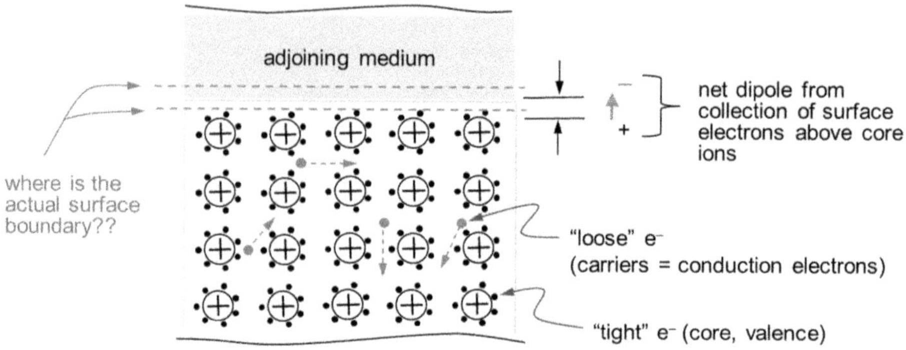

Figure 6.4. Diagram illustrating the basic electrical nature of a solid substance such as a crystalline metal. The positively charged metal ions, termed the core ions, are positioned according to their intrinsic positions and each core ion has surrounding electrons: tightly bound core electrons with large attractive potential energies close to the core and less tightly bound valence electrons in outer shells from the core. At the surface of the solid the top layer of electrons, particularly the outer shell valence electrons, provide a negative surface charge relative to the positive core ions, which then creates a layer of intrinsic surface electric dipoles. In an atomic scale of view, it is not clear where to place the exact surface boundary of the solid. The figure shows an adjoining medium, which can be vacuum, solid or fluid. This medium will interact with the surface dipoles to create an interface different from either the pure solid or adjoining medium.

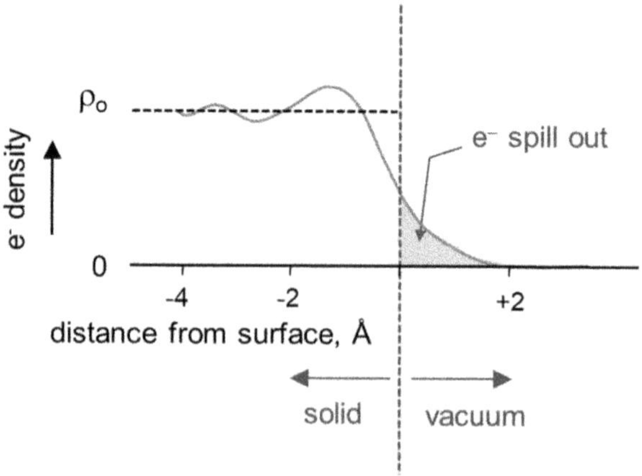

Figure 6.5. A schematic diagram of an arbitrary solid/vacuum interface defined in terms of the decay of the surface electron density probability cloud into the vacuum. The distance scale is shown in atomic units with the exact distances depending on the specific case of the solid. This oscillation is referred to as a Friedel oscillation. Outer features of the curve can be observed by scanning tunneling microscopy (see previous chapters).

6.2.2 The solid/vacuum interface with electric fields applied: the space charge region

Now let's look at a solid with its core ions and electrons and consider the effects of an electric field across the solid, applied between the surface to the bottom of the solid. Here it is useful to consider, in addition to bound core and valence electrons, the case of the presence of conducting (loose) electrons. A schematic diagram is shown in figure 6.6 for the case of a vacuum interface with a planar electrode situated above the

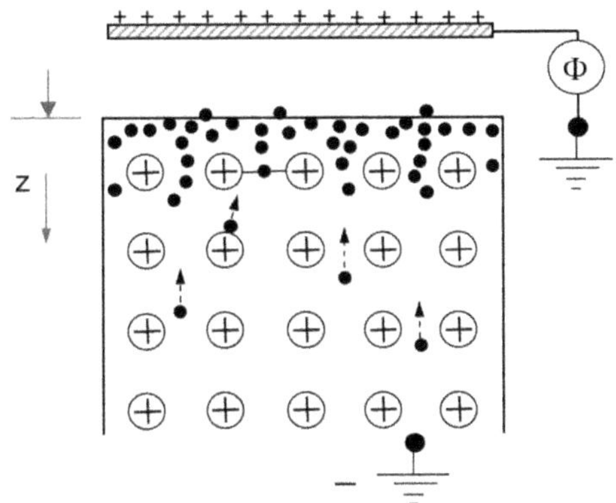

Figure 6.6. A schematic of a hypothetical solid with a positively charged electrode positioned just above the surface at the vacuum interface. The black dots indicate 'loose' or conduction electrons that pile up near the surface. The core and valence electrons are not shown for simplicity. The charge response on the electron depth (z) profile of the electron density tails off exponentially until in a deep region the electrons are no longer affected by the field. If the field is reversed the electron density profile reverses.

surface such that the electrode does not physically contact the solid (a vacuum interface). For example, where the field is applied with a positive charge on the top electrode all electrons in the solid experience a force pulling them towards the surface. Valence and core electrons which are tightly bound to the core ions will only slightly shift their positions a small fraction of their binding distance from the core ions, but any conduction (loose) electrons will be transported by the field to new locations near the surface, as shown by the black dots in the figure. In the case of tightly bound electrons (a dielectric medium) this results in a polarization of the solid. For conducting electrons collections of electrons will start to pile up near the surface. This brings up a very important question of how far the electrons shift their positions with the applied field as a function of depth. The simplest situation has the electron density drop off exponentially from the surface. Beyond some depth the field essentially has no effect on the electron density. This fall off is typically characterized by a convenient parameter, shown as δ in the figure. This effect is treated more quantitatively in exponential decay parameter. The Debye length, d_{Debye}, is the depth over which the charges essentially become free of the applied field, typically termed the screening length, and is given by $d_{\text{Debye}} \propto (n_e^{\text{bulk}})^{-1/2}$ where, n_e^{bulk} = the carrier (conducting electrons) density. Note that the fall off is independent of the applied field and the magnitude depends on the 'tightness' of available e^-. For the case of a large density of conduction electrons $d_{\text{Debye}} \to 0$ and the conduction electrons pile up to the surface. For insulators with $n_e^{\text{bulk}} \to 0$, the effects of the field reach deeply into the solid.

6.2.3 The solid/liquid interface

If the adjoining medium is a liquid the same principles apply for electrical polarization effects. But the presence of liquid interface, particularly with aqueous solutions, can give rise to chemical processes which may produce ions that provide new surface charges. The two main ways to introduce net charge on a surface in contact with solution are discussed below. After that some important associated aspects of charged surfaces at liquid interfaces are discussed.

6.2.3.1 Introduction of surface charge by chemical equilibria with surface groups and by adsorption of ions from an ionic solution as the adjoining medium

For the first phenomenon we use the common example of solids that contain surface –OH groups that act as Brønsted acids or bases which can ionize to provide H^+ or OH^- ions into solution (typically aqueous) leaving charged surface sites. This changes the local pH at the interface. The diffusion of these ions into the solution is controlled by the presence of the remaining oppositely charged ions immobilized at the surface so the concentration of diffusing ions decays with distance by electrostatic forces.

Diffusion of surface-generated ions into the liquid medium is driven by the lowering of their chemical potential by solvation in the liquid (more stable when solvated). But transport into solution causes charge separation of the diffusing ions and the remaining surface counterion charges so a balance arises between solvation stabilization and electrical work of charge separation. These effects are illustrated in figure 6.7. The final ion distribution profile with Z is determined by the Boltzmann energy distribution and the Coulomb forces between surface and solution ions.

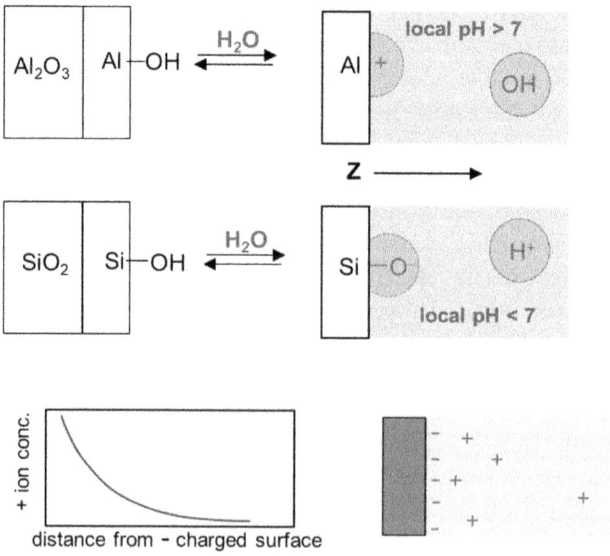

Figure 6.7. Solid surfaces in contact with water often become charged by ionization of surface groups intrinsic to the solid surface. For surface –OH groups this will change the local pH. Typical cases shown at the top are Al_2O_3 and SiO_2, which often behave as basic and acidic surfaces, respectively, under common laboratory conditions. The concentration of the diffusing ions decays into the liquid, as shown in the lower left plot, since the counter ions remain pinned on the surface creating charge separation (bottom-right). Top right: the ions in solution are shown with illustrative solvent shells.

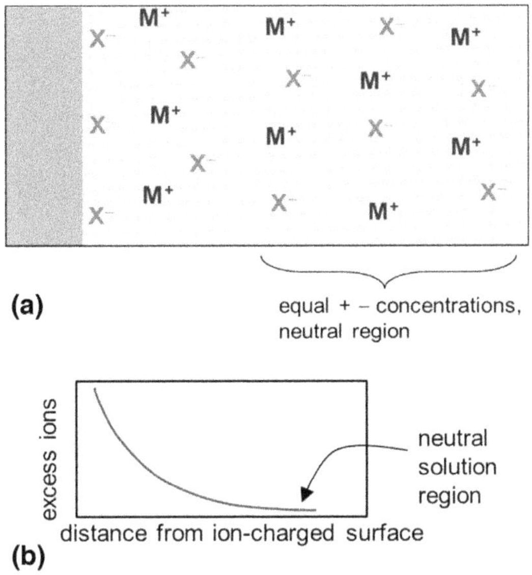

Figure 6.8. (a) Creation of a charged surface later by adsorption of ions of one charge from an originally neutral solution of + and – ions. Near the surface the adsorbed ion layer(s) creates a charge gradient with the solution ions. The gradient tails off with increasing distance away from the surface into a neutral region with balanced ion charges. (b) A hypothetical plot of the ion charge imbalance decay into the neutral solution region.

Typical voltages associated with the potential difference of this charge separation are ~25 mV.

The second surface charge production mechanism involves diffusion of ions from an ionic solution medium onto a neutral surface to create a charge layer. Common cases involve diffusion of Br^- ions onto AgBr nanoparticle surfaces or silver electrodes and adsorption of Cl^- ions on Ag electrodes, common in electrochemical cells. The general case is illustrated in figure 6.8.

6.2.3.2 Definitions of important phenomena associated with charged surfaces at liquid/solid interfaces

6.2.3.2.1 Isoelectric point (IEP) in aqueous media

This state applies to a condition in which the surface has no free charge, termed the point of zero charge (PZC). For the case of surfaces with Bronsted acids or bases this condition arises when the pH is neutral with respect to the (H^+–OH^-) surface equilibria. For surfaces with other ion equilibria (e.g. AgBr with adsorbed Br^- ions) just replace pH with –log(solution ion concentration). The IEP measurements are often used as characterization of nanoparticles. One can measure IEP by titration and watch the motion of nanoparticles between charged electrodes in a cell vanish at the PZC. These measurements, although very useful, are prone to observations of a range of values since particle surfaces often are not well controlled which can give

Table 6.1. Isoelectric point (IEC) for common solids.

Common Brønsted acid–base solids	IEP range
SiO_2	2–6
ZrO_2 (zirconia)	6–10
TiO_2 (rutile or anatase)	5–6.5
Al_2O_3 (α-, γ-alumina)	8.5–9
Al_2O_3 (boehmite)	8–10
ZnO	6–10.3
MgO (magnesia)	~11

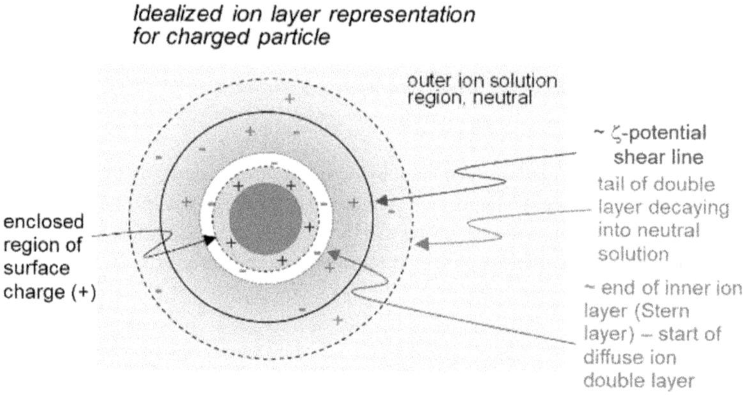

Figure 6.9. Illustration of the basic aspects of ions in a solution pertaining to the measurement of the ζ-potential. The inner Stern layer is tightly bound by electrostatic forces of the ions and the surface charge. Followed by diffuse double layer of ions with more weakly bound ions tailing off into the neutral solution. The shear layer is shown just within in the outer layer.

rise to measurement errors. Table 6.1 gives IEP values for some common Bronsted acid–base types of solids.

6.2.3.2.2 Zeta (ζ) potential (electrokinetic potential)

Charged particles at equilibrium in ionic solutions possess ion shells, as shown in figure 6.9. The inner ion shell consists of ions of opposite charge to the surface and is tightly bound (Stern layer). In the next shell, a diffuse double layer (termed the Guoy–Chapman layer), the ions are more weakly bound and tail out into the neutral region of the solution. When charged particles collectively move through the solution via the force of an applied electrical potential, some of the outer layers of ions past the double layer can be sheared off, typically ~2 layers, which changes the net charge on the particle surface. In this way, measurement of the particle velocity as a function of applied force can be used to determine the ζ-potential (see figure 6.10).

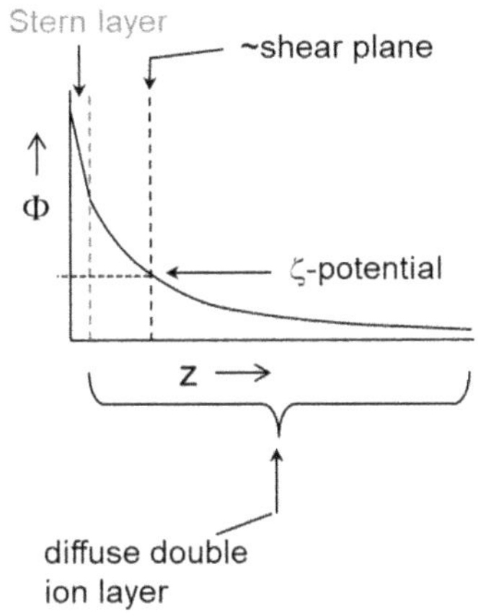

Figure 6.10. Qualitative depiction of the location of the ζ shear plane relative to the different ion layer shells.

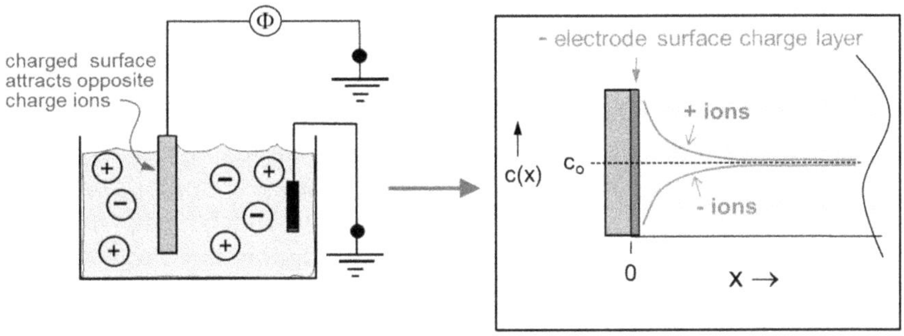

Figure 6.11. Left, ions move toward electrode with opposite charge. The right plot shows the screening of the electrical potential by the accumulation of ions of the opposite charge.

This is a widely used characterization of nanoparticles and colloids and can provide a measure of the particles' ability to not aggregate. Typical values are in the range of 25 mV, depending on the surface charge.

6.2.3.2.3 Ion distribution between two electrodes in an ionic solution with an applied electrical potential

When oppositely charged electrodes are placed at some distance apart in a solution, considerably larger than the intrinsic ion layer shell distances, as shown in figure 6.11, solution ions will move under applied potential to the corresponding oppositely charged electrodes. In that process, ions of like charge will start to bunch up near the appropriately charged electrode. The resulting spatial ion concentration distribution resembles the general one given on the right side of the figure, selected to have a negative charge for illustration; negative ions are driven away and positive ions attracted. This creates a concentration gradient for each ion away from its oppositely charged electrode. At some distance much larger than typical ion shell layer thicknesses, the ion concentrations approach that of the neutral regions in the bulk solution. Recall the

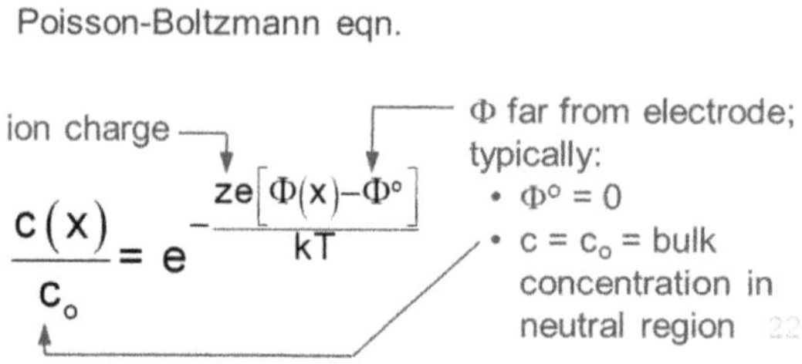

Figure 6.12. The Poisson–Boltzmann equation. The distribution of ion concentration with distance into the solution from an electrode $C(x)$ is given relative to the concentration C_0 in the neutral region of the solution. The quantity e is the electron charge.

concept of the decay depth of electrons in a solid at the electrode–vacuum–solid interface. In that case the estimate of the depth of displace electrons was given by the Debye length. For ionic solutions, we use the known electrostatic behavior of solution ions to provide calculations of the solution ion distribution. Across the solution between electrodes the ions have energies according to the applied potentials of the electrodes but also suffer ion repulsions and attraction depending on the local concentrations of each type of charged ion when they bunch up or separate into different regions. So the two factors adjusting ion concentrations are at play together.

This build up is treated by using a Boltzmann distribution of the potential energy of the ions due to the forces from the electrodes (attractive force) and adding the electrostatic forces between ions as functions of their bunching up in each region (repulsive force). This gives rise to the Poisson (Coulombic ion electrostatic interactions)–Boltzmann (Boltzmann energy distributions of the potential energies away from the neutral center region between the two electrodes) equation, shown in figure 6.12. The Boltzmann distribution part is shown by the kT exponential dependence and the ion electrostatic energies shown in the exponent term. Here Φ° is the electrical potential in the neutral part of the solution relative to a local ion potential at distance x from the electrode. The Poisson–Boltzmann equation is critical in analysing ion distributions near surfaces for solutions with applied electrical potential.

There are numerous familiar applications where solid surface/solution space charge characteristics are important, and can be used to control and/or design, for example:

- Colloid and nanoparticle dispersions, avoiding particle aggregation.
- Functional behavior of proteins with charged surfaces, controlling bio-interfaces.
- Ion transport across biological cell membranes.
- Electrochemistry, electrolytic supercapacitors and batteries.
- Fuel cells.

6.3 Electron energy distributions in solids and their surfaces

6.3.1 Electron bands and the band gap

To this point our main concern has been with the effects and nature of charges and charge layers at surfaces as influenced by electrostatic field and their associated forces on the charges. At this level the simple description of the solid in terms of core ions with surrounded electron shells works quite well. In deeper detail, when the electrons in the outer shells are pushed together at close distances in a solid the electrons on adjacent nearby ion cores interact at a quantum level and cause splitting of their energies and their wave function distributions. So, if any behavior of the solid involves quantum states these aspects must be rigorously treated.

The changes that occur in moving a collection of atoms into close proximity is illustrated qualitatively in figure 6.13 for the case of boron atoms.

In this example the solid-state exhibits bands of energies with lowest energy electrons in very stable core states, tightly bound to the core ions, with the wave functions of each electron undulating along a row of repeating core ions in the associated crystal plane. These wave functions are spread among all the core ions to provide maximum electron density around each atom, with strong binding energy to the cores. For the lowest energy core states the wave functions essentially undulate around each core ion. At higher energy levels the wave functions undulate with wider and wider undulation distances, thereby creating undulating density to the core ions along the lattice row which results in weaker binding. A simple analog to this can be given by the periodicity or undulation of atom or particle positions in phonons (see chapters 5 and 6). Eventually, just below the band gap, higher energy

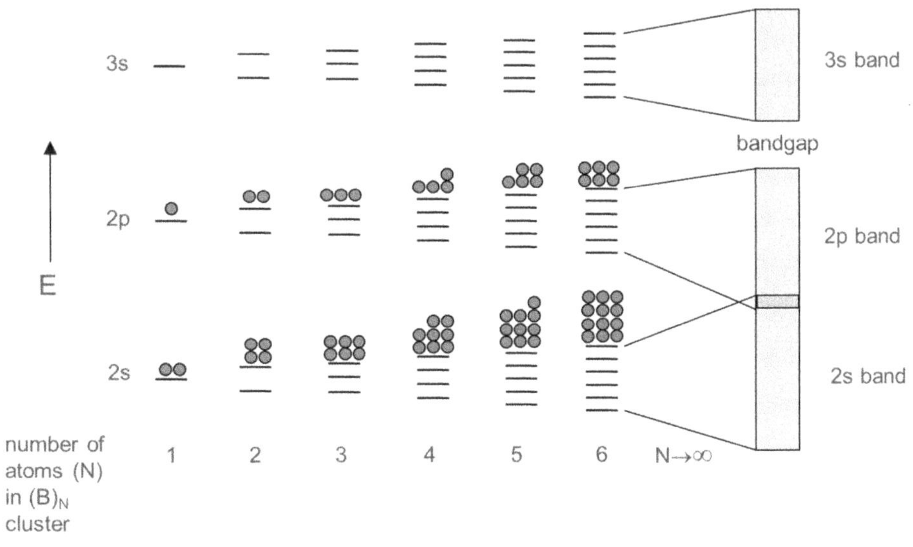

Figure 6.13. Illustration, for the case of boron atoms, of the formation of discrete quantum states when the atoms are increasingly brought close together to the eventual bonding distance in the solid state as required by the Fermi statistical character of electrons. Note that the original atomic s and p levels increasingly split into more states according to the number of atoms involved. Finally, in the formation of solid boron the original isolated atom electron states form near continuous bands of closely spaced energy states, described according to the quantum numbers of the isolated atom states. Note also that as each of the bands widen in terms of the energies of electrons contained, the bands start to overlap and at higher energies form a separation between the highest filled states of the solid and higher energy unfilled states. This latter gap is called the band gap.

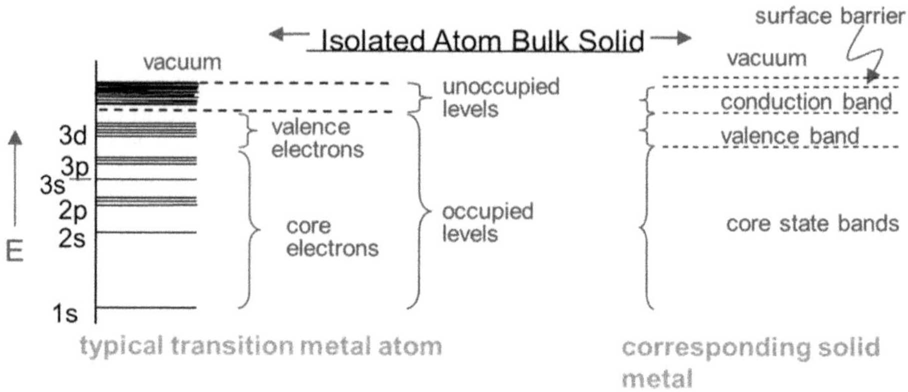

Figure 6.14. Schematic illustration of the differences in electron energy states in a single atom and a bulk solid formed from the atom.

valence level states begin to appear with more weakly bound electrons. For a solid surface the very highest energy states in the surface region exhibit a wave function extending just beyond the surface, described earlier in figure 6.5 as an electron spill-out state which is part of the contribution to the surface dipole (see figure 6.4).

Once above the band gap the electron wave functions are no longer associated with overlap around core ions and are free to move across the crystal as charge conduction carriers when excited by some process such as application of applied electrical potential or impact with energetic particles or photons. Further details on wavefunctions in solids can be found in various solid state physics texts.

A simple schematic illustrating the energy levels in an isolated atom versus its solid state is shown in figure 6.14. This helps one to think about the definition of the bands and the characteristics of the associated electron states. As a note on energy scales, E (or often U) typically refers to the work required to excite an electron in the given energy level to a vacuum state with no kinetic energy (essentially 'hovering' over the surface) and as such is a positive energy. The negative value is the potential energy. The term binding energy is often reserved for a specific meaning and defined later in the section on surface spectroscopies. We will see in this later section the role of the work function in interpreting surface spectroscopy measurements.

6.3.2 Distribution of numbers of energy states per energy increment

A common and very useful description of the 'energy stacking' of states in each of the energy bands is the density of states (DOS), defined as the number of energy states per energy interval. So if there are a large number of electron energy states appearing within small increment of increasing energy, this region of energy has a great many states and the DOS is large. The DOS can be important in determining several properties of the solid, including heat capacity and chemical reactivity. Figure 6.15 illustrates an example of the DOS for the number of valence electron energy states appearing in an energy increment. If we take small increments of energy around each specific energy and count the electron states in each increment as

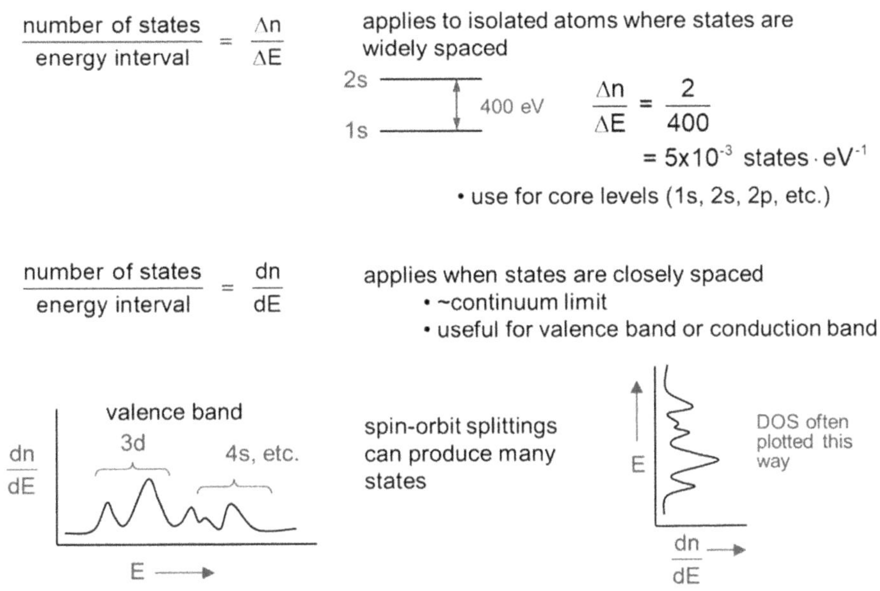

Figure 6.15. An example of the DOS associated with different valence band energy states per energy increment. The valence states have typical distributions of energies around the peak due to local effects that broaden the energy envelop. The number of distinct states for a given electron valence state can be large due to phenomena such as spin–orbit splitting.

a function of the energy we arrive at plots such as in the figure. The DOS of solids can be measured by various types of spectroscopies using x-rays, ultraviolet light and electron impact. These will be discussed in further sections.

6.3.3 Classification of energy bands structures in terms of associated electrical properties

There are three common classes of energy band structures, each distinguished in terms of their electrical behavior: conductor (usually a metal), semiconductor and insulator, as shown in figure 6.16. In the case of a conductor, the highest energy filled state has no band gap, e.g. it is located just below the lowest energy unfilled state of the conduction band; thus, very little energy is needed for excitation into the conduction band. At 0 K there are no filled states in the conduction band but as the temperature rises or as excitation energy is supplied (such as an applied voltage) some electrons move into the conduction band. The semiconductor case shows a finite band gap and the energy of the gap can be overcome by applied potential, elevated temperature or other means, e.g. photons, to excite electrons into the unfilled conduction states. Finally in the case of an insulator, the band gap is relatively large and population into the conduction band requires significant energy (higher applied voltages, etc) but often this increase in required excitation energy may be sufficient to start breakdown of the insulator material before conduction

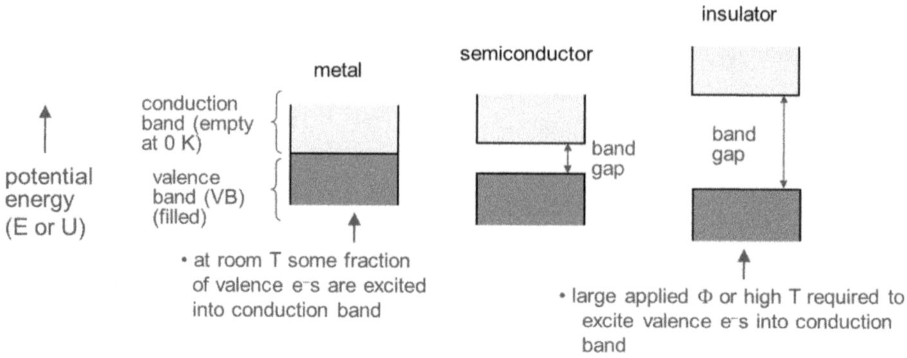

material	band gap	carrier conc. at room T	δ(space charge depth)
metal	~0	large	small (atoms)
semiconductor	~0.5 - 2 eV	small	medium (many atoms)
insulator	~2 - 6 eV	~0	large (~μm)

Figure 6.16. Diagrams describing general classes of band structures in terms of the types of electronic properties they represent. Conducting metals have no band gap and any excitation of lower-level electrons creates conduction electrons. Semiconductors have a small band gap and applied energy (electrical potential, etc) is needed for excitations to create conduction electrons. Insulators have large band gaps and significant energy is needed for conduction. Also shown are typical band gaps energies (in eV), associated conduction carrier densities and associated space charge depth for each of these cases.

ensues in an otherwise unperturbed material. Note in figure 6.16 the magnitude of energies (in electronvolts, eV) required for band gap excitation. Note also, the depth of the space charge region (see figure 6.6), the average depth over which an applied electrical potential at the surface no longer influences buried charge in the solid (refer to figure 6.6 and the Debye screening depth,d_{Debye}).

6.3.4 Fermi levels and electron distributions—common definitions

A widely used way to characterize the electron energy in a solid is the Fermi level. The value of this definition is useful in understanding and predicting how electrons redistribute between two media in electrical contact to balance the overall energy of the system. As shown in figure 6.17, the Fermi level energy (E_F) is drawn at the halfway energy in the band gap, regardless of its magnitude, for the condition of $T = 0$ K. As T increases electrons are excited from the top of the filled state band and redistribute to populate the lowest states of the filled band, in which case E_F acts as a balance point around which the electron flow between filled and unfilled states shifts. The newly formed empty states in the filled level are termed 'holes'. At a given T this amounts to a chemical potential shift in the electrons and thus E_F is a reference point for the chemical potential of the system. For convenience, here E_F^0 denotes the Fermi level and E_F the uppermost energy filled level. Typically, and later on we just use E_F for the Fermi level.

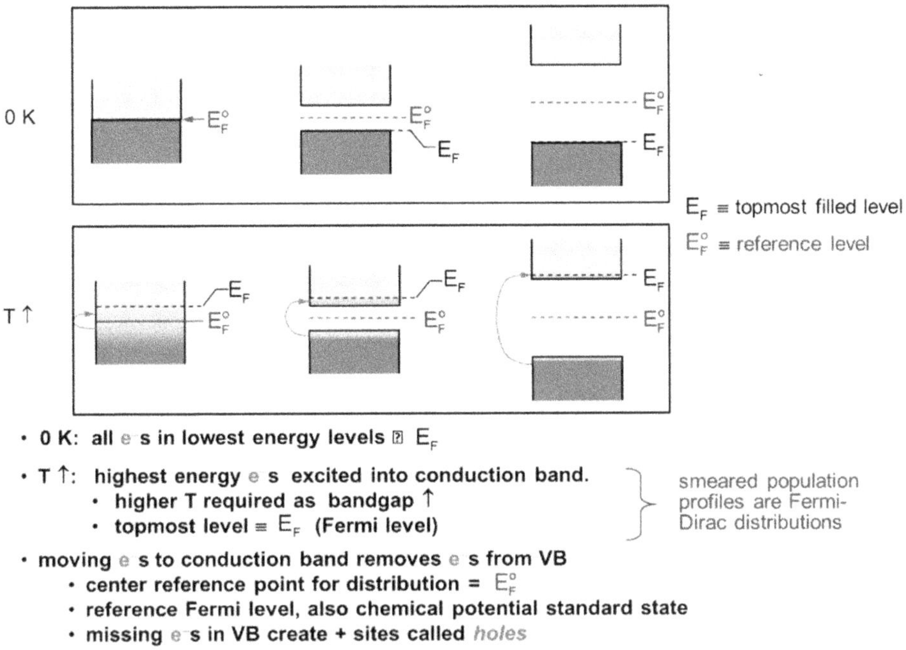

Figure 6.17. Definition of the Fermi energy level in terms of the band gaps for conductors, semiconductors and insulators. As T increases from 0 K, electron populations shift from unfilled states to filled states. The Fermi level serves as a balance point between newly formed unfilled band states (holes) and newly populated unfilled band states. The energy re-distributions in each band after shifting are given by Fermi–Dirac statistics. The population shifts are illustrated schematically in terms of the gradients in the red color density.

6.3.5 Conduction mechanisms with electrodes contacting a solid

Now that we have established the general structure of the electron energies in a solid, we can describe the essential behavior of conducting charge through the solid. When an electrical potential is applied by contacts across a solid with a population of conducting electrons, charge will flow via the transport of conducting electrons injected at the negative electrode to the positive electrode where they are collected and flow through the circuit, as depicted in figure 6.18.

A reverse case of carrier type can occur when the solid has no conduction electron states available as charge carriers but where application of a potential across the solid can excite valence electrons from the top of the conduction band to leave an unfilled conduction band state (hole), as illustrated in figure 6.19. This allows charge to flow when electrons are injected into a hole at the negative electrode and ‘hop’ through the solid to the positive electrode to be collected. This still involves electron transport through the solid but with the electrons ‘hopping’ from hole to hole.

6.3.6 The work of removing an electron from a surface at the solid/vacuum interface.

When an electrical potential or other form of energy (particle or photon collision) is applied to a solid, surface electrons can be ejected. The energy required not only

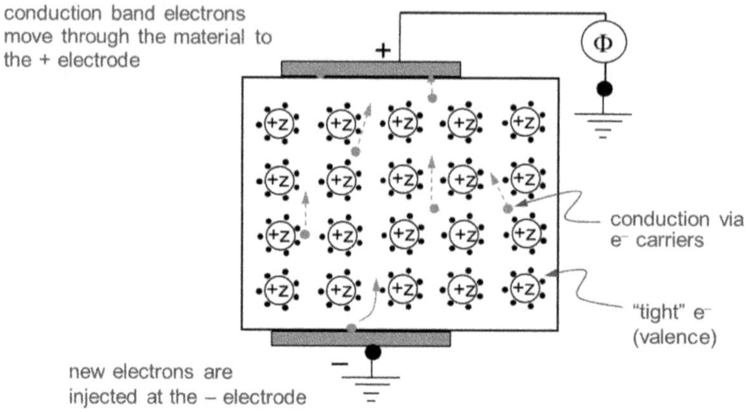

Figure 6.18. Schematic illustration of conduction through a solid with potential applied at two sides. Electrons in the conduction band move through the solid under applied potential with injection at the negative electrode and collection at the positive one.

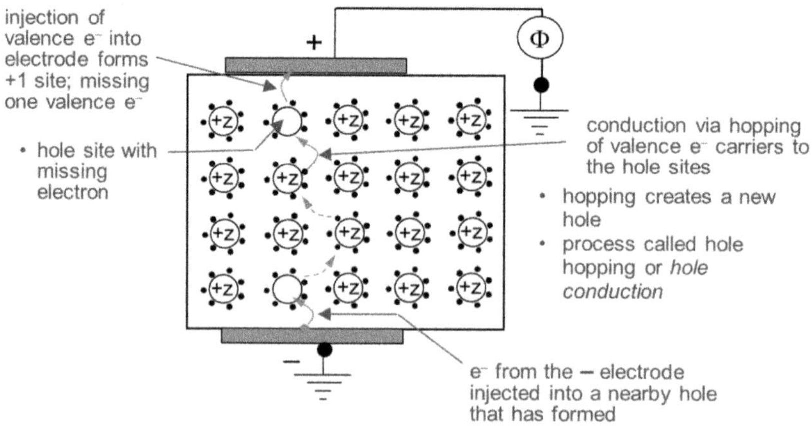

Figure 6.19. Schematic illustration of hole conduction. In this case, compared to conduction electron flow between electrodes, the injected electrons from the negative electrode hop from hole to hole to be collected at the positive electrode.

involves supplying the energy to overcome the binding energy of the electron in its associated band but further involves supplying an energy to overcome an inherent electrostatic barrier that exists at the surface/vacuum interface. This barrier is associated with the electron density at the surface edge and is associated with the surface dipole and the electron 'spill out' cloud which places electron density just above the surface. This is illustrated in figure 6.20. Even for the highest lying energy state at the surface, removal of the electron requires work to overcome this 'electrostatic penalty', which can be thought of as an 'electron jump' barrier. This jump energy is termed the work function, $\Delta\phi$, with units given typically in electronvolts (eV). This is worth more detail. When the electron at the surface has

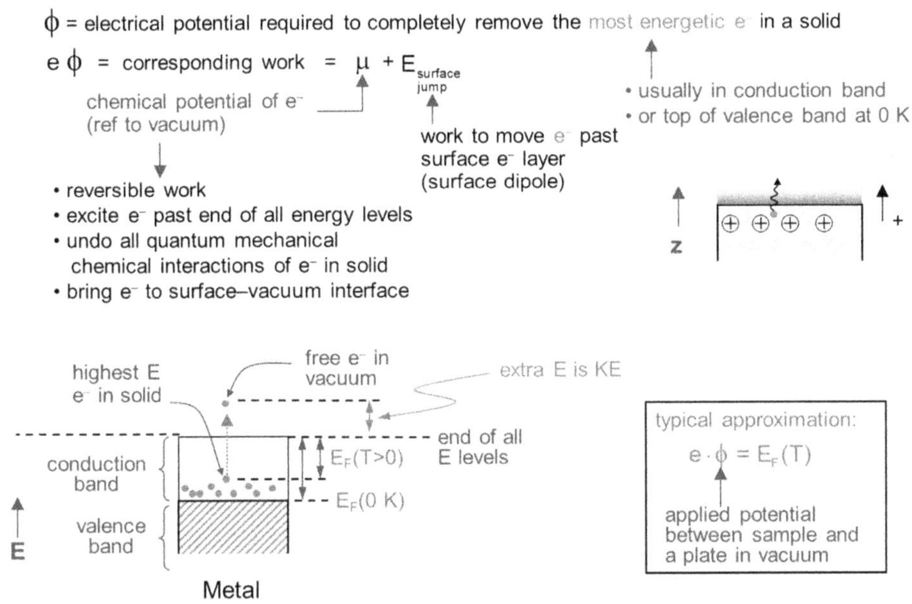

Figure 6.20. Schematic depiction of the surface work function and its basis in terms of the extra layer of electron density above the surface at the vacuum interface. The work function represents the extra work required to remove a surface electron, which has become uncoupled from its quantum chemical interaction with the underlying solid, out into the vacuum.

no more quantum chemical interaction with the solid there still is this work penalty to release the electron into the vacuum

6.3.7 Measurements of work functions at surfaces

Over the years, extensive measurements of work function values have been measured for a variety of surfaces. There are several ways to measure $\Delta\phi$. The most typical way is by several types of electron and x-ray surface spectroscopies, as described in a later section of this chapter. These spectroscopies require a high vacuum condition for the sample and thus are desired methods since the surface to be probed is under clean conditions with minimal surface contamination in the measurement environment. A simple method often used involves capacitance measurements with an electrical probe touching the surface under an applied potential to determine the electrostatic potential required to move the outermost surface electron off the surface to the probe tip. This type of probe is typically vibrated to change its distance from the surface in an oscillating way.

Work functions for a given solid are very dependent on the detailed nature of the surface such as roughness, including steps and kinks, and chemical states, so are an important characteristic of each type of surface. Work functions also can exert control on charge transfer processes from the surface and on chemical interactions, as will be discussed shortly.

Table 6.2. Work functions for different crystal faces of tungsten, molybdenum and tantalum. The values are given in electronvolts (eV) [1, 2].

Crystal face	W [1]	Mo [2]	Ta [2]
110	4.68	5.00	4.80
112	4.69	4.55	—
111	4.39	4.10	4.00
001	3.56	4.40	4.15
116	4.39	—	—

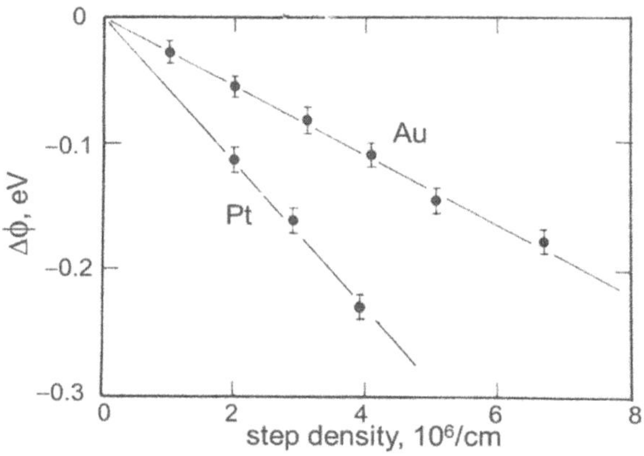

Figure 6.21. Work function data for Au and Pt crystal faces of differing roughness as characterized by step density. Rougher surfaces clearly exhibit lower $\Delta\phi$ values.

Some values of measured work functions for different crystal faces of three metals are given in table 6.2.

Thought questions:

1. Why might $\Delta\phi$ depend on crystal face structure?
2. Take noted that W, Mo and Ta are all bcc unit cells. You might try to see if the surface density of the faces correlates with the $\Delta\phi$ trend.

Work functions for Au and Pt crystal faces with varying step densities are shown in figure 6.21.

Thought questions:

1. Why might steps have lower $\Delta\phi$ values than terraces?
2. Is this trend consistent with the data in table 6.1?

6.3.7.1 Work function versus ionization potential—what is the difference?
Ionization potentials (I_P) are well known characteristics of atoms and chemical compounds. Whereas I_P represents the work required to remove an electron from a

Table 6.3. Comparison of work functions and ionization potentials for different alkali metals.

Metal	$\Delta\phi$, eV	I_P, eV
Li	2.9	5.392
Na	2.75	5.139
K	2.30	4.341
Rb	2.16	4.177
Cs	1.81	3.894

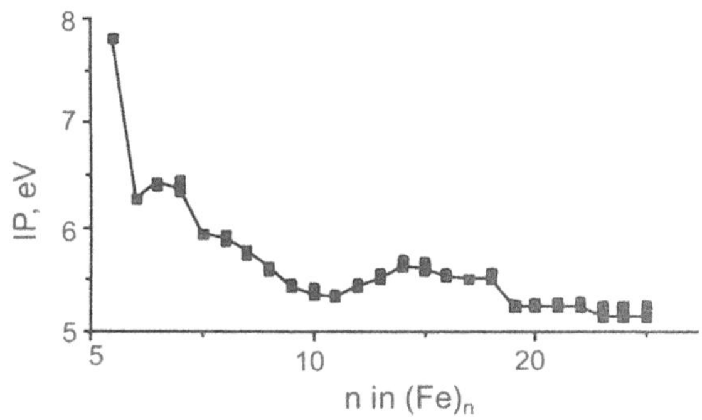

Figure 6.22. Correlation of the ionization potential of iron clusters as a function of the number of atoms in the cluster.

single species, typically a gaseous atom or molecule, $\Delta\phi$ involves removal from a solid surface. Representative values for gaseous alkali atoms are given in table 6.3. It is of interest to compare values of I_P with $\Delta\phi$ since both are energies of electron removal into vacuum but differ since the removal is from a different form of a material entity.

One can immediately see the differences between the two sets of values: $I_P > \Delta\phi$ by a substantial energy in all cases. A further insight into the differences is given by the correlation in figure 6.22, where I_P varies with the number of atoms in iron atom clusters. These data show that the work function of a surface seems to develop as the number of atoms in a solid increase from a few, such as in a cluster of a few atoms to a solid with a large number of constituent atoms with formation of a repeating lattice.

Thought question:
How might you explain the comparisons in table 6.3 and the connection to the correlations in figure 6.22? You might try to think about the trend in terms of charge stabilization by a polarizable medium.

6.3.8 Charge transfer at solid/solid interfaces

We now look at charge transfer in the case of two solids placed in intimate contact, as illustrated in figure 6.23. Upon contact electrons can flow across the interface

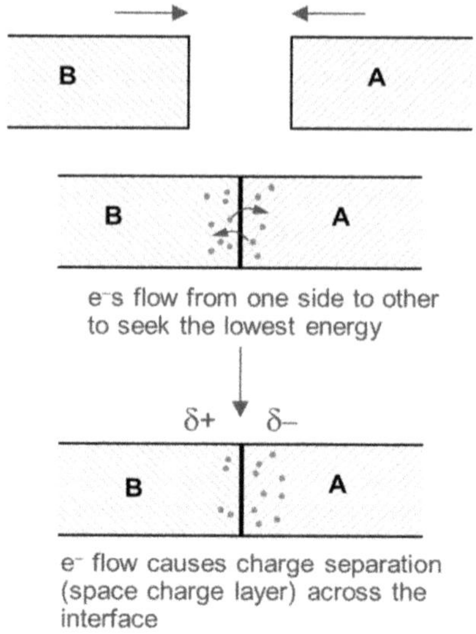

Figure 6.23. Schematic of two solids, A and B, placed into intimate contact at their surfaces. If A and B have different values of their Fermi levels (E_F) electrons will flow across the contact point to balance the E_F values. As a result, a charge imbalance occurs which creates a space charge layer at the A/B interface.

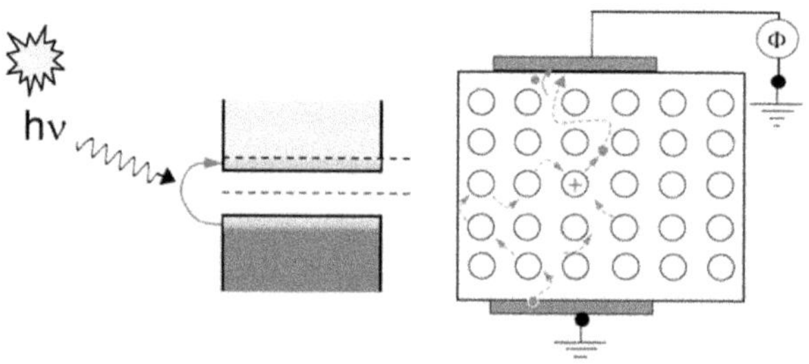

Figure 6.24. A schematic illustration of the basics of a photovoltaic cell. The cell is made by sandwiching the semiconductor between suitable electrodes. Upon exposure to visible light high lying valence band electrons are excited into the conduction band to create a space charge layer which causes a potential difference between the electrodes; thus current can flow to provide useful work from the circuit.

according to their relative energy levels. The solid with a higher Fermi energy (E_F) will act as a source to transport electrons into the solid with the lower E_F. This process continues until the Fermi levels are equal. This flow creates a space charge layer at the interface.

6.3.8.1 *A useful example: photoexcitation and charge conduction in photovoltaic cells for solar energy conversion*

Photovoltaic cells can be constructed by placing suitable electrodes to sandwich a photosensitive semiconductor which allows excitation of electrons in the high lying valence band into the conduction band at common frequencies of light in the solar spectral range. A schematic diagram is shown in figure 6.24. This process produces holes (h^+) in the semiconductor and drives electrons (e^-) into one of the electrodes. Current then can flow through the circuit with e^- injected back into the semiconductor to complete the circuit by hopping across h^+ locations. This process can provide useful work to be done in the external part of the circuit.

However, the process is not completely efficient. While some e^- and h^+ each reach the electrodes to provide a voltage difference, a fraction wander in a random walk fashion from site to site through the solid to meet up again and recombine, releasing energy (just heating the sample) and lowering the efficiency of the process.

Not all semiconductor materials will work as photovoltaics since only a few types are capable of photoexcitation to excite across the band gap. The photoexcitation process requires an internal coordination of lattice motion involved in phonon vibrations (see chapter 5, especially section 5.1.2 on surface vibrations.) so that the lattice structure is arranged during an excitation event to allow wavefunctions of states in the valence and conduction bands to be compatible for transfer.

Only a limited number of semiconductor material match this requirement. The most common materials include Si, GaAs, CdSe, $CuIn_x$, and $Ga_{(1-x)}Se_2$ (variable x). In practice, additional efficiency is gained by adjusting the work functions of the semiconductor surfaces to lower the barrier for charge transfer across the interface. This can be done by chemical or physical treatment of the surfaces to reduce impurities and create full contact.

Thought question:
It is common to adjust the work functions of the two electrodes to produce surface dipoles that help drive e^- and h^+ to the desired opposite electrodes. Explain how this might work.

6.4 Adsorption of molecules and atoms at surfaces

In this section we consider the interaction of chemical species, in particular molecules and atoms with solid surfaces from the point of view attachment to the surface. The most well understood cases involve the solid/gas interface. Solid/liquid interfaces provide complexities from the solvent medium so for simplicity we restrict the discussion to solid/gas interfaces. Reflecting back to cases of the solid/solid interface, there we just discussed the transfer of charge. To be complete we would need to consider transport and diffusion of atomic species, which brings in the complexity of solid-state chemistry, but for simplicity we also leave that aside.

There are two common types of adsorption phenomena. First, physical adsorption (or commonly physisorption), where the adsorbed species (adsorbate) does not form chemical bonds with the surface (more explicitly, overlap of the adsorbate wave functions with surface wave functions). Second, chemical adsorption (or commonly chemisorption) where chemical bonds form; at one extreme weak bonding and at the other extreme vigorous reaction which can eventually disrupt the chemical structure of the substrate surface and even rip a few layers below the original solid surface.

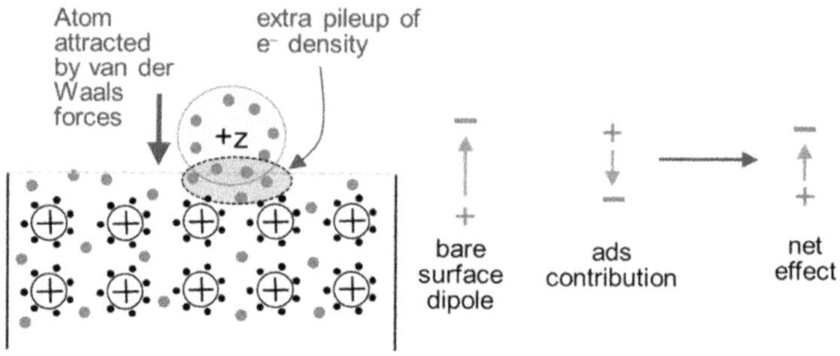

Figure 6.25. Schematic of a non-chemically bonding adsorbate attached to the substrate surface by vdW polarization forces. As the adsorbing molecule or atom approaches the surface, electron density from the outer orbitals of the adsorbate form a negative charge layer. This layer creates a surface dipole that is inverted from the dipole of the bare surface. The net effect is to create a reduced surface dipole which leads to a lower value of the surface work function. Pauli exclusion forces set a limit to how close the adsorbate can approach the surface.

6.4.1 Physisorption

Physisorption arises when a molecule or atom approaches a solid surface and is attracted only by simple electrical polarization forces. The latter are due to electron redistribution in the local region of the surface and within the adsorbate. The forces are almost always due to so-called van der Waals (vdW) forces and increase only slowly when the adsorbate approaches the surface in comparison to chemical forces which, when operating, lead to sharp onsets of attractive forces with approach of a molecule or atom. A detailed discussion of these forces will be encountered later in chapter 9. The simplest example of vdW forces is given by the attractions between the chemically inert rare gas molecules (e.g. Ar, etc) that give rise to condensed phases at sufficiently low temperatures.

Compared to a clean, bare surface in vacuum with some associated work function $\Delta\phi_o$, coverage of a surface by physisorption of an adsorbate typically will lead to a value of $\Delta\phi < \Delta\phi_o$. This effect is due to a 'pile up' of electron density at the adsorbate/solid interface, as shown in figure 6.25. The adsorbed species is driven closer to the surface by vdW forces until electron repulsion from the Pauli exclusion forces start to arise to repel the adsorbate.

6.4.2 Chemisorption

Chemisorption can involve a wide variety of adsorbate species, from atoms to complex molecules. A large number of fundamental studies have been done over the years with atom chemisorption on bare single crystal metal surfaces at the solid/vacuum interface. These studies have served to reveal the basic chemical types of interactions involved in chemisorption.

For example, electropositive atoms such as alkali metals donate electrons into empty surface states of the substrate which results in a decrease in the value of $\Delta\phi$ for the metal surface as would be expected from an increase in the electron density at the surface with a tendency for an electron to be more easily removed under an applied potential. In contrast, electronegative atoms such as O will accept electrons from the substrate surface making the substrate surface charge more positive with a tendency to make transfer of an electron into the vacuum require more work and lead to increase in $\Delta\phi$ values. Other examples include aromatic molecules, which leads to decreases in $\Delta\phi$ because of the donation of electrons from the aromatic pi-cloud. Some cases involving molecules with acceptor orbitals such as CO cause increases in $\Delta\phi$ and H atoms, which can chemically bond to many reactive transition metals to form surface hydrides by H–metal bonding cause increases in $\Delta\phi$. Taken together, these data indicate many types of surface chemical interactions. Overall, the data make a good case for using measurement of $\Delta\phi$ before and after chemisorption as a diagnostic for the bonding mechanisms. The schematic in figure 6.26 illustrates these effects. In many cases of rough surfaces in terms of the presence of steps and kinks, measurements of $\Delta\phi$ versus coverage can be a good diagnostic of when the most active defect sites are populated with adsorbate before adsorption occurs on less energetic surface sites.

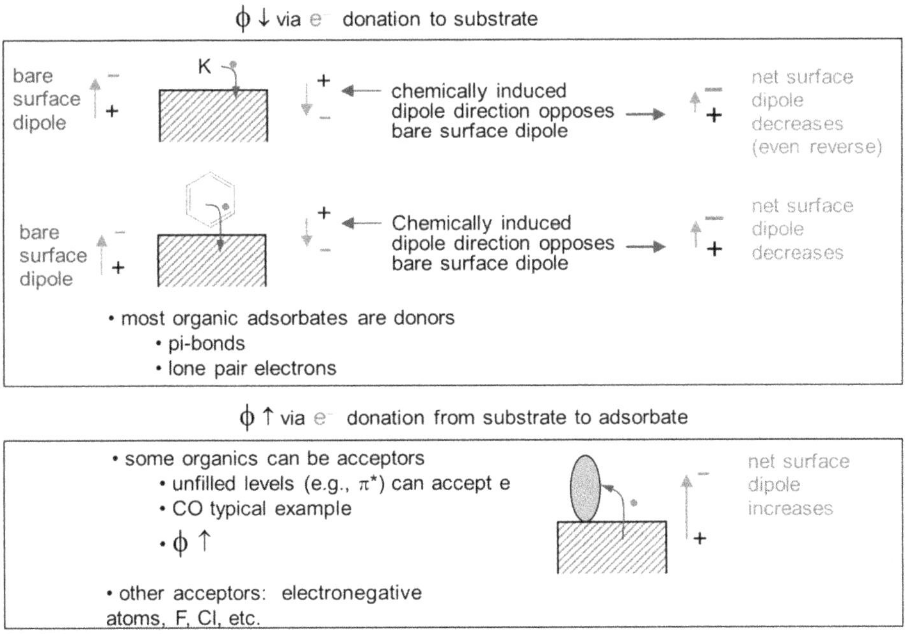

Figure 6.26. Depiction of the charge transfer directions of chemisorbed species at the solid/vacuum interface. Two general cases are: (a) adsorbates that direct electron charge into the substrate and (b) those that accept electrons. Typical donors are the obvious cases of highly electropositive alkali metal atoms but also include aromatic and related molecules that are rich in electron density due to multiple bonds such as benzene and conjugated species. Some acceptor molecules typically include those with unfilled pi-electron levels such as C≡O which can bond with electron rich donor species. The effect on the surface work function (here denoted as ϕ) is indicated in the diagrams.

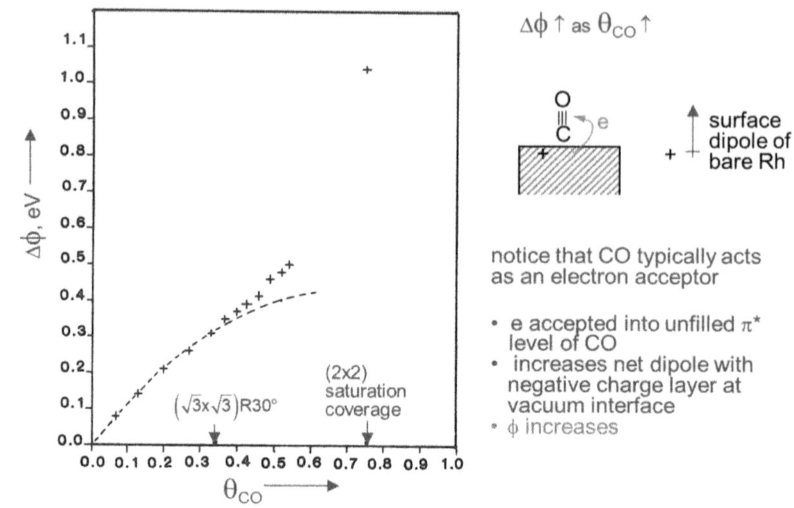

Figure 6.27. Plots of the change in work function versus CO coverage on a rhodium surface at the solid/vacuum interface. Notice that $\Delta\phi$ continues to increase as the coverage approaches 1.0. The dashed line is a prediction of what might be expected for the work function behavior at approach to high coverage. At low coverage a $(\sqrt{3}x\sqrt{3})R30^{\circ}$ exists. Near saturation coverage adsorbate structure shifts to a (2 × 2)(3 CO) overlayer. The schematic shows how electron donation to CO molecules will increase the surface dipole which results in a decrease in the work function.

A specific example of the effect of C≡O chemisorption on a rhodium surface is given in figure 6.27 where the change in work function is plotted against the CO coverage (given as Θ, where 1 is full coverage). In this case CO is an acceptor with donation from the surface state electrons in the metal. This changes the surface dipole which is reflected in the shift of the work function.

Finally, we note that for studies of electronic conduction in nano-objects such as molecules and nanoparticles, the method for making contacts involves chemisorption to the solid state contact. This allows maximum interaction of the nano-object electron states and the contact states. Examples of this commonly involve organo-sulfur groups on gold. In these cases, the chemisorption bonding is critical to the conduction efficiency.

Thought questions:

1. Draw the electron 'dot' (Lewis) structure of CO and show why the C atom might be negative and thus be a σ-electron donor to unfilled states in the metal surface.
2. Saturation coverage has a (2 × 2)−3CO structure. Starting with a $(\sqrt{3}x\sqrt{3})R30^{\circ}$ structure, why might the change in adsorbate structure cause the continuing increase in the work function rather than the predicted one (dashed line)?

6.5 Measurement of surface region electronic and chemical characteristics

The surface region of a solid has slightly shifted electronic states compared to the deeper region of the bulk, as would be expected from the abrupt termination of the solid lattice. Clearly the existence of inherent electric dipoles and electron spill-out clouds at the surface underlie some differences in electronic behavior with the bulk solid. For the chemistry aspect you can think of the roles of molecular frontier orbitals in dictating the chemical reactivity of molecules. Surface region electronic properties are fundamental to important behavior such as charge transport across interfaces and surface chemistry, so we need to dig deeply to understand these behaviors.

In chapter 1 we made comparisons of properties of different sized nanoparticles as a function of surface/volume (S/V) ratios. Many of the differences were explained quite well by looking at the simple missing atom model which focused on the effects of a missing outer layer (or a few layers) of atoms at the surface of the nanoparticle. The main structural effect is the absence of attractive bonding forces at the outer terminal layer of atoms that would otherwise come from the missing layer above, which results in compression of the outer three layers or so. This was seen to affect properties such as melting points, for example. With a change in the packing density of this outer surface region one also can expect shifts in the associated electronic states. One example was revealed by shifts in the optical spectrum of CdS nanoparticles as a function of size (figure 1.11). This can be understood in a simple way by assuming that the compression of the surface layer has an analogous relationship to the increased splitting of energy levels and changes in band gaps as a function of assembling atoms into a dense solid, as shown in figure 6.12. Overall, however, the missing atom model only weakly applies to understanding changes in the electronic properties. In particular, electrons are confined by an electric potential force field from local nuclei, with their properties distributed like a fluid or cloud and valid predictions require a quantum mechanical approach.

Other than doing quantum chemical calculations, which are outside the scope of this book, the most straightforward approach is to make experimental measurements of the surface region characteristics. In particular, we now discuss several common methods that analyse the chemistry and electronic properties of solid surfaces: primarily those that can non-destructively characterize electronic states, including mapping states from core level to valence band, and chemical species. In these methods, measurement involves interrogating the surface by impact of particles, ranging from electrons to photons. The response of the impact is ejection of the same type of particle or a different type, but with different energies or momenta, and this change is analysed for surface region characteristics. It is most useful when the method does not involve destruction of the sample, e.g. by sputtering or etching material from the surface region. The most important methods involve ejection of electrons under non-destructive conditions. We will focus on these methods.

A summary of methods to stimulate emission of electron from solid surfaces is shown in figure 6.28. Of the methods shown only photoemission and electron impact are useful for surface analysis.

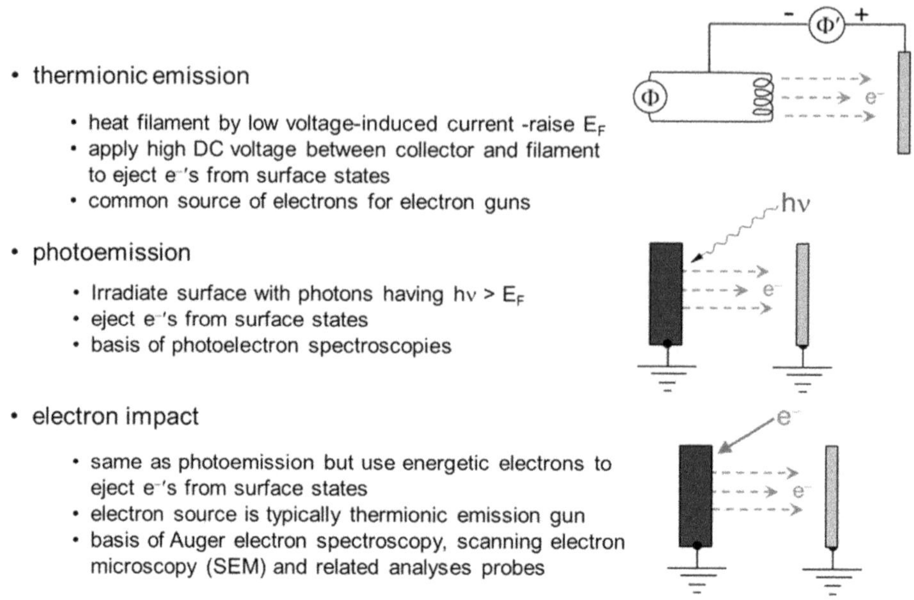

Figure 6.28. Three main ways to cause ejection of electrons from a solid surface. (A) emission from a heated sample raised to a temperature which causes electrons in the highest energy levels to be excited into the vacuum state under applied potential. This method is the basis of traditional vacuum electronic tubes used for electronic devices, however, because of the high temperatures of the emitting surface it is not useful for surface analysis. On the other hand, it does provide a highly controllable method for providing electron beams for impact on another surface to be studied. (B) Emission by a flux of photons, typically selected from convenient wavelengths in the UV or x-ray region where energies are sufficient to cause electron emission from a variety of energy states in the solid, from core to valence band levels. This method is the basis for several types of surface spectroscopies. (C) Electron impact in which a beam of energetic electrons impacts the sample and creates electron emission which under the optimum conditions provides a basis for surface analysis. Electron beams offer the advantage that the energies, flux and sharpness can be controlled by electron optics methods (see A).

Since the main basis for surface region spectroscopies all involve electron emission, and sometimes injections, it is useful to take a quick look at the mechanisms involved in electron scattering processes within solids. Upon impingement of a beam of primary electrons upon a solid surface, the electrons start a cascading series of scattering processes off the electron clouds around core ions of the solid. The scattering branches out in different directions with deeper and deeper penetration, as shown in figure 6.29. The main point to be made here is that the process of secondary ion scattering also is a fundamental aspect of the electron emission in photoemission spectroscopy and needs to be considered in detail for any surface spectroscopy which depends on analysing the energies of electrons emitted from the solid surface.

The eventual result of the internal scattering is that the electrons are slowed and eventually leak out through the ground on the sample. The average depth to which the electrons penetrate before colliding away and losing energy is important in an inverse way when an energetic electron is generated deep within the solid and one

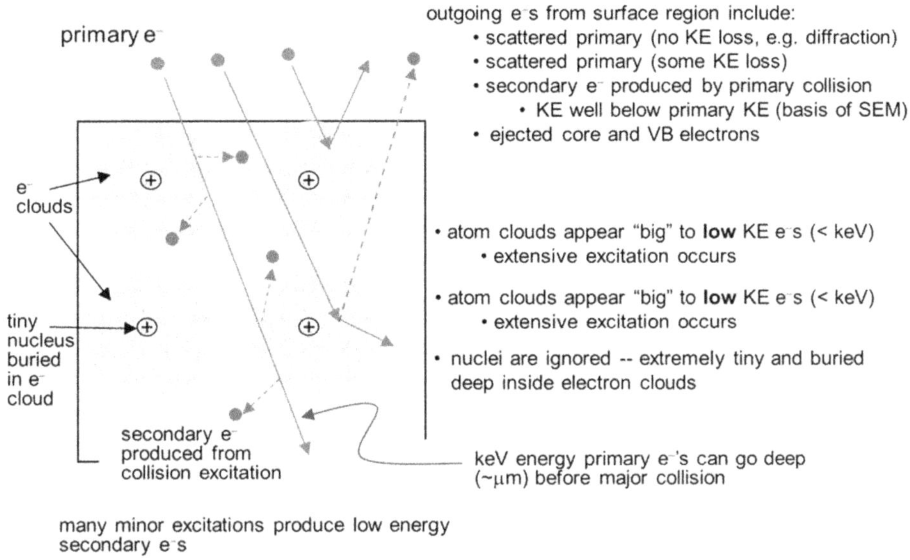

Figure 6.29. Schematic illustration of electron scattering processes within a solid after impact from a primary electron beam for a solid electrically grounded to avoid net charge build up. When the primary electron has keV energies or higher, penetration can go deep into the solid many micrometers. After a number of collisions each of the secondary scattered electrons dissipate energy heating up the solid and are eventually captured at the grounding electrode.

Table 6.4. Representative values of inelastic mean free paths for different materials for an x-ray energy of 1 keV. (Data from [6]).

Material	IMFP (nm)	Typical range of values (nm)
Graphite	2	1–2.5
Diamond	1.5	1–2.2
Aluminum (Al)	2.6	2.4–2.9
Silicon (Si)	2.5	2.4–2.6
Gold (Au)	3.0	2.9–3.2
Silver (Ag)	2.7	2.6–2.8
Copper (Cu)	2.4	2.2–2.5
SiO_2 (quartz)	3.5	3.2–3.6
Al_2O_3 (sapphire)	3.8	3.5–4.0

can ask what the probability is that the electron will escape the surface into vacuum. This depth is often termed the electron escape depth. Since the actual process involves the average or mean free path to escape it is more properly termed the inelastic mean free path (IMFP) which is the proper term for the probability an electron with a specific kinetic energy will traverse a specific distance in the solid. Representative values are given in table 6.4. The aspect of secondary electron

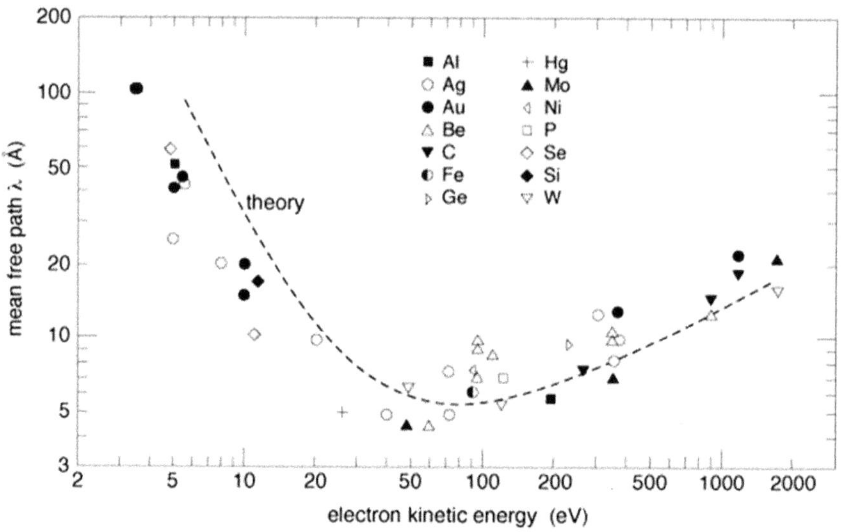

Figure 6.30. A widely useful correlation of the energies of the probabilities of an electron scattering within a solid to escape from the surface as a function of the electron depth and the specific kinetic energy. At 50–100 eV the electron has an associated most probable escape depth of 5–100 Å. Depths below this value are associated with a low probability of the electron escaping into the vacuum and thus being available for analysis. Reprinted from [3], Copyright (2015), with permission from Elsevier.

scattering and IMFP correlations with electron energy and the specific solid material has been studied extensively and the results often summarized in a 'universal' correlation plot as shown in figure 6.30. Theoretical methods and equations have been developed and are available for accurate calculations. These data are extremely useful for analysing electron surface spectroscopy data as will be seen in the next sections.

6.6 Electron spectroscopies—tools for surface/interface characterization

6.6.1 Photoemission spectroscopy, overall summary

The major tool for surface analysis at the solid/vacuum interfaces is photoemission spectroscopy with the impact photons ranging from ultraviolet to x-ray frequencies, which these frequencies ranging from soft x-rays (below 5–10 keV) to hard x-rays (>5–10 keV and sometimes beyond for certain types of analyses). Most chemical and electronic state surface analysis is done with soft x-rays, which have energies that can probe core level and lower lying valence band levels. For probing higher energy valence bands closer to the Fermi level energy, UV photons are used.

Table 6.5 below gives a summary of the electron spectroscopy methods. The methods are characterized by photon energies, electron energy states excited, the parameters measured (intensity of the photo-emitted electrons or the flux of emitted electrons as a function of the photoelectron kinetic energy) and gives the common abbreviations for the technique.

Table 6.5. Summary of common photo emission spectroscopies as a function of the input photon energy, the energy states probed, the type of information gathered and the abbreviations for the methods.

hv		e^- excitation to vacuum state	Measurement	Abbreviation
UV	Fixed	VB $\rightarrow \infty$	Intensity versus photoelectron KE	UPS
X-ray	Fixed	Core $\rightarrow \infty$	n_e versus photoelectron KE	XPS
X-ray	Fixed	VB $\rightarrow \infty$	n_e versus photoelectron KE	VBXPS
X-ray	Scan	Core + VB $\rightarrow \infty$	n_e versus *hv*	NEXAFS

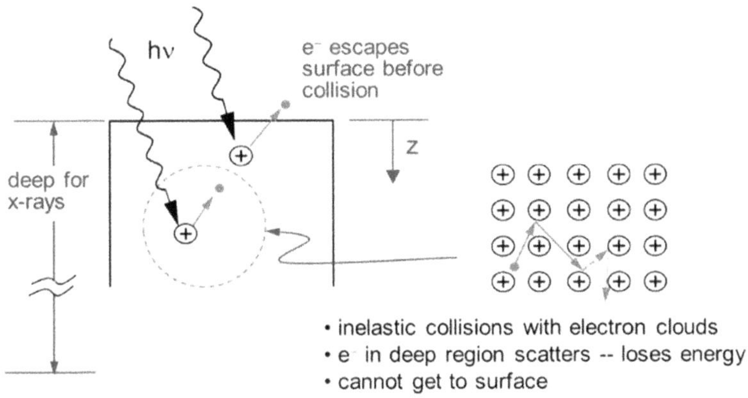

Figure 6.31. Illustration of the role of the IMFP in determining what excited electrons make it past the surface as determined by inelastic scattering processes.

Generation of the impinging photons is generally done with standard sources such as UV lamps and electron irradiated Al or Cu anodes. For highly controlled beams over a range of energies, synchrotron sources are used at facilities built for this purpose at different locations around the world. The methods involving scanning input photon energies require the use of beams at a synchrotron facility and as such are more difficult to employ. Most methods involve fixed photon energy [4], and we discuss these first.

6.6.2 XPS, UPS and VBXPS

In all these methods high energy photons impinge on the sample surface. X-rays penetrate and have sufficient energy to excite electrons from their energy levels in the sample. The excited electrons scatter through the sample and some fraction is directed towards the surface. This is where the character of the inelastic mean free path (often termed α) becomes important. Only a fraction of the excited electrons escapes to the surface, as depicted in figure 6.31 to be analysed, and this depends on

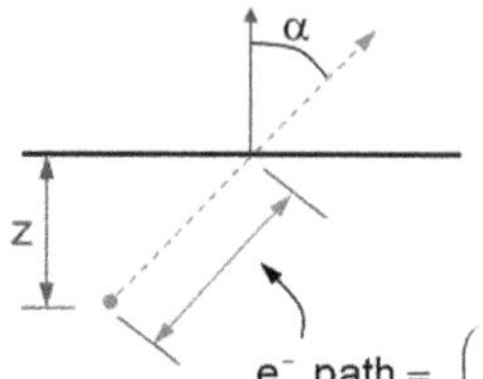

using I = intensity of the photoelectron count (n_e) at the detector

$$\frac{I(z)}{I_o} = \frac{e^{-}\text{ emission from depth } z}{e^{-}\text{ emission from } z = 0} = e^{\left[-\frac{1}{\lambda}\left(\frac{z}{\cos\alpha}\right)\right]}$$

$$e^{-}\text{ path} = \left(\frac{z}{\cos\alpha}\right)$$

example: for $\alpha = 0$, $\frac{I(z)}{I_o} = e^{\left(-\frac{z}{\lambda}\right)}$ and $\frac{I(\lambda)}{I_o} = \frac{1}{e}$ or ~37% attenuation

Figure 6.32. Schematic of the geometry of the take-off angle for capture of electrons arriving at the surface from different depths z on average. The formula gives the probable intensity I of photoelectrons coming from a depth of z as a function of the takeoff angle α and the IMFP (λ). In the example, setting z = IMFP for this solid and electron energy, there is a 37% attenuation in the count of escaping electrons at angle α compared to normal to the surface.

the energy of the excited electrons, as show earlier in figure 6.30. A further important aspect is the angle from the surface normal used to orient the detector electron optics. This limits the analysed electrons to those that follow this trajectory. This geometry is shown in figure 6.32. Shallow takeoff angles associate with long trajectories of the electrons in the sample that escape the surface and steep, near normal, angles associate with short trajectories. Accordingly, shallow angles probe escaping electrons very near the surface and thus probe the shallower surface region states. The controlled variation of takeoff angles allows probing different depth regions of the surface and is a standard variation used widely in practice often termed angle resolved photoemission (ARPES).

Once photoelectrons are collected and analysed for their number at each kinetic energy (KE) interval [$n(E)$], the binding energy of the electron state probed is calculated from the conservation of energy equation, as shown in figure 6.33. Notice that the work function of the surface appears in the calculation. In fact, using photoelectron spectroscopy one can readily measure the work function.

A very instructive example for the information that can be obtained from an XPS spectrum is shown in figure 6.34. The figure shows several different energy states probed: core level, valence band levels, Auger electron peaks (see next section) and the work function offset from the KE = 0 photoelectron (the vacuum energy of the electron with no kinetic energy).

Further analyses can be done using different photon energies in two ways: (i) looking at the depth dependence on photon energy in figure 6.30 and table 6.4, one can choose the depth region of interest and (ii) for a higher resolution spectrum one can use UV photons since these will probe the valence band states in much greater detail.

One of the main values of photoemission spectroscopy for surface analysis is using the detailed binding energy (BE) peak positions of different elements as a method of identifying their chemical states at the surface and in the surface region.

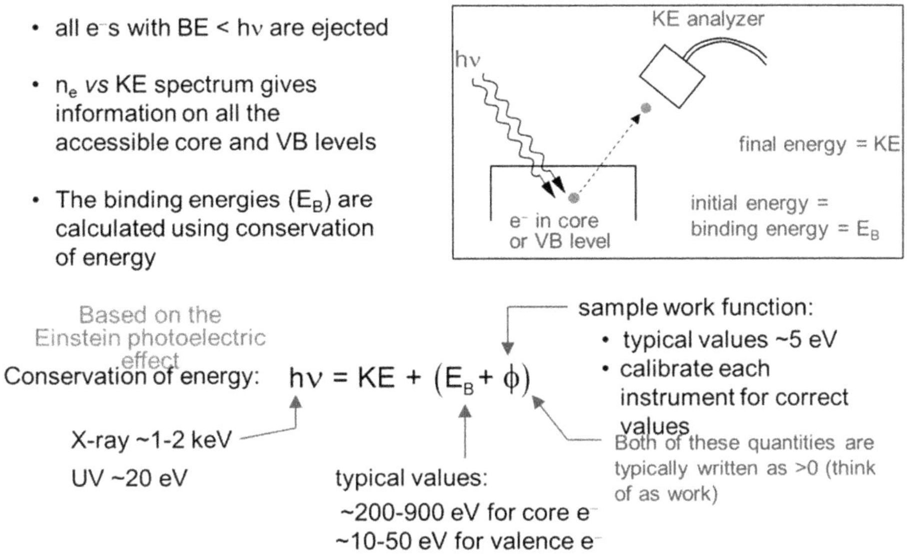

Figure 6.33. The binding energy of the state from which the photoelectron was ejected is done using conservation of energy in the formula above. The KE of the emitted electron must be the sum of the binding energy, KE, and the work function for the solid/vacuum surface. Typical values are shown in the figure.

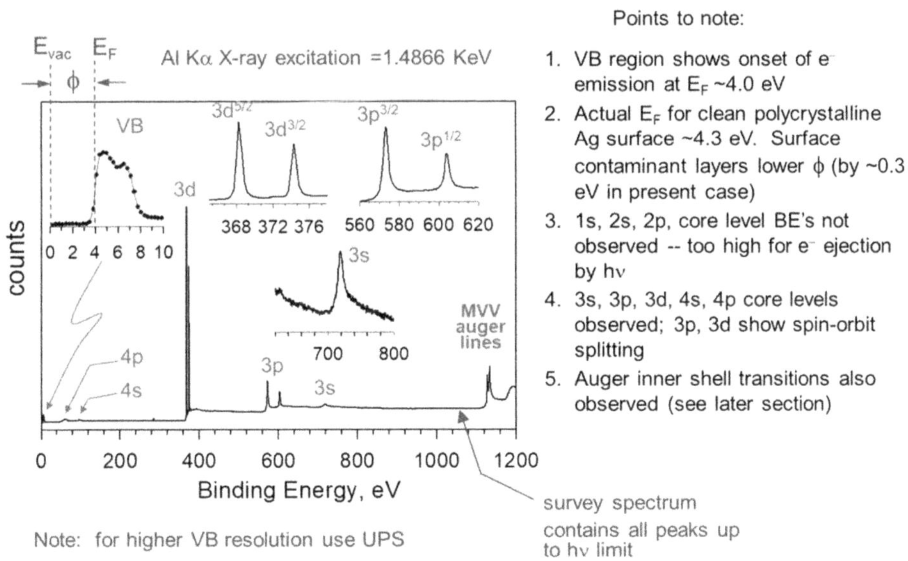

Figure 6.34. Example photoemission spectra for a Ag surface, which shows the various types of electron states in the solid that can be probed. Ag is often used as a calibration for setting up XPS instruments and these peaks are standard for this process. In the upper left figure, notice the onset of the valence band peak, set off from zero energy by the work function. Also seen there are the low binding energy 4p and 4s core level peaks. The spectrum also shows Auger peaks (see section 6.6.4).

Table 6.6. Selected C (1s) chemical binding energy shifts.

Functional group	Structure	BE, eV (± 2 eV)
Hydrocarbon	C–H, C–C	285.0
Amine	C–N	286.0
Alcohol, ether	C–OH, C–O–C	286.5
Cl bound to C	C–Cl	286.5
F bound to C	C–F	287.8
Carbonyl	C=O	288.0
Amide	N–C=O	288.2
Acid, ester	O–C=O	289.0
Urea		289.0
Carbamate		289.6
Carbonate		290.3
2F bound to C	$-CH_2-CF_2-$	290.6
C in PTFE	$-CF_2-CF_2-$	292.0
3F bound to C	$-CF_3$	293–294

This is a common approach used for a variety of materials ranging across metals and their compounds, semiconductors, insulators, and organic materials such as polymers. In general, each element has distinct BE values for core electrons: e.g. the C 1s, O 1s, N 1s, and Si 2p BE peaks are all measurably different and can be used for identification of elemental content in surface region. Careful measurement of peak areas at each BE are done. The observed peak areas are compared against reference material surfaces; then after corrections for the known cross sections for excitation of the elements involved, the areas are used to compute the relative composition of the surface region.

The chemical states of elements can be inferred from chemical shifts (as in nuclear magnetic spectroscopy). Local e^- density around each element in a solid affects the BE of a given transition. The observed BE values are compared against reference materials. The shifts are used to infer the chemical states and compositions of the analysed material surface. With careful analysis one can calculate accurate (±5% or so) chemical composition ratios.

Representative C(1s) BE shifts are shown in table 6.6.

6.6.3 NEXAFS—variable x-ray energy spectroscopy

Although near edge absorption fine structure spectroscopy can be a very incisive surface spectroscopy it is not widely used because variable x-ray photon energy sources are needed to scan the specimen, and these are only available at special synchrotron facilities. A diagram of the energy levels involved in the photon excitation process and the associated electron emission is given in figure 6.35. The figure summarizes some of the requirements and associated details needed to obtain this spectrum. An example spectrum is given in figure 6.36 of a self-assembled

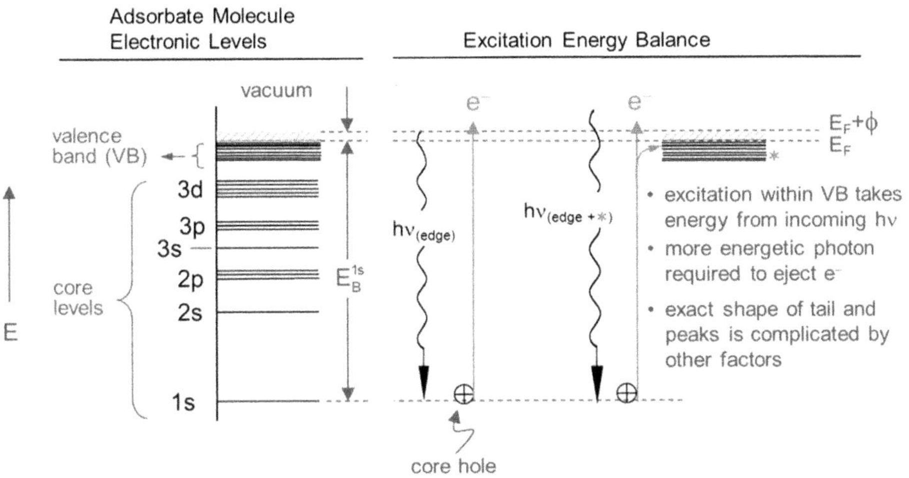

Figure 6.35. An energy diagram, photon absorption processes and electron emission schematic for a NEXAFS experiment. The left-hand side shows the core and valence level energy states of the hypothetical sample. Once the photon is absorbed an electron is ejected to form a core hole. In the diagram the absorption and emission are shown involving the lowest lying 1s state. Two possibilities are shown: (i) an electron is ejected out into the vacuum while (ii) the input *hν* energy is partly used to excite a VB electron into a high lying VB level.

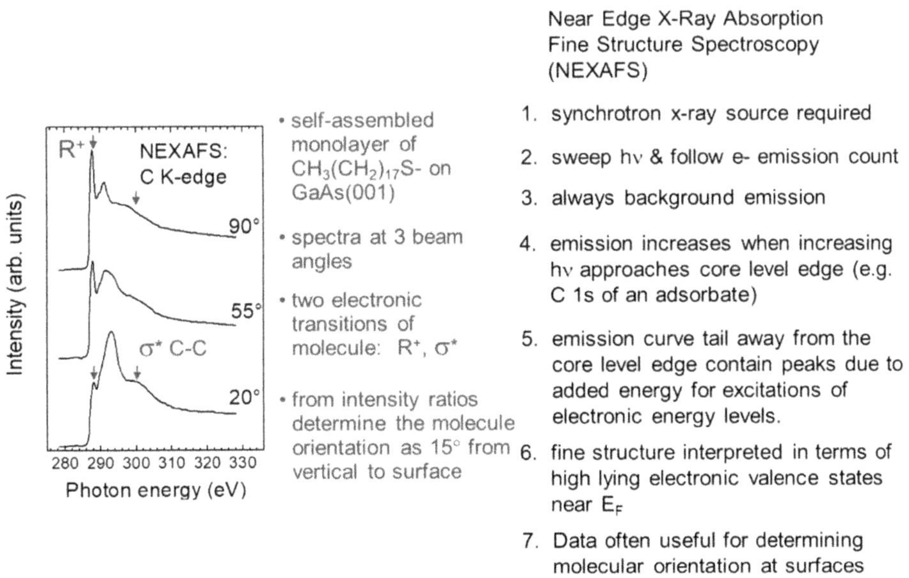

Figure 6.36. NEXAFS spectra of a SAM of octadecane thiol on a GaAs(001) surface taken at different takeoff angles. The tailing energy edge slopes down from a sharp *R** high energy excitation at 285 eV then further reveals sigma energy levels of the C atoms of the molecule. The use of variable angles allows some interpretations of the tilt angle of the C–C–C molecular axis on the GaAs surface.

Table 6.7. Summary of photon in–electron out spectroscopy methods and their uses.

Method	Input photon	Primary capabilities	Disadvantages
XPS	X-ray (∼1–10 keV), fixed energy (monochromatic; anode or synchrotron)	• Surface atomic and chemical composition • Core-level spectroscopy • Variable takeoff angle for depth profile analysis	• Possible x-ray damage for organics • Sample charging for insulators
NEXAFS	X-ray (∼1–10 keV), tunable energy (synchrotron)	• VB spectroscopy • Work function and Fermi edge measurements	• Same as XPS • Synchrotron required
UPS and VBXPS	UV (∼10–100 eV), (He arc lamp or synchrotron)	• VB spectroscopy • Work function and Fermi edge measurements	• Possible UV damage for organics • Sample charging for insulators • Energy resolution is poor for VBXPS

monolayer (SAM) of octadecane thiol on a GaAs(001) surface. The spectrum shows a sweep of photon energies from 280–340 eV.

The lowest energy x-ray photons excite electrons from a K (s electron) shell with appearance of excitation from different K shell states of the C atoms in the alkyl chains of the SAM. With increasing energy electrons lose energy by causing excitations within VB states, as shown in figure 6.33. With decreasing photon energy other C atom peaks appear. Since these peaks have intensities that change with the orientation of the chains, using changing take-off angles these peaks can be diagnostic for the average chain orientation. In the example spectrum the chain orientations were determined to be 18° from normal incidence.

A summary of the use of different types of photon in–electron out techniques is shown in table 6.7. The capabilities and liabilities for each technique are noted.

6.6.4 Core hole spectroscopy—the Auger electron process

Auger electron spectroscopy (AES) is a rather unique spectroscopy which involves inner electron state relaxations synchronized following emission of an electron by a photon or electron beam. Once the core hole is created, a higher energy electron relaxes to fill that hole and, to conserve energy, another electron is emitted (see figure 6.37). The energy of the emitted electron is independent of the energy or form of the incident radiation, as long as a core hole is created.

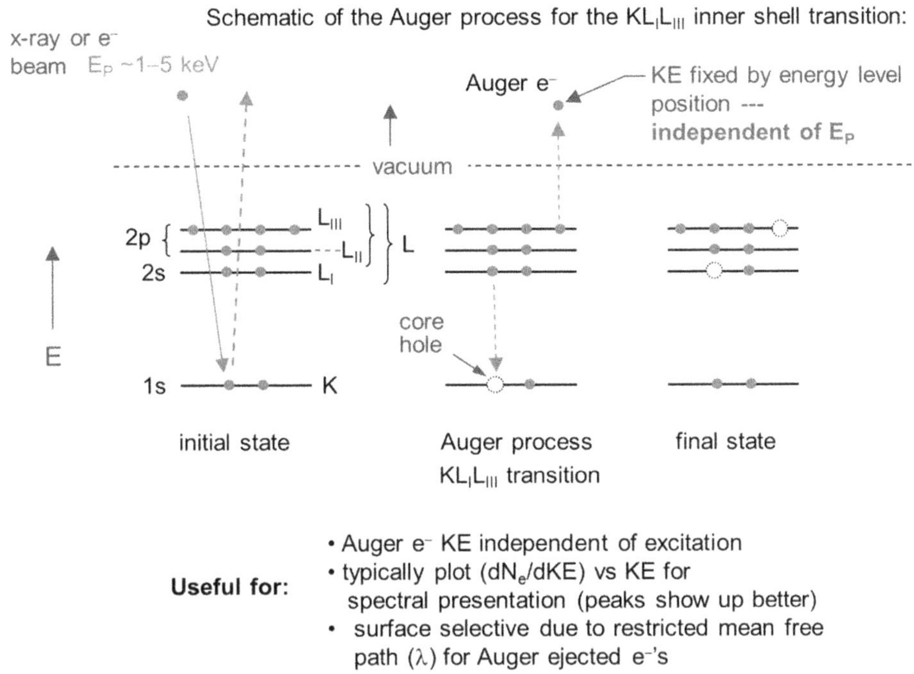

Figure 6.37. Schematic diagram of Auger electron spectroscopy involving states in the KLL energy levels. Deep level electrons from the 1 s K shell of the sample surface region are excited into the vacuum by incident electrons or x-rays with energies in the 1–5 keV region to create a core hole. Electrons from a level in the L or M shells (only L transitions are shown) can decay into the K shell states with the synchronous emission of an L shell into vacuum (the Auger electron which is detected) and creation of a final state with two empty states in the L shell.

An example of Auger peaks with x-ray irradiation was shown earlier in the photoemission spectrum in figure 6.34 (look at the MVV peak in that figure). In that case the Auger peak appeared with the x-ray energy used for excitation but was not the main information needed from the spectrum. By using high energy photons or electrons the Auger spectrum can be made to be the dominant feature. Often high energy electron beams are used to generate Auger electron spectra since these are more convenient to generate and offer easily tunable energy compared to x-rays. Further, e-beams can be focused to ~20–50 nm diameters while it is extremely challenging to focus x-rays to these dimensions. This advantage allows spatial imagining of surface for composition mapping, etc, by rastering the focused electron beam across the surface.

The intensity of Auger peaks is a function of the specific element being probed; light atoms have higher probability of Auger emission than heavy atoms. This is in general contrast to other surface spectroscopies which are often limited to elements heavier than the first few in the periodic table. AES is often used to provide maps of the composition of elements on a surface. This contrasts with standard secondary emission scanning microscopy (SEM) in which the electron beam is rastered across the surface and the scattered electrons are used to simply map the general topographical features on the surface. In SEM the incident electrons generate x-rays, whose energy is characteristic of the elements. Since these x-rays are detected from as deeply as 1 μm below the surface, this technique is not a very surface sensitive technique.

6.6.5 Inelastic electron surface/interface scattering spectroscopy—high resolution electron energy loss spectroscopy (HREELS)

In this spectroscopy a controlled narrow energy electron beam of energy E_P is directed at a sample surface where collision occurs. The scattered electrons that arise from a near specular reflection (such as reflection from a mirror) are collected and analysed for their energies. The energy of the electrons in this outgoing beam and compared to the input energy ($E_P - E_{out}$). Some fraction of the scattered electrons can lose energy in specific collisions with adsorbates that excite vibrational modes. These 'loss' peaks are interpreted in terms of the adsorbate structures. Further, since the electric fields at the surface for metals are oriented perpendicular to the surface the loss peak intensities can be interpreted in terms of the adsorbate orientation, similar to how infrared spectroscopy is used. The spectroscopy process is represented in figure 6.38.

A summary of the use of different types of electron in–electron out techniques is shown in table 6.8. Note that for a given primary electron KE range, several

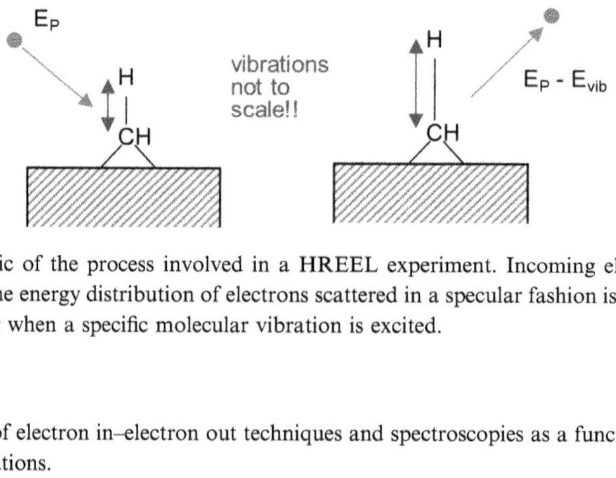

Figure 6.38. Schematic of the process involved in a HREEL experiment. Incoming electrons of energy E_P impact the surface. The energy distribution of electrons scattered in a specular fashion is analysed. Loss peaks with $E_P - E_{vib}$ appear when a specific molecular vibration is excited.

Table 6.8. Summary of electron in–electron out techniques and spectroscopies as a function of energy ranges and associated applications.

Emitted e^- KE range, eV	Technique and source of emitted electrons	Application
$\sim 0 \rightarrow 50$	SEM—secondary excitation with new emitted e-s	Imaging of surface topography

(*Continued*)

Table 6.8. (*Continued*)

Emitted e^- KE range, eV	Technique and source of emitted electrons	Application
$\sim 100 \rightarrow 300$	AES—discrete core-level excitation + re-emission via Auger process (+ x-ray excitation also)	Chemical and atomic composition
$\sim (E_P$—$50) \rightarrow E_P$	EELS (electron energy loss spectroscopy)	Characterization of electronic states; states probed depend on input energy
E_P—δ ($\delta \sim$ few eV)	HREELS—vibrational transitions	Vibrational spectra
Emitted at E_P	LEED—elastic scattering (no KE loss)—diffraction	Surface ordering

scattering and spectroscopic processes can occur simultaneously, and E_p is chosen to avoid unwanted processes, e.g. diffuse electron scattering. What may look like noise or side effects when doing one specific type of measurement may be another process that can provide useful information.

6.6.6 Inelastic electron tunneling spectroscopy (IETS)

This type of experiment is based on the same principles as scanning tunneling microscopy (STM) that was described in detail in chapter 3. In the case of IETS the interest is in using electrons that tunnel through a molecule or nano-object (particle or nanowire) in contact between two macroscopic contacts, typically gold nanoprobes, as a probe of the energy states intrinsic to the object. The quantum states typically involve electronic energy states and vibrational mode states. The mechanism of IETS involves interactions with the tunneling electrons at applied potentials swept over a range of values. When the applied potential approaches a quantum state excitation energy, the electron can scatter from the state to excite it with a corresponding loss in kinetic energy. The excitation of the nano-object state opens a new channel for tunneling with a corresponding increase in current which then continues for all higher applied potential values.

A schematic of the experimental data is shown in figure 6.39. The experiment typically follows the pattern described below and as shown in figure 6.40:

1. Sweep (increase) V across junction.
2. Measure current (I) of electrons that tunnel across the gap and through the nanowired-in nano-object.
3. When the electron energy matches the energy E_{vib} a small fraction of electrons excite a vibration and lose E_{vib}.
4. The excitation opens new tunneling channels (like the fifth lane in a four-lane highway) for the electrons that lost energy.
5. The tunneling current rises slightly, but measurably.
6. Extra channels remain open thereafter as V increases further.

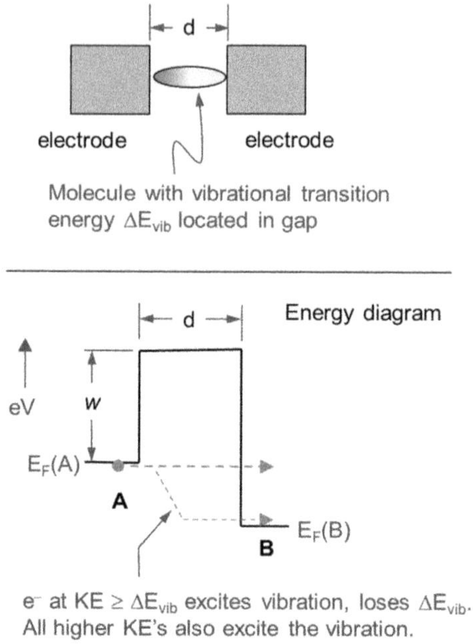

Figure 6.39. Diagram of the configuration of a molecule bonded between two electrodes for an IETS experiment. The electrical potentials are kept as low as possible to allow tunneling and avoid any damage to the molecule. The interest here is in matching the energy of the tunneling electrons to the molecular electronic and vibrational mode energies to produce a tunneling current with peaks appearing at the excitation energies. The lower diagram illustrates the energetics of the tunneling process.

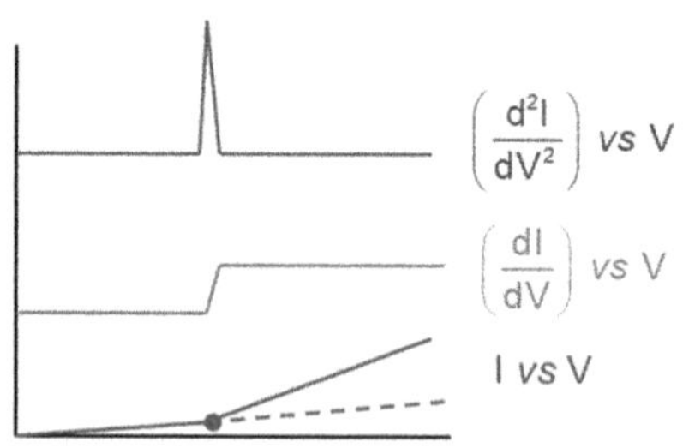

Figure 6.40. Hypothetical current (I) versus applied voltage (V) curves and derivatives for an IETS experiment. As the potential reaches an energy for an excitation the current increases and continues to increase thereafter. To bring out this onset a plot of the second derivative shows it as a peak.

7. Plot the second derivative of the current with respect to voltage to get the excitation peak.

There are several types of junction configurations that have been used which involve bundles of molecules arranged between counter electrodes:

1. Large area contacts with SAMs between the contacts. These have included superconducting arrangements with vapor deposited Pb top electrodes with

the measurements done below the critical temperature (T_C) for the onset of superconduction in the Pb electrode.
2. SAMs arranged between crossed Au nanowires with the electrode contacts configured by contact along the nanowires.
3. SAMs deposited between vertical arranged nanowire contacts.
4. SAMs deposited in nanopores with a Au bottom contact and a vapor-deposited top Au contact.

The ideal experiment with a single molecule set between two opposing electrodes has been challenging and awaits a favorable experimental that has less noise in the measurements, e.g. STM measurements of vibrations in a single molecule is extremely challenging, even at the lowest cryogenic temperatures. Note though that this field is changing rapidly as workers have been progressively able to fabricate nano gaps from conductors such as Au that allow thiol-based SAMs to be placed between the contacts.

An example of above experimental configuration [3] is given in figure 6.41. The second derivative of the current shows peaks characteristic of the molecular

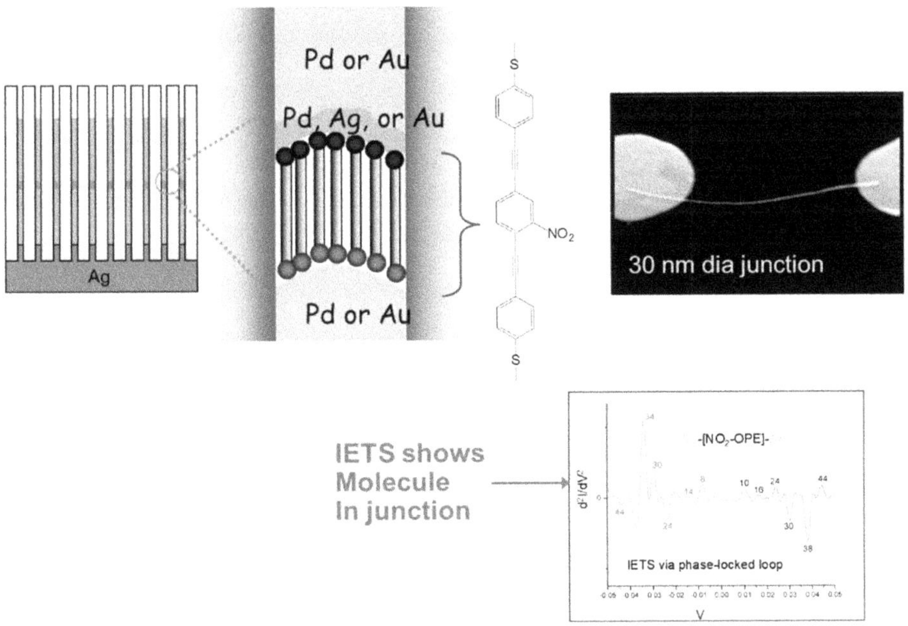

Figure 6.41. An example of IETS using a bundle of molecules that have been configured to be contacted between Pd and/or Au contacts in a nanowire configuration [5]. The nanowire was laid across two macroscopic contacts to form a ~30 nm junction. The molecule in the figure is a dithiol which allowed bonding to the opposing contacts, either Pd or Au. The voltage scan at low temperature was done to obtain the second derivative of current versus voltage, shown in at bottom-right. The peaks that appear correspond the known molecular vibrations. Further examination of the spectra revealed modes that indicate geometric changes in the molecule with applied voltage, which indicates a switching behavior in the junction. Reprinted with permission from [5]. Copyright (2005) American Chemical Society.

vibrations which were used to interpret the changes in the molecular geometry during excitation [4].

Overall, these types of experiments point to the end goal of studying the electrical properties of single molecules and other nano-objects, which involve quantum states of these nano-objects and their environment.

Chapter 6 Problems

1. When K is deposited on the (111) face of Rh the work function of the surface decreases remarkably. At 20% monolayer coverage, the work function change attains its minimum value of −1.8 eV. If the Rh−K interatomic distance is 0.12 nm, what is the charge transfer at the adsorption site of the alkali metal?
2. What is the Gouy–Chapman solid/liquid double layer? Discuss the experimental evidence for its existence.
3. Why is the energy of the Auger electron independent of the energy of the incident electron? Do you expect its relative intensity to be independent of the incident energy? Why or why not?
4. Discuss some of the challenges of inelastic electron tunneling spectroscopy through a single molecule. Find an experiment in the literature where this was done and discuss it.
5. Assume octanedecane thiol is self-assembled on an atomically smooth surface. If the S 2p and C 1 s electrons are collected, do you expect the ratio of the S 2p photoelectron intensity to C 1 s photoelectron intensity to be greater at electron collection angles normal to the surface or at glancing angles? Derive an expression for the relative S 2p and C 1 s intensities as a function of collection angle.

Further reading

General reference texts

1. Woodruff D P 2016 *Modern Techniques of Surface Science* (Cambridge University Press). An intermediate to advanced level text describing characterization of surfaces.
2. Somorjai G A and Li Y 2010 *Introduction to Surface Chemistry and Catalysis*, 2nd edn (Wiley). This text provides general advanced undergraduate and beginning graduate level material and chapters 5 and 6 are focused primarily on the electrical properties of surfaces and the surface chemical bond.

Specialized reference article

1. Kosmulski M 2009 Compilation of PZC and IEP of sparingly soluble metal oxides and hydroxides from literature *Adv. Coll. Int. Sci.* **152** 14–25 10.1016/j.cis.2009.08.003

References

[1] Kaminsky M 1965 *Atomic and Ionic Impact at Metal Surfaces* (Academic)

[2] Protpopov O D *et al* 1966 Emission parameters of tantalum and molybdenum single crystals *Sov. Phys. Solid State* **8** 909

[3] Adams F and Barbante C 2015 Spatially confined analysis *Comprehensive Analytical Chemistry***vol 69** ed F Adams and C Barbante (Elsevier) ch 2 pp 29–66

[4] Stevie F A and Donley C L 2020 Introduction to x-ray photoelectron spectroscopy *J. Vac. Sci. Technol.* A***38*** *063204*

[5] Cai L T, Skulason H, Kushmerick J G, Pollack S K, Naciri J, Shashidhar R, Allara D A, Mallouk T E and Mayer T S 2004 Nanowire-based molecular monolayer junctions: synthesis, assembly, and electrical characterization *J. Phys. Chem.* B **108** 2827–32

[6] Powell C J and Jablonski A 2010 *NIST Electron Inelastic-Mean-Free-Path Database* (National Institute of Standards and Technology)

Chapter 7

Surface chemical bonding

In this chapter the details of chemisorption to surfaces are discussed. We go from van der Waals attraction of inert gases to decomposition of ethylene on metal surfaces. The relative reactivities of different metal surface are compared. Reactions with metal, oxide, semiconductor and organic surfaces are considered. The use of various characterization techniques to monitor the mechanism of these chemical reactions is also reviewed.

7.1 Introduction

The surface of an object in vacuum is occupied by an outermost cloud of distributed electron density with associated electronic states and local electrostatic structure—this layer controls chemical bonding of adsorbate species.

- For a pure solid, adsorbate bonding is controlled by the surface electrons which are distributed according to the structure of the surface atoms.
- For a functionalized or chemically modified surface (such as native oxide-covered silicon or silanol-covered SiO_2), adsorbate bonding is primarily controlled or mediated by the electronic and electrostatic character arising from the surface groups.

A molecule might bond weakly with a surface. This is termed physisorption, and takes place through van der Waals bonding, weak hydrogen bonding, dipole or weak donor–acceptor interaction. Typical bonding energies, $U_{\rm ads}$ might range from 40 to 60 kJ mol^{-1}. In the case of weak bonding the surface and molecule are relatively unperturbed.

A molecule might chemisorb with a surface through covalent, ionic, or strong donor–acceptor bonding. This bonding will generally be greater than 60 kJ mol^{-1}. *The bonding may be dissociative or non-dissociative.* Often the bonding includes molecular distortion, bond breaking, and surface reconstruction.

doi:10.1088/978-0-7503-4916-1ch7

Before bonding the surface may be 'clean' as in ultrahigh vacuum conditions. The surface may be an abrupt termination of the bulk material or may be reconstructed. In this case the adsorption will directly bond to the substrate material; this is the most reactive form of the surface.

On the other hand, the surface may be exposed to the ambient of another exposure. In this case, there will be an overlayer present; for example, and the surface may be oxidized, the oxides may have reacted with H_2O to form –OH. Other contamination may also be present. These previously exposed surfaces generally have lower reactivity than clean surfaces. Also, the reaction chemistry with the exposed surface will be different than that seen with clean surface.

Typical metals such as Fe, Ni, and Al readily form oxides. The oxides will further react with H_2O to form surface –OH groups. Precious metals such as Au will not form oxides but are readily contaminated with hydrocarbons. Non-metals such as Si, Ge, Ga, As will also readily form oxides. The oxides are likely to react with H_2O to form surface –OH groups. Bulk oxides will also react with H_2O to form surface –OH groups.

7.2 Fundamentals of physisorption

Physisorption bonding between an adsorbate and the surface is primarily through van der Waals (vdW) attractive forces. Between inert atoms vdW interactions are driven primarily by electron fluctuation forces (London dispersion). That is, a momentary dipole on one atom induces a dipole on a nearby atom. The attraction energy increases with r^{-6}. As the atoms approach each other, they repel each other with r^{-12}. As an atom approaches a surface, the attractive potential is usually approximated as r^{-3} and the repulsive potential is taken as r^{-9}. The vdW forces tend to scale with atomic polarizability (α) or the solid refractive index at visible frequencies.

The are some general trends in vdW attraction. Metals attract more strongly than semiconductors, which attract more strongly than insulators. As the atomic number of the substrate proceeds down a periodic table column the vdW attraction increases (see figure 7.1, for the adsorption of He onto bare metals). Similarly, the vdW attraction increases for a given substrate as the atomic number of the adsorbate proceeds down a group column of the periodic table. The vdW attraction also increases for organic molecules as the π-electron density increases. The strength of the vdW attraction is related to the number of polarizable electrons—that is, the number of electrons that can be induced to form a dipole.

The bonding of inert gases to metals Cu, Ag, and Au are displayed relatively in table 7.1. The bonding of He to Cu is taken as 1.00. For reference, the binding energy of Xe on Au is approximately 19 kJ mol^{-1}. Again, the more polarizable the adsorbate and substrate the stronger the bonding.

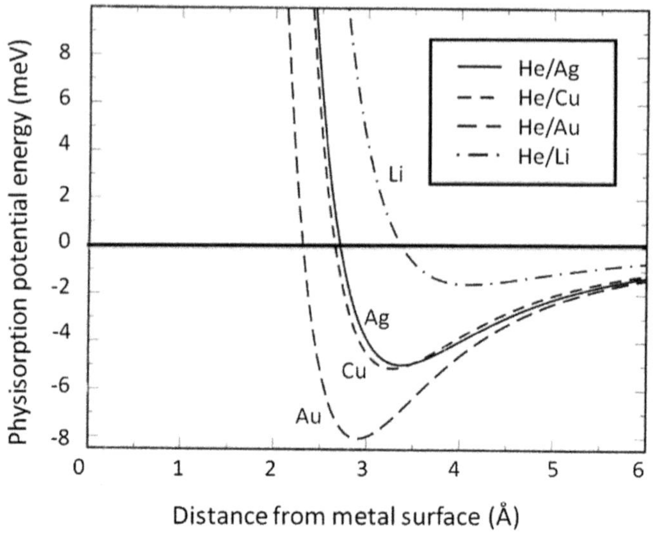

Figure 7.1. Van der Waals attractive potential of He to various metals. (This Physisorption potential image has been obtained by the author(s) from the Wikimedia website where it was made available by Ironck13 under a CC BY 3.0 licence. It is included within this book on that basis. It is attributed to Ironck13 [11].)

Table 7.1. vdW binding of inert gases to metals.

Metal	Relative vdW adsorption energy				
	He	Ne	Ar	Kr	Xe
Cu	1.00	2.01	6.67	9.38	13.6
Ag	1.11	2.23	7.21	10.1	14.6
Au	1.22	2.46	7.86	10.9	15.7

7.3 Bonding to clean transition metal surfaces

To understand basic surface chemistry, a variety of adsorbates on transition metal surfaces have been studied. They vary from simple diatomics such as H_2 and Cl_2, to electron-rich molecules with pi-bonds such as O=O, C≡O, $RHC{=}CH_2$ ($R{=}CH_2$, C_2H_5, …), HC≡CH, N≡N, toluene or pyridine. Pyridine is interesting because it has high electron density of the aromatic ring, but also includes a lone pair of σ-electrons on the N. NH_3 also includes a long pair of σ-electrons on the N.

Types of bonding include ligand donor–acceptor as in RS bonded to Au, OC on Cu, and π-electron bonding from toluene to Pt. Most often electrons are donated to empty metal states, for example d-bands. Occasionally backbonding to empty adsorbate states, as is the case for molecular adsorption of CO onto metal surfaces. Covalent bonding occurs between H-atoms and Pd surfaces, N atoms and Fe surfaces and RC forming three bonds with Pd surfaces. Ionic bonding is seen for adsorption of Cl^- on Na.

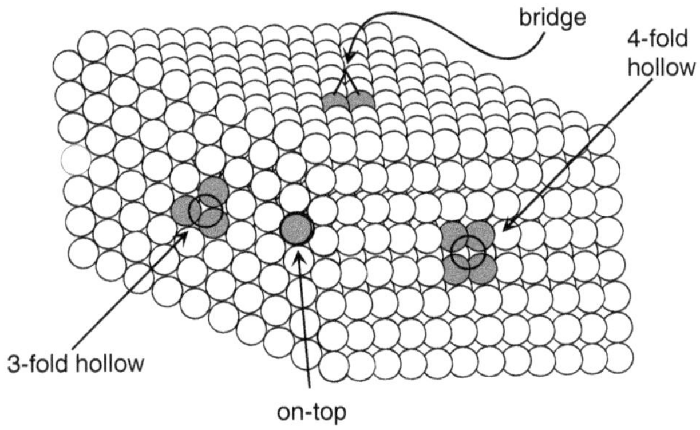

Figure 7.2. Four possible bonding sites for an adsorbate on a metal surface.

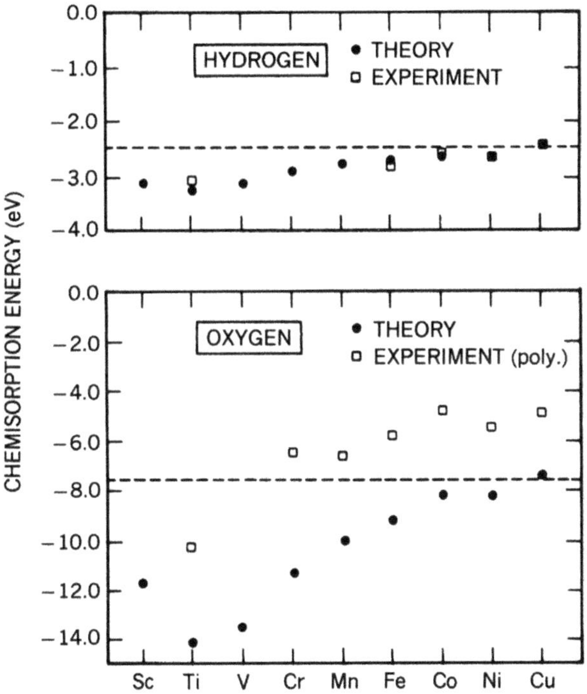

Figure 7.3. Chemisorption energy in eV, for H (top) and O (bottom) on transition metal surfaces. Reproduced from [1].

Of course, molecules and atoms chemisorb at specific surface lattice sites, as seen in figure 7.2, showing bonding to one, two, three and four substrate metal atoms are all possible

For single atom adsorbate bonding on transition metal surfaces, ΔH_{ads} of adsorbate–metal at the surface tends to decrease as the atomic number of the metal increases across the row. Bonding can occur via e^- to a neutral, zero-valent metal surface.

However, as shown in figure 7.3(top), for hydrogen, which donates e^- to the surface, ΔH_{ads} increases as atomic number, Z, decreases. The is because there are

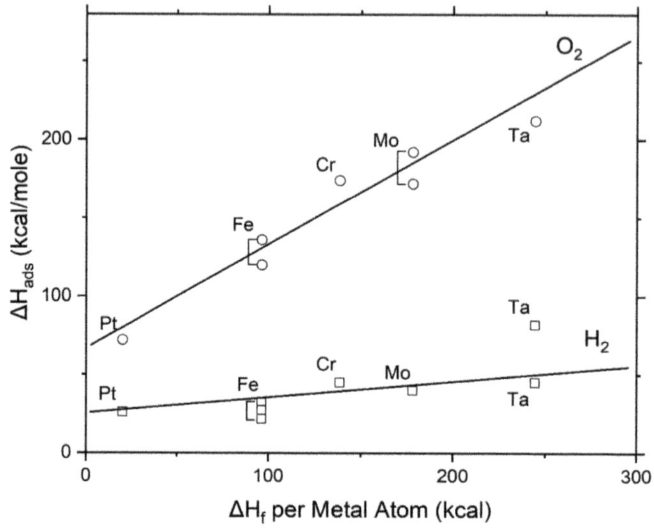

Figure 7.4. Heats of adsorption of O_2 and H_2 on various transition metals as a function of the ΔH_f of the corresponding oxides and hydrides, per metal atom [2].

more available empty states at the surface leading to stronger A–M bonding. This mechanism is true for mid-transition metals to high Z metals.

Similarly for the case of the adsorption of O, bonding tends to be via e^- donation from a neutral atom to a metal surface, ΔH_{ads} also increases as Z decreases. Metals with lower Z are more electropositive. This mechanism dominates with mid-row to lower Z metals, as shown in figure 7.3 (bottom).

In general, for single atom adsorbate bonding on transition metals, ΔH_{ads} of A–M at the surface scales with the bulk A_xM_y lattice energy, because the bonding at the surface is similar to bonding in the bulk compound. Figure 7.4 is plot of ΔH_{ads} for O_2 (top) and H_2 (bottom) bulk oxide or hydride ΔH_f on transition metals. Note that the transition metals that have different valence states available (e.g. Fe^{+2} and Fe^{+3}) will have more than one adsorbate bonding energy depending on the valence.

An interesting case is the adsorption of C≡O on metals. As figure 7.5 shows, that on surfaces where C≡O adsorbs molecularly the ΔH_{ads} is relatively constant (the left side of the plot), independent of the metal surface. On those surfaces where C≡O adsorbs dissociatively, (the right side of the plot) the ΔH_{ads} follows the $\Delta H_{formation}$ (per metal atom) of oxide formation.

Problem: Guess the dissociation energy based on the lattice energies of bulk compounds.

For molecular adsorbate bonding at transition metal surfaces, a good general rule is: *Transition metal surface bonding follows the bonding in organometallic cluster compounds of the corresponding adsorbate molecules and the zero-valence metal atom.*

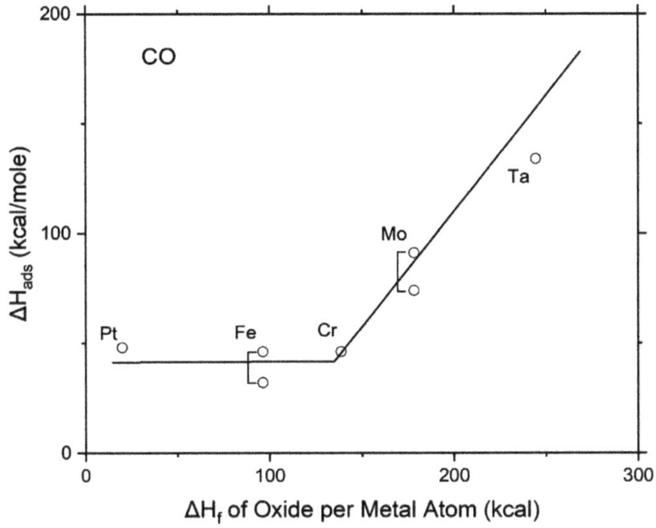

Figure 7.5. Heats of adsorption of CO on various transition metals as a function of the ΔH_f of the corresponding oxides per metal atom [2].

Common modes of olefin and aromatic molecule surface bonding include:

- Decrease in electron density at the adsorbate.
- Decrease in π bond order in the adsorbate.
- Distortion of bonds in the adsorbate.
- New surface dipole decreases the work function, ϕ.

For example, ethylene and benzene both have great deal of electron density in their respective π-systems. They tend to bind parallel to the surface and donate electrons to the surface. Since there is less electron density remaining in the π-system, the bond weakens and lengthens, as observed in the infrared spectrum.

Bonding can also involve bond scission and rearrangements. For example, ethylene can bond end on to a surface, lose a H that is now bonded to the surface. The molecule $H_2C{=}CH$- is bonded to the surface along with the H atom. $H_2C{=}CH$- might further rearrange to form $H_3C{-}C{\equiv}$, where the terminal carbon may from one bond to each of three separate metal atoms.

Common names for chemisorbed hydrocarbon species include:

- Methyl, for CH_3 singly bonded to the surface.
- Methylidene, for CH_2 either double bonded to the surface or twice singly bonded.
- Methylidine, for CH triply bonded to the surface or three times singly bonded.
- Ethylidyne, for CH_3C triply bonded to the surface or three times singly bonded.

An interesting case is demonstrated by the adsorption of benzene on Rh(111). The benzene lies flat on the Rh surface. The center of ring is centered over a three-fold hollow, and due to the donation of electrons to the surface, the C–C bonds weaken, and two C–C vibrational frequencies are seen, one consistent with bond lengths of 1.56 Å and another corresponding to a C–C bond length of 1.46 Å. In benzene the C–C bond distance is 1.41 Å. This separation into two bond lengths suggests that the adsorbed benzene has distorted into a boat-like structure as the result of electron donation. Interestingly, the benzene will not order on the Rh(111) surface unless it is coadsorbed with CO [3].

The bonding of molecules to a metallic surface is consistent with the bonding of the molecule to a metal cluster. A famous case compares the binding of ethene ($H_2C{=}CH_2$) to metal surfaces. For example, ethene bonding to some metal surfaces rearranges to form ethylidyne triply bonded to the surface:

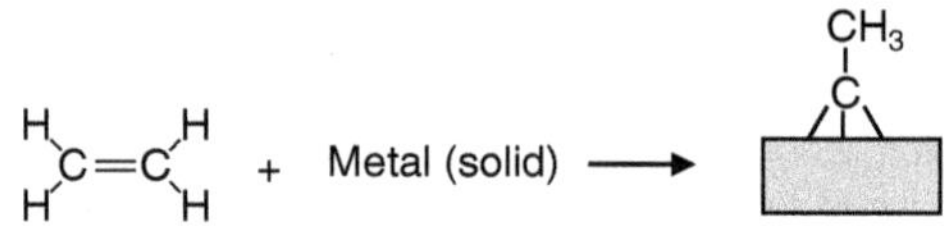

For comparison consider $Co_3(CO)_9CCH_3$, where the CCH_3 is ethylidyne with one bond between the carbon and each of the Co atoms in the cluster (see figure 7.6). Low energy electron diffraction shows the C–C bond length for the surface adsorbate on Pt(111) is 0.150 nm while x-ray diffraction shows the C–C bond length on the Co cluster is 0.153 nm. For reference, the $H_2C{=}CH_2$ bond length is 0.133 nm and the C–C bond length in $H_3C{-}CH_3$ is 0.154 nm. The C–M bond length is 0.200

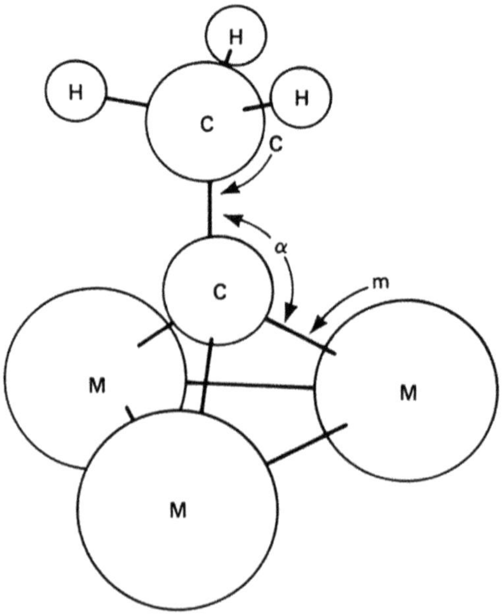

Figure 7.6. Structure of ethylidyne wherein the C closest to the surface is bonded to three metal atoms.

for ethylidyne on Pt(111) is 0.200 nm and the C–M bond distance is 0.190 nm on the Co cluster [4].

CO bonding on a metal surface is very interesting. C≡O is triple bonded. Charge can flow from the metal surface into the CO $2\pi^*$ antibonding level, creating a strong C–M σ bond, but weakening the CO bond in the process. For an example, Consider the case of CO bonding to the Ir(100) surface.

What characterization probes could be used to check the weakening of the CO bond?

In light of the previous discussion comparing inorganic clusters to surface bonded CO, we can directly relate the bonding of CO on Ir(100) to $Ir_4(CO)_{12}$. The ultraviolet photoelectron spectra, of CO(gas), $Ir_4(CO)_{12}$ and CO on Ir(100) are compared in figure 7.7. Ultraviolet photoelectron spectroscopy (UPS) and valence bond x-ray photoelectron spectroscopy can be very powerful tools in determining the details of adsorbate bonding. The similarity of the cluster and the adsorbate spectra imply similar, but not identical, bonding. The CO (gas) photoelectron spectrum shows three peaks, The highest energy occupied molecular orbital has the lowest binding energy and is the 5σ level and corresponds to the lone pair of electrons primarily situated on the C atom. This level is not seen on for the $Ir_4(CO)_{12}$ and CO on Ir(100)

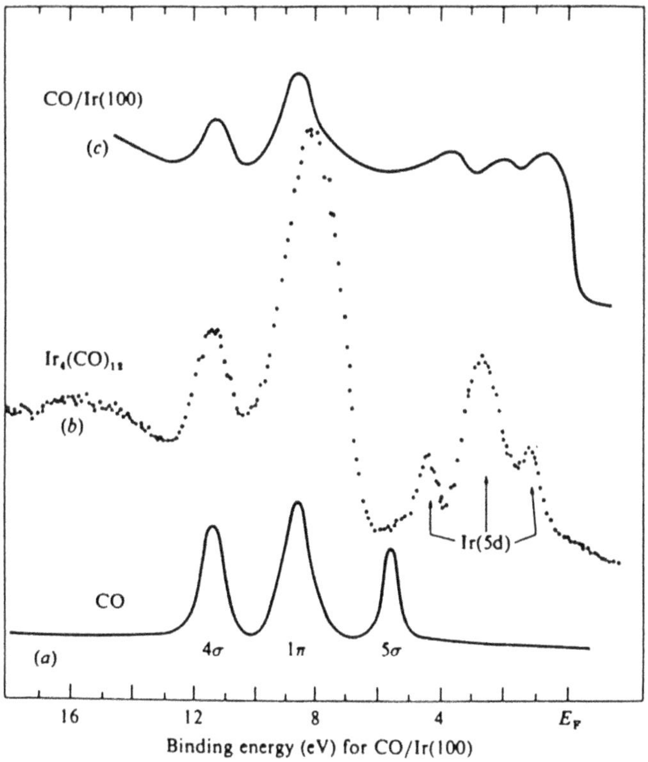

Figure 7.7. Photoelectron spectra of gaseous CO, the complex $Ir_4(CO)_{12}$ and CO adsorbed onto Ir(100). The number of photoelectrons emitted are plotted as a function of their binding energy (eV). Reprinted from [5], Copyright (1976), with permission from Elsevier.

because those electrons have been delocalized into the substrate Ir. The triple bonds in the 2 1π and 1 4σ for all three species occur at roughly the same binding energy. The conclusion for the adsorbate and the cluster is that the 5σ electrons are involved in bonding via electron donation to the empty states in Ir 7d valence band [5].

A last aspect to make for clean transition metal surface chemisorption is the case of organic molecules that contain highly polarizable atoms or are highly polarizable groups that can form strong bonds to generally inert metals such as the coinage metals, Ag, Cu and Au. This type of bonding has been very important to the development of self-assembled monolayers in which molecules with the above groups can organize into periodic assemblies on the clean metal surfaces. The typical molecules involved include organosulfur molecules as well as organ isocyanates. In these cases, the highly polarizable atoms or functional groups are able to bond to the substrate via strong polarizability interactions. In the case of Au, in particular, there are no empty substrate orbitals to accept electron donation from the adsorbing functional groups and the major interaction appears to be from polarizability effects, although the exact bonding is not well defined.

In the case of organothiols chemisorbed on Au, which has been extensively studied, the bonding is localized in three-fold hollows on a Au(111) surface. This infers that the maximum exposure of the S atoms to neighboring Au atoms is optimum for bonding strength. Examples of the surface structure of these monolayers is given earlier in chapter 3. For organothiols on Ag the strength of the more covalent Ag–S bond tends to cause reconstruction of the Ag lattice, whereas with Au–S bonds only slight reconstruction occurs. In the case of organoselenols the reconstruction effects are notable given the strong Au–Se bond strength.

7.4 Bonding to semiconductor surfaces

A good general rule in considering bonding to semiconductor surfaces is that bonding follows the bonding of the corresponding adsorbates with the nonmetal atom. Common elemental semiconductors are Si and Ge, and typical bonding is covalent. Examples include Si–Cl, Si–H, Si–CH_2R. Si–C bonds are very strong. The covalent bonding occurs with dangling surface bonds, as shown for Si in figure 7.8.

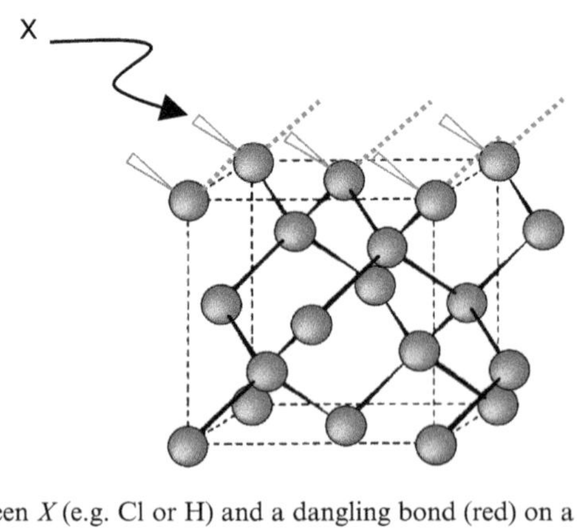

Figure 7.8. A reaction between *X* (e.g. Cl or H) and a dangling bond (red) on a Si surface. A similar reaction could take place on a Ge surface.

Typical compound semiconductors include GaAs, InP, InAs, GaN and SiC. The crystal structure of GaAs is similar to Si, the atoms have tetrahedral symmetry. The bonding to adsorbates varies depending upon the species but is typically covalent. For example, RS–(InAs) and Cl–(GaAs) are largely covalent bonds.

Controlled adsorbate bonding to compound semiconductors is often done to overcome the defects inherent in the formation of a surface. This is referred to as 'surface passivation'. Surface defects can 'pin' the bulk Fermi lever (E_F) so it cannot change when a voltage is applied to a device. Surface defects can also trap (or donate) conduction electrons; this residual charge lowers device conductance. Also surface oxides on the surface of these compound semiconductors contain many defects leading to poor electrical characteristics. Adsorbate bonding may preferentially react with some of these surface defects thereby removing them. Similarly bonding of adsorbates are often proposed to prevent oxide formation.

7.5 Bonding to oxide surfaces

As a general rule oxide surface chemistry follows trends of the corresponding bulk oxide chemistry. The surface chemistry of oxides is dominated by reaction with water. For nonmetal oxides (e.g. SiO_2) act as acid anhydrides and react with H_2O to form Brønsted acids ($SiO_2 + 2H_2O \rightarrow H_4SiO_4$). They often dissolve in aqueous base. Terminal –OH groups can react with organics, e.g. $(EtO)_3P = O$ and $(EtO)_4Si$.

For metal oxides, electropositive metal oxides typically hydrate to form bases and dissolve readily in aqueous acid, e.g. $CaO + H_2O \rightarrow Ca(OH)_2$ and $Na_2O + H_2O \rightarrow 2NaOH$. Some metal oxides are amphoteric and act as both acids and bases, e.g. $Al_2O_3 + H^+(aq) \rightarrow Al^{+3}(aq)$ and $Al_2O_3 + OH^-(aq) \rightarrow Al(OH)_4^-(aq)$. Hydrated forms can also react with organics, e.g. $(iPrO)_4Ti$.

Typical oxide surface chemistry involves surfaces exposed to ambient conditions. These might be water vapor present in the humid ambient atmosphere or after cleaning in aqueous media. These conditions produce surfaces with –OH groups which follow Brønsted behavior. The groups can be slightly acidic, neutral or slightly basic. Surfaces may hold many layers of adsorbed H_2O, through hydrogen bonding. SiO_2 is used a desiccant because of its ability to adsorb water. Surface bonding can occur through acid-base chemistry or through covalent bonding. These reactions tend to follow typical organic alcohol chemistry to form ethers, esters, etc.

When surfaces are dehydrated by, for example baking the surface at high T, the chemistry switches to typical Lewis acid interactions. The active sites are typically acidic sites and are often useful in catalysis.

7.6 Bonding to surfaces of carbon materials and organic polymers

A good general rule is that pure C surfaces are functionalized by making new C-organic group bonds which require vigorous reactions. Carbonaceous materials can vary from sp^2 to sp^3 bonding. On the sp^2 end of the spectrum we have graphene and graphite stacks. At the sp^3 end there is diamond. In between, starting with materials that have largely sp^2 bonding are nanotubes and fullerenes, glassy vitreous

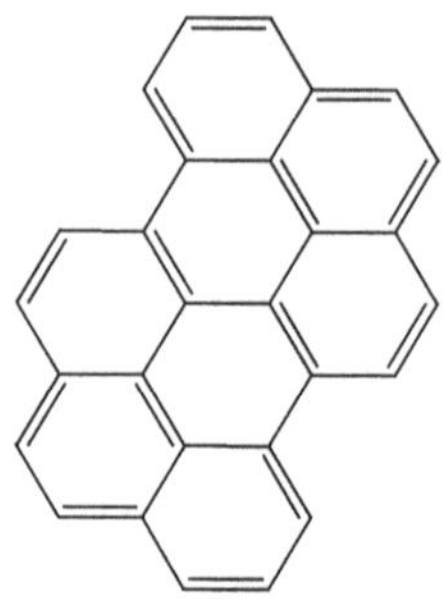

Figure 7.9. Chemical bonding in graphene.

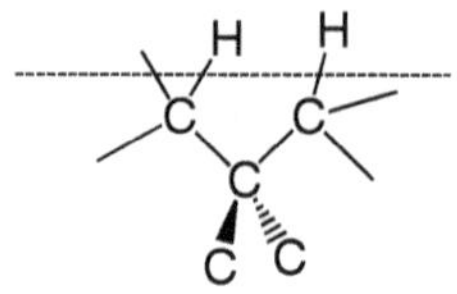

Figure 7.10. Hydrogen terminated surface of sp^3 carbon.

carbon, carbon fibers, amorphous carbon including soot, carbon black, hydrogenated amorphous carbon, and the diamond-like carbon.

Figure 7.9 shows the chemical bonding in graphene. The face, an unreacted aromatic graphite ring, tends to be highly stable. The edge, on the other hand is highly reactive due to the presence of C–H and C–OH groups. In graphitic mixtures (glassy carbon, etc) have both faces and edges exposed at the surface.

In figure 7.10 we show the surface of a material like diamond. The terminal sp^3 bonds either terminate with C–H or C–OH or remain as dangling bonds. In this aspect the surfaces of diamond are similar to those of clean silicon.

All forms of C surfaces are typically functionalized by organic reactions that convert C=C, C–C, and C–H bonds to functional groups. The typical ease of conversion C=C < C–C < C–H. Graphite containing mixtures tend to functionalize at the exposed edge regions first.

Typical functionalization processes include oxidation. For example, exposure of C=C to oxygen results in –OH, phenols, etc, and rupture of the surface plane. Exposure to oxygen is usually quite aggressive, including air roasting, ozone, peroxides, chemical oxidizing agents such as chromate, and others. Exposure of C–H termination of sp^3 to oxidation creates –OH, $-CO_2H$, ketones, and even CO_2. It is difficult, especially in the case of sp^3 materials to limit the reactions to effective surface functionalization and not actually etch the material. There is a huge literature on the functionalization of carbonaceous materials, but few details are known.

Application of controlled carbon materials surfaces include:

- Glassy carbon electrodes for electrochemistry.
- Graphene for novel, advanced electronic devices.

- High surface area supports for transition metal catalysts.
- Nanoporous carbon electrodes for batteries, capacitors, and fuel cells.
- Activated carbon particles for gas adsorption.
- Diamond-like coatings for tools and medical devices.

Molecules can also adsorb onto polymer surfaces. A general rule is that polymers are typically functionalized by organic reactions to make covalent bonds. These reactions depend on the surface characteristics. In the first case, the polymer might be characterized by having inert monomers, for example, polyethylene $–(CH_2–CH_2)_n–$. These surfaces are functionalized with vigorous reagents such as CrO_3, oxygen plasma, MnO_4^-, forming $–CO_2H$, $–C{=}O$, $–OH$ at the surface.

A second group of polymers might contain reactive side groups such as poly (acrylic acid), with a backbone of $–(CH_2–CHCO_2H)_n–$. The carboxylic acid side chains would be very reactive and could be targeted with a variety of synthetic organic reagents including bio-active reagents. A common surface stability problem with polymer surfaces is that they often relax, and polar groups roll over into the bulk to remove the effect of functionalization.

One application is the conversion of low cost, nonpolar polyolefin surfaces to polar surface to enhance inking of the surface, adhesion to the surface, wetting of the surface and polymer blending. Inexpensive, mechanically robust polymers can be converted to bio-compatible polymers for implants. Surfaces of silicone polymers that are used for soft lithography stamps and channels can be oxygen functionalized to produce SiO_2 surfaces.

7.7 Using temperature programmed desorption (TPD) to probe surface bonding

In a temperature program experiment a substrate is loaded into an ultrahigh vacuum chamber. It is cleaned, typically through sputtering and annealing at elevated temperature. The structure of the clean substrate is checked with LEED, IR, XPS, or other techniques. The substrate is then dosed with the adsorbate to the desired coverage. The structure of the adsorbed covered substrate is checked with LEED, IR, XPS, or other techniques. Then the temperature is ramped up and the desorbed species are monitored with a mass spectrometer (see figure 7.11). TPD can be applied to a variety of practical surfaces, including particles. The interpretation of the results from powders and non-uniform surfaces is much more difficult.

Temperature program desorption can provide an estimate of relative adsorption bond strengths. The simple interpretation for a single mass species desorbing at different T_P values is that the surface is heterogeneous with different bonding sites. Higher T_P values represent desorption from strong bonding sites, for example terraces versus steps, versus kinks. Alternatively, it might represent bind in different modes; for example, on-top versus bridge versus three-fold hollows. Desorption at different T_P might also represent desorption from various chemical and physical defects. Note that TPD is a kinetic process, and one cannot obtain true thermodynamic data. In fact, TPD only represents the state of the surface at the desorption

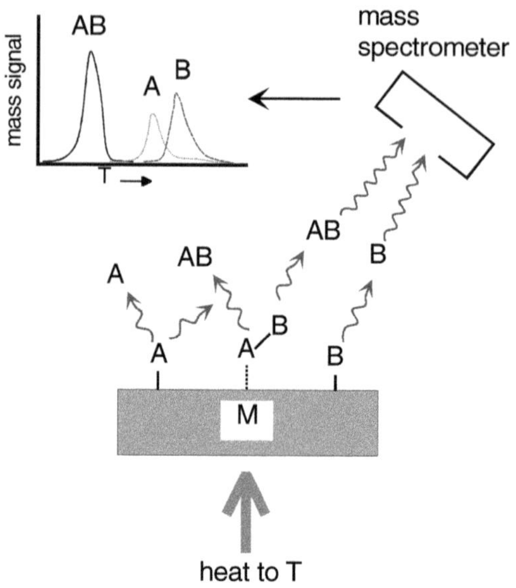

Figure 7.11. A sample substrate with adsorbate AB is heated to increasingly higher temperature in UHV. As the temperature increases, the adsorbates desorb and are detected in a mass spectrometer. The mass signals are plotted as a function of time.

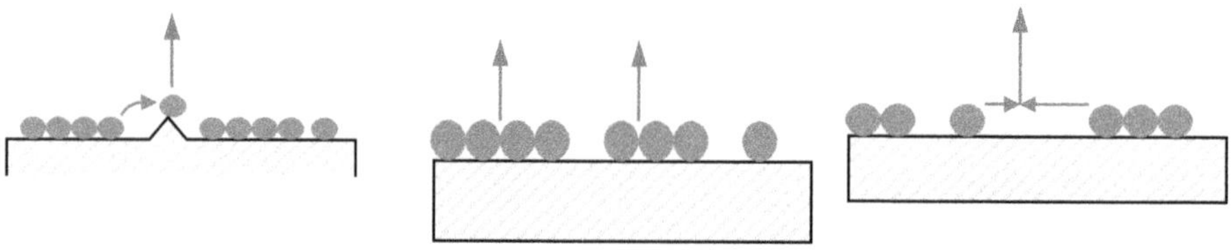

Figure 7.12. Three possible desorption mechanisms for adsorbates on a solid substrate, as the substrate is heated.

temperature. Species may diffuse from one site to another, where they more easily desorb.

TPD may distinguish between molecular and dissociative chemisorption. In figure 7.11, the desorption of the species AB only probably means that AB was adsorbed molecularly. Observation of only A and B peak likely means that there was dissociative adsorption (at least at the desorption temperature). A mixture of A, B and AB is difficult to interpret. It may mean that there was a mixture of binding states, due for example a mixture of binding sites, or that some AB adsorbates dissociate immediately prior to desorption.

In figure 7.12 we see three proposed mechanisms for desorption. On the left, adsorbates diffuse around on the surface until they find a weak binding site; from that site they quickly desorb. The rate of desorption is then limited by the number of weak binding sites, not the number of adsorbates. As a result, the rate of desorption

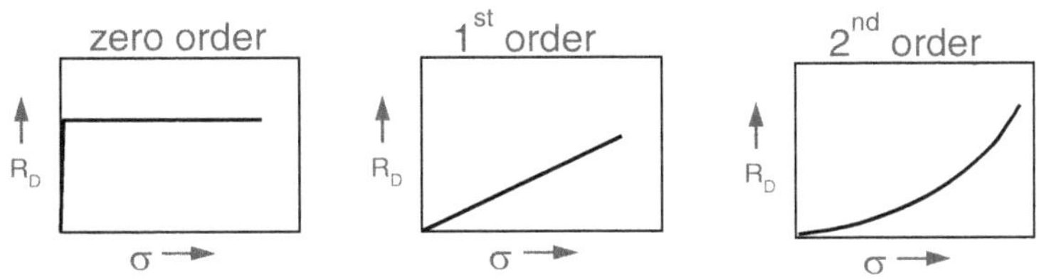

Figure 7.13. Plots of the rate of desorption of adsorbates as a function of the amount of adsorbate remaining. These desorption experiments are performed isothermally.

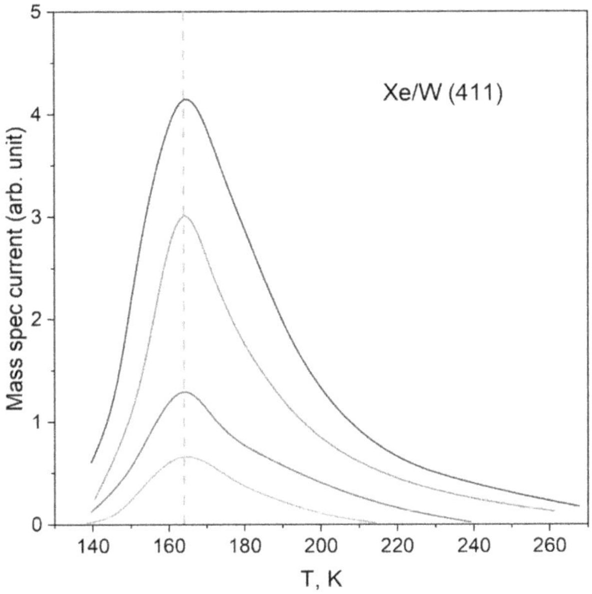

Figure 7.14. A plot of the thermal desorption of Xe from W(411) as a function of T (in K).

will not depend upon the coverage of adsorbates, but only the number of weak binding sites that does not change with coverage. The kinetic order of desorption with respect to coverage will then be zero. See the left panel in figure 7.13. The peak of desorption temperature, T_P, increases with increasing coverage, σ.

In the center of figure 7.12, the rate of desorption just depends upon how many adsorbates are there. If there twice as many adsorbates present, twice as many will desorb. The kinetic order of desorption for this process is first order. See center figure below in figure 7.13. The maximum desorption rate, T_P, will not depend upon coverage.

On the right of figure 7.12 two adsorbates must find each other and form a dimer to desorb. The rate of desorption will increase with the square of the coverage. This is second order desorption with respect to coverage of adsorbate. See right figure below in figure 7.13. Perhaps not surprisingly, T_P will decrease with increasing σ.

A good example of first order desorption is the TPD of Xe from W(411). As is evident in figure 7.14, a plot of Xe mass spectrometer intensity as a function of

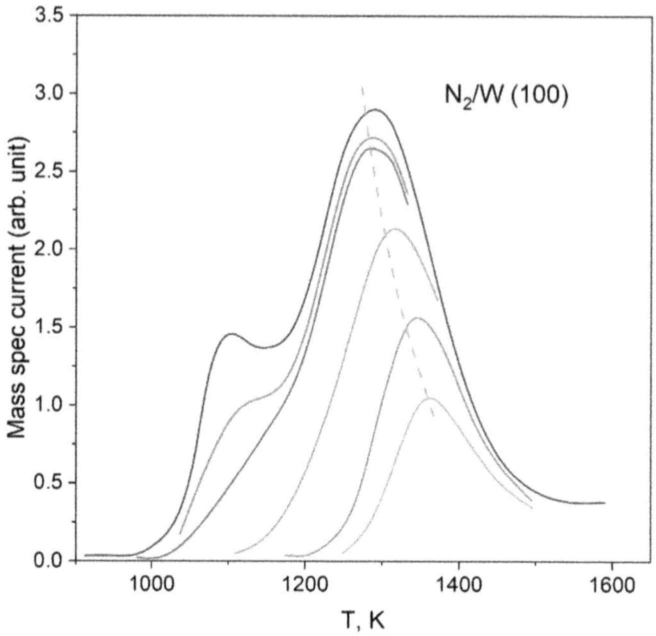

Figure 7.15. A plot of thermal desorption of N_2 from W(100) as a function of T (in K). Desorption suggests recombination of N before desorption.

temperature, T_P does not depend upon the coverage of Xe. Each Xe atom is desorbing independently of the others [6].

The temperature program desorption of N_2 from W(100) as a function of coverage in figure 7.15 shows recombination of dissociated N_2 at low coverages [6]. Figure 7.15 plots the mass spectrometer signal for N_2 as a function of surface temperature in kelvins. The N_2 desorption is second order because two dissociated N atoms must recombine to form N_2. As expected, the T_P moves to lower temperatures with increasing coverage. As the highest coverages studied, there is some desorption of molecular nitrogen, presumably due to the presence of a weak chemisorption site.

The example of TPD of CO from stepped Pt(533) as a function of coverage shown in figure 7.16 is an interesting case [6]. At low coverage the CO is relatively strongly bound in stepped sites. It desorbs from these sites with first order kinetics; the T_P remains constant as the coverage increases. Once the stepped sites are filled, the CO fills the more-weakly binding terrace sites. As the coverage increases the T_P moves to lower temperatures suggesting second order kinetics. It is unlikely that the second order kinetics are due to recombination of the C and O, but more likely due to some repulsive interaction between CO on the terraces.

In summary, temperature programmed desorption is a kinetic phenomenon, not thermodynamic. It usually has non-ideal, complex shapes. It is a good 'fingerprint' of specific bonding configurations. However, it is not usually incisive by itself—TPD

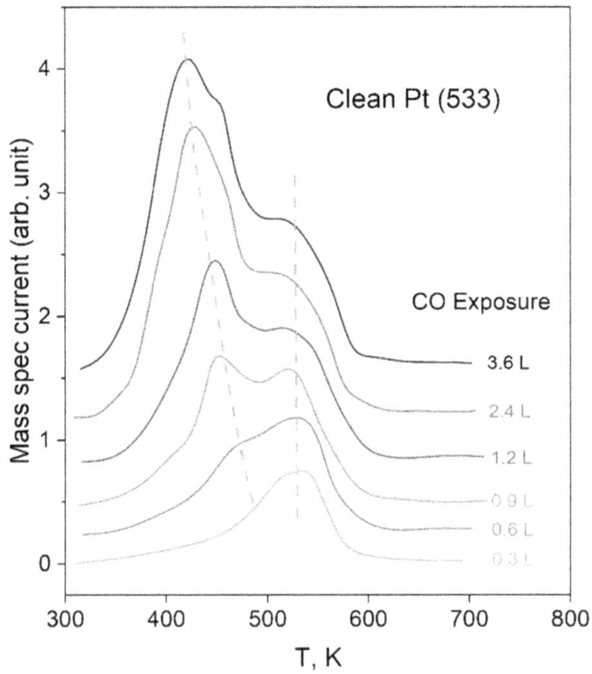

Figure 7.16. Temperature programmed thermal desorption of CO adsorbed on Pt(533). Plots suggest two binding states with first and second order desorption.

Figure 7.17. Two surface chemical reaction mechanisms following Langmuir–Hinshelwood kinetics. Left, the red atom diffuses until it finds a green atom and reacts. On the right, the red atom diffuses until it finds an active site and reacts. In both mechanisms, the molecules desorb immediately after reacting.

works best combined with other surface-active analyses such as IR, photoemission, ToF-SIMS, and theory.

7.8 Surface chemical reactions

Two model surface reactions are used to describe chemical surface reactions, the Langmuir–Hinshelwood and the Eley–Rideal mechanisms. These are considered in more detail below.

Two mechanisms can be described by the Langmuir–Hinshelwood mechanism. In the mechanism, two species diffuse on the surface until the encounter each other (see figure 7.17, left). Once the encounter each other they have a probability of reacting to a product. The adsorbates are randomly distributed and do not interact except for reaction. The reaction is $A + B \rightarrow C$, and the reaction rate, $r_R = k_R \theta_A \theta_B$, where θ_i is the fractional coverage of species, i, and k_R, is the rate constant.

The diffusion of a species to an active site (designated by * in the figure above to the right) where the product is formed and desorbs follows the same kinetics. Here, the reaction can be written $A + * \rightarrow D$, and follows the kinetics $r_R = k_R \theta_A \theta_*$. For both mechanisms the rate of reaction follows the activated form $k_R = Ae^{-(\frac{\Delta E^*}{k_B T})}$, where A is the pre-exponential that contains factors accounting for orientation, collision rates and others, ΔE^* is the activation energy, k_B is the Boltzmann constant, and T is the absolute temperature.

The major steps in Langmuir–Hinshelwood reactions follow

- Surface collisions of reactants from the gas phase.
- Accommodation at the surface with bond formation.
- Diffusion to reactive sites or encounters with reactants.
- Reaction.
- Desorption of the product.

Another surface reaction mechanism is the Eley–Rideal mechanism. Basically, the surface has coverage of A adsorbates. The reactant B only sticks to surface when it reacts with the A that is already present and forms a product, C, which desorbs. This reaction rate, $r_R = k_R \cdot \theta_A P_B$. If the reaction rate depends upon the sticking coefficient of B, S_R, the that rate constant $k_R = S_R = (\frac{a_S}{\sqrt{2\pi M_B RT}})Ae^{-(\frac{\Delta E^*}{RT})}$, where A is the pre-exponential term, ΔE^* is the activation energy, R is the gas constant, a_s is the area per reaction site, T is the absolute temperature. Recall that the flux of B molecules striking the surface is $S_R = \frac{N_{Av} P_B}{\sqrt{2\pi M_B RT}}$, when N_{Av} is Avogadro's number and M_B is the molecular weight of B. Eley–Rideal processes are relatively rare; most surface reactions follow Langmuir–Hinshelwood mechanisms.

Surface chemical processes are generally activated. That means that there is an energy input required to activate the reaction, and the source of energy might be thermal ($k_B T$), photochemical ($h\nu$), electrochemical (electrical potential). In the dynamics section we discussed the Boltzmann probability factor. All other factors being equal, the rate of a process will vary with temperature the as rate $\propto A \cdot e^{-(\frac{\Delta E^*}{RT})}$. Here activation energy is expressed as J mol^{-1}. Alternatively, activation energy could be express as eV molecule^{-1}. In that case, rate $\propto A \cdot e^{-(\frac{\Delta E^*}{k_B T})}$, where the gas constant, R, is replaced with the Boltzmann constant, k_B.

Figure 7.18 shows the calculated energies as a function of the progress of the reaction [7]. It shows the progression of the reaction starting with gas phase H_2O and CO and a clean Cu surface progressing to H_2 plus CO_2 and a clean Cu surface. Minima in the energy diagram are obtained when H_2O is adsorbed on the surface and CO remains in the gas phase. Then the H_2O splits into adsorbed OH plus adsorbed H. Then at considerable energy penalty the system switch to gas phase H_2 plus adsorbed CO and O. Finally, the O plus CO combine and desorb as CO_2. All

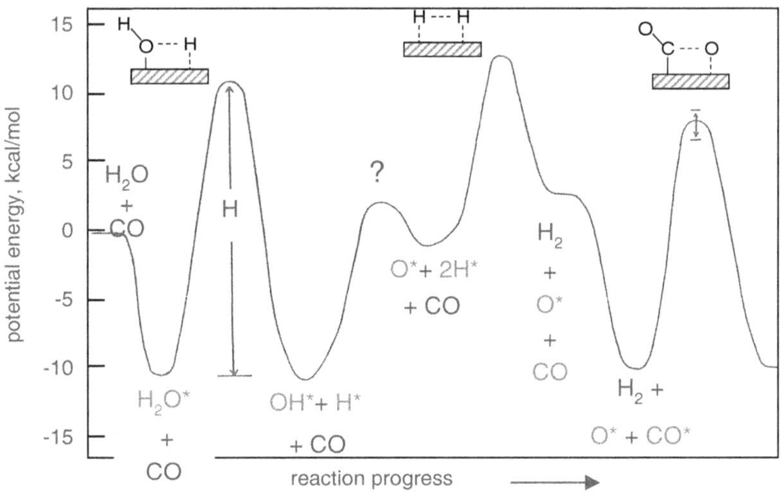

Figure 7.18. Complex potential energy diagram for the reaction $H_2O + CO \rightarrow CO_2 + H_2$ on Cu. Species in red with asterisk are adsorbed. Blue species are gas phase.

along the way there are large activation barriers for the reaction to proceed. For that reason, it is important to provide energy to surface throughout the reaction.

Note that the reaction intermediates (displayed along the top of the energy diagram) are high energy states as the electrons formerly forming bonds within the molecule now form bonds with the surface.

The reaction could proceed in the reverse direction—it is admittedly endothermic in that direction. However, to limit the reverse reaction it is important to remove the H_2 and CO_2 after formation to limit the probability of the reverse reaction.

Temperature affects the formation of different species on a surface according to the different activation energies. Thus, a species stable at low temperature may not be the lowest energy state available. A good example is the adsorption of C_2H_4 on Rh(111) as a function of temperature [8]. In figure 7.19 we show the high-resolution electron energy loss spectra (HREELS) as a layer of adsorbed C_2H_4 is heated from 77 K (bottom spectrum) to 310 K (middle spectrum) to 450 K (top spectrum). The HREELS spectra show the vibrational features of the adsorbed species.

At 77 K the C_2H_4 is adsorbed molecularly. It is primarily bound to the surface through van der Waals interactions. As the temperature in increased the molecule forms three σ-bonds with the surface through one of the carbon atoms. One hydrogen atom on that carbon atom is lost, and the other is transferred to the carbon atom above the surface. The π-bond is broken, forming ethylidene. Finally, on heating to 450 K two of the hydrogens on the terminal carbon atom are lost forming a mixture of methylidyne and C_2H with five bonds to the surface.

Can you guess the approximate ΔE_A values from this information?

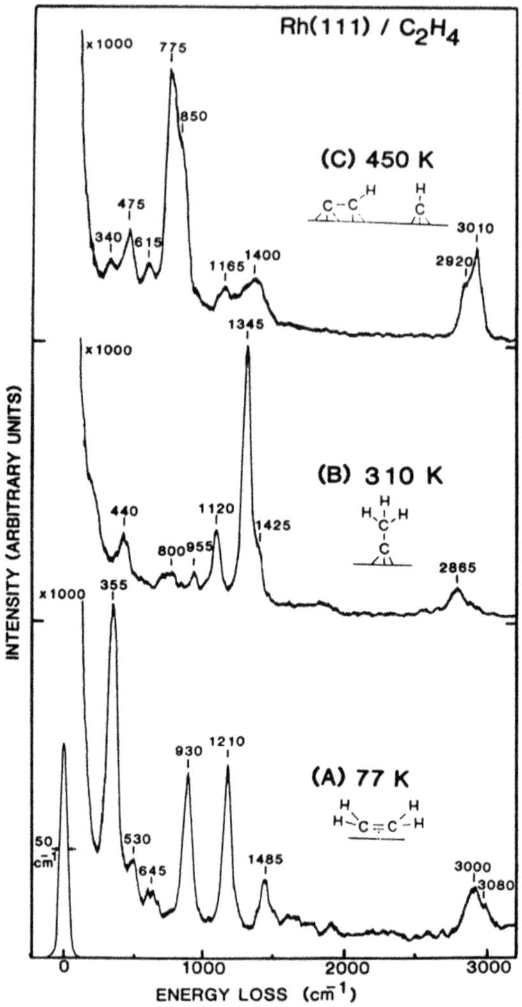

Figure 7.19. HREELS spectra of C_2H_4 adsorbed on Rh(111) as a function of temperature. Reprinted from [4].

Chapter 7 Problems

1. Nitrobenzene appears to be chemisorbed in a disordered layer at 300 K on the Pt(100) and Pt(111) surfaces. List those parameters that are characteristic of the the surface that control ordering. What kind of experiments would you perform to aid the ordering of the chemisorbed monolayer? See [9].
2. The average heat of adsorption of hydrogen and krypton on tungsten can be estimated to be 45 and 2 kcal mole^{-1}, respectively. Estimate the residence times of these two species at 3, 300, and 1500 K, if $T_s = T_g$.
3. Adsorbate-induced restructuring of surfaces could explain the formation of cluster-like bonding of adsorbates on metal surfaces. Discuss how the

strength of the chemisorption bond is likely to influence the restructuring of metal surfaces.

4. When C_2H_4 chemisorbs on Rh(111), it lies with its C=C bond parallel to the surface at low temperatures, forms ethylidyne (C_2H_3–) at 300 K, and dissociates to C_2H– and CH– groups at 410 K. Find similar organometallic multinuclear cluster compounds with similar organic species attached and discuss their bonding behavior include bond distances, binding sites, and bond angles. See [10].
5. Surface defects, steps, and kinks dissociate molecular bonds more readily and exhibit higher heats of adsorption for the chemisorbed atoms or molecular fragments. Find two examples of such chemical behavior and discuss the possible relationship between the electronic structure and the atomic structure of the defect and its reactivity to break adsorbate bonds.

Further reading

1. Somorjai G A and Li Y 2010 *Introduction to Surface Chemistry and Catalysis* 2nd edn (John Wiley). This text provides general advanced undergraduate and beginning graduate level material and chapter 46 is focused primarily on the surface chemical bond.
2. Somorjai G A 1972 *Principles of Surface Chemistry* 1st edn (Prentice-Hall). This text provides general advanced undergraduate and beginning graduate level material and chapter 5 concentrates on the surface chemical bond.
3. Woodruff D P 2016 *Modern Techniques of Surface Science* 3rd edn (Cambridge University Press). This book emphasizes characterization techniques many of which are relevant to characterization of the surface chemical bond.

References

[1] Norsko J K 1990 Chemisorption on metal surfaces *Rep. Prog. Phys.* **53** 1253

[2] Toyoshima I and Somorjai G A 1979 Heats of chemisorption of O_2, H_2, CO, CO_2, and N_2 on polycrystalline and single crystal transition metal surfaces *Catal. Rev.* **19** 105–59

[3] Koel B E, Crowell J E, Mate C M and Somorjai G A 1984 A high-resolution electron energy loss spectroscopy study of the surface structure of benzene adsorbed on the rhodium (111) crystal face *J. Phys. Chem.* **88** 1988–96

[4] Bent B E, Mate C M, Kao C-T, Slaving A J and Somorjai G A 1988 Molecular ethylene adsorption on Rh(111) and Rh(100): estimation of the C–C stretching force constant from the surface vibrational frequencies *J. Phys. Chem.* **92** 4720–6

[5] Brodén G and Rhodin T N 1976 Chemisorption of CO on Ir(100) studied by photoemission *Solid State Commun.* **18** 105–9

[6] King D A 1978 Kinetics of adsorption, desorption, and migration at single crystal metal surfaces *Crit. Rev. Solid State Mater. Sci.* **7** 167–208

[7] Nakamura J, Campbell J M and Campbell C T 1990 Kinetics and mechanism of the water-gas shift reaction catalysed by the clean and Cs-promoted Cu(110) surface: a comparison with Cu(111) *J. Chem. Soc. Faraday Trans.* **86** 2725–34

[8] Bent B E 1986 Bonding and reactivity of unsaturated hydrocarbons on transition metal surfaces: spectroscopic and kinetic studies of platinum and rhodium single crystal surfaces *PhD Thesis* University of California, Berkeley, CA https://escholarship.org/uc/item/9d85175q
[9] Somorjai G A and Szalkowski F J 1971 Simple rules to predict the structure of adsorbed gases on crystal surfaces *J. Chem. Phys.* **54** 389–99
[10] Muetterties E L, Rodin T N, Band E, Brucker C F and Pretzer W R 1979 Clusters and surfaces *Chem. Rev.* **79** 91–137
[11] Ironck13 2008 File:Physisorption 2.jpg *Wikimedia Commons* https://commons.wikimedia.org/wiki/File:Physisorption_2.jpg

IOP Publishing

Surface and Interface Science

David L Allara and Robert L Opila

Chapter 8

Catalysis

In a surface catalytic process, the reaction occurs through a sequence of elementary steps involving surface chemistry. These include adsorption, diffusion, chemical rearrangements (including bond breaking bond-forming, and molecular rearrangement) and finally desorption. The basic principles are described and a real-world example of the Born–Haber synthesis of ammonia from hydrogen and nitrogen is described in detail.

8.1 Definitions and backgrounds

A catalyst speeds a given chemical reaction without itself being altered or consumed and without changing the reaction thermochemistry. For example, consider the desired reaction sequence

$$A + C \rightarrow D$$

with undesired side reactions

$$A \rightarrow B$$

$$A + D \rightarrow E.$$

A good catalyst would accelerate the formation of D from undesirably slow to acceptable. It would also accelerate the formation of D relative to side products, in the case above, B and E.

There are two types of catalysts, *homogeneous* and *heterogeneous*. Homogeneous can be in solution or the gas phase. Typical examples include H^+, OH^-, and M^{+n} (transition metal ions). Heterogeneous or surface catalysis is a bulk material, although only the surfaces are involved in the reaction. Common examples include transition metals and their oxides.

There are some terms useful when considering processing using a heterogeneous catalyst. The *catalyst turnover frequency*, f, is the product molecules second^{-1}. $1/f$ is the time per product molecule. The *turnover rate*, R, is f/A, and is the product

doi:10.1088/978-0-7503-4916-1ch8

molecules $cm^{-2} \cdot s$, where A is the catalyst area (often taken as the BET area). *Turnover number* is $R \cdot \delta t$, which is product molecules cm^{-2}. δt is the reactor run time. Assuming an area/site (e.g. ~25 Å^2), $R^{site} \cdot \delta t$ is the product molecule/site. A practical catalyst should work for run time such that $R^{site} \cdot \delta t > 10^2$. *The reaction probability* (RP) is the (product formation rate) divided by the (reactant incidence rate). RP is equal to R/F, where F is the gas flux.

We show a summary of reaction parameters for hydrocarbon conversion reactions on Pt in figure 8.1. As you proceed from low temperatures to higher temperature there is an increase in reaction complexity. At the lowest temperatures Pt catalyzes the hydrogenation of π bonds. As the temperature increases, it becomes possible to isomerize and cyclize olefins. Around 400 K Pt catalyzes the dehydrogenation of hydrocarbons. And at temperatures greater than 500 K alkanes can undergo hydrogenolysis, isomerization, and cyclization. Notice that the reaction rates (displayed on the left axis) can be quite low. This is because when hydrocarbons adsorb on Pt surfaces, many intermediate species can be formed, and some of these will block reaction sites.

Consider the uncatalyzed reaction between hydrogen and oxygen to form water. If the free energy is less than zero, it is a spontaneous process:

$$H_2 + 1/2O_2 \rightarrow H_2O \Delta G^0 = -229 \text{ kJ mol}^{-1} \quad \Delta H^0 = -242 \text{ kJ mol}^{-1}.$$

With no catalyst the only energy available is thermal energy. For H_2 and H_2O at room temperature (~300 K) in translation, rotational, and vibration modes is $6RT \sim 15 \text{ kJ mol}^{-1}$. If the following reaction sequence is followed to form $2H_2 + O_2 \rightarrow 2H_2O$:

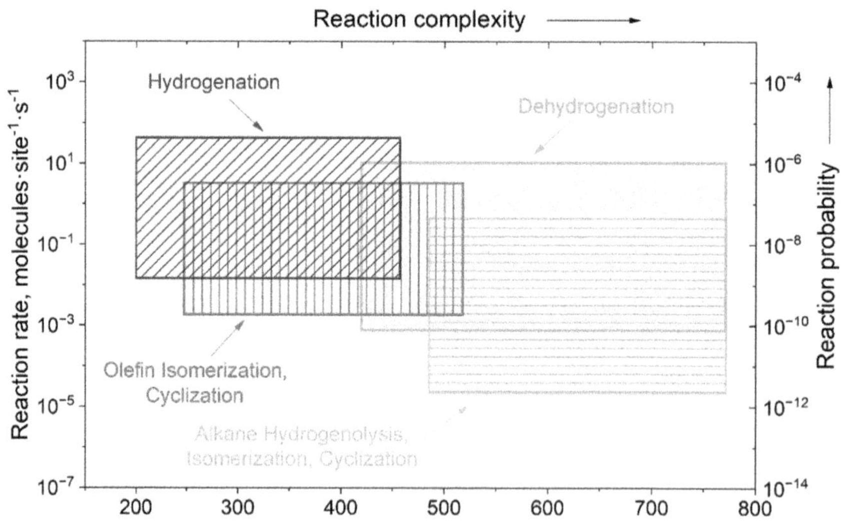

Figure 8.1. A block diagram of hydrocarbon conversion over Pt catalysts plotted as a function of reaction rate as a function of temperature as discussed in [1].

$$H_2 \rightarrow 2H$$

$$H + O_2 \rightarrow HO_2$$

$$HO_2 \rightarrow HO + O$$

$$H + HO \rightarrow H_2O$$

$$O + H_2 \rightarrow H_2O.$$

The first reaction in the sequence requires about 436 kJ mol^{-1} activation energy, significantly more than the 15 kJ mol^{-1} thermal energy available. Each of the subsequent reactions in the sequence require a small activation energy, but slightly more than thermal energy available. We can estimate the rate of reaction with the following approximation, rate $\sim Ze^{-\frac{\Delta E_a}{RT}}$, where ΔE_a is the activation energy and Z is the collision frequency ($<10^{12}$ s^{-1}). Using 436 kJ mol^{-1} as the activation energy results in a rate of $\sim 10^{-64}$ reactions s^{-1}. The reaction would take centuries at room temperature.

With a Pt catalyst the energetics change considerably. If the following reaction pathway is followed to form $2H_2 + O_2 \rightarrow 2H_2O$

$$2H_2(g) + 4\,Pt(sites) \leftrightarrow 4Pt - H\,(H^*)$$

$$O_2(g) + 2Pt(sites) \leftrightarrow 2Pt - O\,(O^*)$$

$$4H^* + 2O^* \rightarrow 2H_2O^*$$

$$2H_2O^* \rightarrow 2H_2O(g).$$

The chemisorption in the first two steps and the surface diffusion in the third step are exothermic. The last step occurs very quickly at room temperature (300 K). The rate $\sim Ze^{-\frac{\Delta E_a}{RT}}$, where the activation energy is close to 0, and z is the surface collision rate, means that we have essentially an explosion! This is all done by the catalyst removing the activation energy to dissociate H_2.

8.2 Kinetic laws and mechanisms

We will recall a few basic concepts in reaction kinetics. First, a *reaction step* is the simplest chemical reaction even involving one single activation step. For example, in a unimolecular reaction there is only one reactant, $A \rightarrow B + C$. In a bimolecular reaction, there are two reactants (they can be the same), $A + B \rightarrow C$.

A *rate limiting step* is the slowest step in a reaction sequence. It limits the product formation rate. It might be considered a bottleneck. For example, in the reaction $A + B \xrightarrow{k_1} C \xrightarrow{k_2} D$, if one step is the slowest, it is the rate limiting step.

Then we have *rate expressions*. If we have a first order reaction, $A \rightarrow B$, the rate of formation of B, $\frac{d[B]}{dt} = k_1[A]$, where $[I]$ represents the concentration or pressure of I, and k_1 is the first order rate constant. In the event of a second order reaction,

$A + B \rightarrow C$, the rate of formation of C, $\frac{d[C]}{dt} = k_2[A][B]$, and k_2 is the second order rate constant.

Typically, for simple reaction steps Arrhenius behavior is followed. That is the rate constant, $k \sim Ae^{-\frac{\Delta E^*}{RT}}$, where the A factor includes reactant collision rate multiplied by and steric effects on collision, and ΔE^* is the activation energy (or energy barrier height). The relationship is very general, and for surfaces includes processes ranging from adsorption to reaction between adsorbates.

Global rate expressions incorporate multiple reaction steps. Often the rate, R, $R = k \cdot f(P_j)$, where k is the global rate constant, and the $f(Pj)$ is a function of the reactant and product concentrations or pressures. Often the global rate constant, even for multiple step reactions, will follow Arrhenius behavior, but ΔE^* and A are usually constant over a limited T range.

Examples of kinetic parameters exist for hydrocarbon catalytic reactions [2]. Olefin hydrogenation has an ΔE^* of 25–50 kJ mol^{-1} with a useful temperature range of $\sim$300 K. Isomerization, cyclization, dehydrocyclization, and hydrogenolysis have ΔE^* of 150–190 kJ mol^{-1} and a useful temperature range >400 K. The reaction of n-hexane with excess H_2 over a Pt catalyst is shown in figure 8.2.

For simultaneous parallel reactions:

$$A \xrightarrow{R_1} P_1$$

$$A \xrightarrow{R_2} P_2$$

$$A \xrightarrow{R_j} P_j$$

$$A \xrightarrow{R_N} P_N.$$

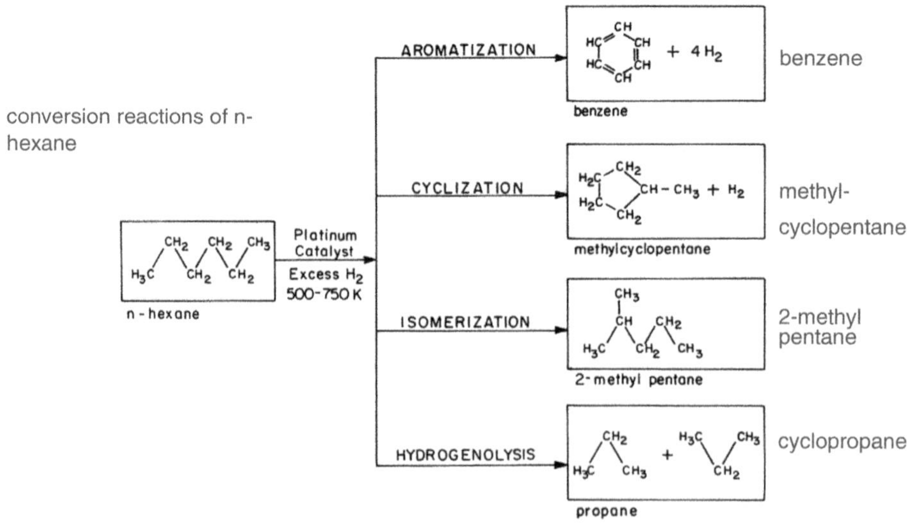

Figure 8.2. Organic species that can be produced by catalytic conversion n-hexane. Reproduced from [2], with permission from Springer Nature.

The fractional catalytic selectivity of path j is $S_j = \frac{R_j}{\sum_{i=1}^{n} R_i}$. Figure 8.2 is a case where multiple, parallel reactions can occur with the same reactants and the same catalyst. It is a challenge to arrange conditions so that one reaction pathway is maximized to make a selected product.

8.3 Catalyst preparation

A typical configuration for transition metal catalysts has metal particles (nano-particles) dispersed on a high surface area support. The supports are commonly SiO_2, Al_2O_3, C, MgO, ZrO_2, or TiO_2. They have large surface area, $\sim 100 - 400\ m^2\ g^{-1}$. The supports are often porous and can have surface areas up to $10^3\ m^2\ g^{-1}$.

A common prep is to impregnate the supports with metal salts, reduce the metal ions to metal with, for example, H_2, and then heat to form metal clusters. The structure then is a support with large surface area containing small metallic clusters. A bimetallic catalyst contains two different metals. Often ionic additives are added. For example, K can act as an electron donor and Cl as an electron acceptor.

8.4 Catalyst life cycle

The life cycle of a catalyst is shown in figure 8.3. After preparation of the catalyst, it is used until it is spent. Often the catalyst is poisoned. Typical poisons include S, C, trace metals, salts. These poisons take strong chemisorption active sites that interfere with the intended reaction. The poisons can also restructure the surface of the catalyst. The poisons might be inherent in the process chemicals or initial impurities in the formulation that segregate to the surface.

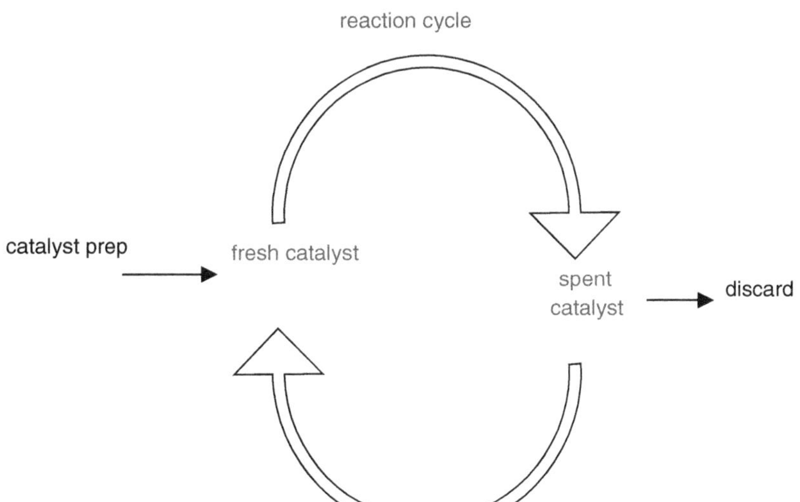

Figure 8.3. The life cycle of a catalyst, described in detail in the text.

Once the catalyst is spent it can either be discarded or regenerated. Regenerate might include oxidation of the poisons, sintering the metal, volatilizing the poisons. Sintering is a process of heating to allow metal atoms to move and reform a stable structure.

8.5 Choosing metal catalysts

Generally, catalytic processes have three steps. First adsorption of an inexpensive reactant such as O_2, H_2, N_2, CO, followed by reaction with an adsorbed molecule such as an aromatic or olefin. Finally, the reacted species desorbs. The series of reactions is shown below with X_2 representing the reactant, A the adsorbed molecule, and M is the metal:

$$X_2(g) \overset{M}{\rightarrow} 2M - X$$

$$M - X + M - A \rightarrow M - (XA)$$

$$M - (XA) \rightarrow M + XA\,(g).$$

There are optimal bonding characteristics for efficient catalysis. For the first reaction above, the M–X bond should be strong to facilitate dissociation. For the second reaction the M–X bond should be relatively weak so the X could rapidly diffuse and form M–(XA). Finally, the M–(XA) bond should be weak so that XA could quickly desorb from the surface. Note that some of these requirements are apparently contradictory, so optimal balance of M–X bond is needed for optimal reaction to take place.

Following the Sabatier principle, the interactions between the catalyst and the reactant should be neither too strong nor too weak. If the interaction is too weak, the reactant will not bind to the catalyst, and if the interaction is too strong, the product will fail to desorb. Very often if a property such as relative reactivity is plotted against chemical number (Z), the reactivity will peak at one element or another. Because the plots increase monotonically as a function of Z to the optimum value, and then decrease monotonically, these plots are called volcano plots (see below for reductive bond scission in the presence of H_2).

Figure 8.4 shows the rate of C–C bond breaking as a function of Z (for Re, Os, Ir, Pt, and Au) [2]. The peak of the volcano is at Os. For the breaking of C–N bonds the peak is at Ir, and for the breaking of C–Cl bonds the peak is at Pt. Recall that as Z increases the ΔH ads decreases. In these plots, the ability of the catalyst to dissociate bonds is balanced against the ability of the catalyst to release product.

8.6 Useful concepts and definition for catalytic processes

Some reactions are *structurally insensitive.* That means that reaction rate, R, is constant with catalyst dispersion or size. Recall that R is normalized to the area available for reaction. For a *structurally sensitive* reaction, R changes with catalyst dispersion. This is not surprising because in general there is a correlation with reactivity as one goes from large particle to cluster to atom.

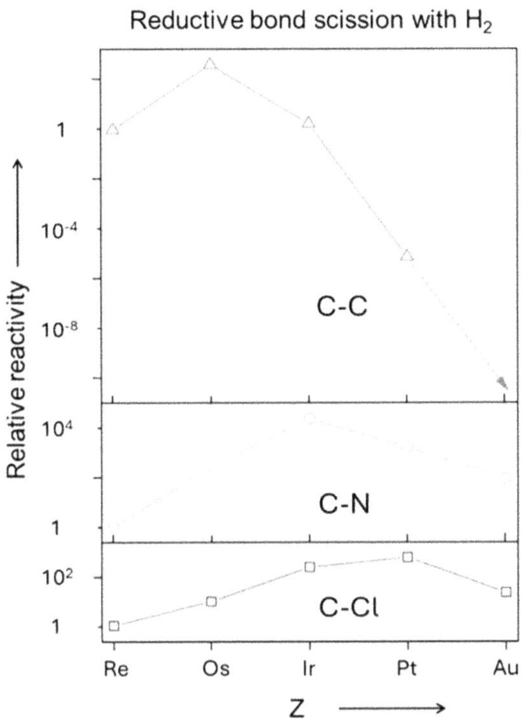

Figure 8.4. The catalytic activities of transition metals for the hydrogenolysis of the C–C bond in ethane (C_2H_6), the C–N bond in methylamine (CH_5N), and the C–Cl bond in in methyl chloride (CH_3–Cl).

Surprisingly there is often a *compensation effect* when comparing similar catalytic systems. Occasionally one finds for variations in catalyst conditions a, b, c, ... that $k_a \sim k_b \sim k_c \cdots$ at some particular run condition (T, P). If the overall rate constant for each of these systems is described by Arrhenius behavior, i.e. $k = Ae^{-\frac{\Delta E^*}{RT}}$, or $\ln k = \ln A - \frac{\Delta E^*}{RT}$, it might be found that while the rates are relatively constant, $A_a \neq A_b \neq A_c \cdots$ and $\Delta E^*_a \neq \Delta E^*_b \neq \Delta E^*_c \cdots$. For constant rate constants then $\ln A$ and ΔE^* must compensate. It has been found that

$$\ln A = \alpha + \frac{\Delta E^*}{R\theta},$$

where θ is the isokinetic temperature, that is, the temperature at which the rates for the catalysts are the same.

Hydrogenolysis reactions show good examples of compensation effects. Below, in figure 8.5, $\ln A$ versus ΔE^* is summarized for a series of hydrogenolysis reactions, including $H_2 + H_3C - CH_3 \rightarrow 2CH_4$, $H_2 + H_3C - NH_2 \rightarrow CH_4 + NH_3$, $H_2 + H_3C - Cl \rightarrow CH_4 + HCl$ over a series of catalysts. These are the same rates that we showed in figure 8.4 [2].

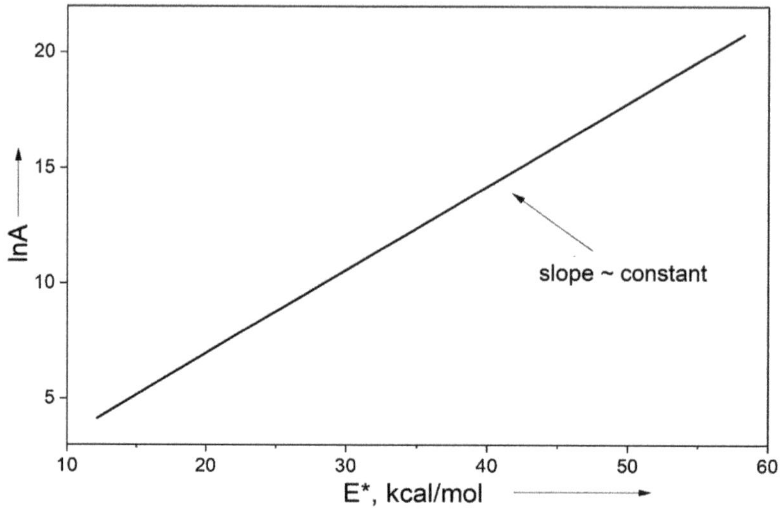

Figure 8.5. Compensation effect for hydrogenolysis reactions. The logarithm of the pre-exponential factor ($\ln A$) is plotted against the apparent activation energy (ΔE^*) for several different transition metal catalysts from reference [2].

The compensation effect can be understood by looking at the underlying thermodynamics. If we re-express the rate constant in terms of free energy of activation (ΔG^*),

$$k = A\mathrm{e}^{-\frac{\Delta E^*}{RT}} = c\mathrm{e}^{-\frac{\Delta G^*}{RT}}.$$

Knowing that $\Delta G^* = \Delta H^* - T\Delta S^*$, and taking $\Delta H^* \sim \Delta E^*$, we have

$$k \sim c\mathrm{e}^{\frac{\Delta S^*}{R}}\mathrm{e}^{-\frac{\Delta E^*}{RT}}.$$

To keep k constant in a variety of situations we need to balance $T\Delta S^*$ and $-\Delta E^*$.

Here is how this might come about. Consider the case where there is a low activation energy. The transition state will appear conformably very similar to the initial state. Since there is minimum bond distortion, there are minimal new ways for the transition state to rearrange. Thus, both the activation energy, ΔE^*, and entropy of activation, $T\Delta S^*$ are low.

Compare this to the case where the energy of activation is high. The transition state structure will be shifted toward products, with extensive bond distortion and multiple ways to arrange. As a result, $T\Delta S^*$ will be high. Examining the equation above shows that in the presence of the compensation effect an increase in energy of activation will be compensated by an increase in the entropy of activation (in the prefactor), yielding little change in the overall rate.

Finally, there a few more catalytic terms. A *structure modifier* is an agent that favorably restructures the substrate surface. It may, for example, block unfavorable sites, e.g. sites that cause side reactions. A *bonding modifier* may tune the strength of the adsorbate bond by donating electrons, e.g. potassium, or withdraw electrons, e.g. chlorine. Often there is a *strong metal support interaction*. In this case, the active sites might be at the support/metal interface. Having the adsorbate bond simultaneously to both the metal and support may have pi-electrons donated to the metal, or an acid–base interaction with the oxide support.

8.7 Acid–base catalysis

Acid-base or donor-acceptor interactions involve charge transfer, either an ion, such as H^+, or an e^-. No change in redox states is preserved. In Lewis interactions and electron is transferred, as in the case of NH_3 and its extra lone pair of electrons interacting with BF_3, missing a lone pair. A Brønsted interaction involves H^+ transfer, as in the case of transfer to or from H_2O.

Dehydrated MgO and Al_2O_3 are good example of solid Lewis acids—they would like to accept an electron. Solid Brønsted acids included hydrated surfaces. Hydrated surfaces of SiO_2 are slightly acidic, and hydrated surfaces of P_2O_5 are strongly acidic. Hydrated surfaces of Al_2O_3 are weakly basic—they would like to donate OH^-. Surfaces of hydrated CaO are moderately basic.

A good example of acid–base catalysts are zeolites. These are an arrangement of tetrahedral Al and Si oxides in an aluminosilicate material that contain a cubo-octahedral cavity. In figure 8.6 the Si or Al atoms are depicted as solid circles and O is the open circles. The alumina moieties act as Lewis acids and the hyrated silica are Brønsted acids.

In figure 8.7 we show the crystal structure of mordenite zeolite assembled from corner-sharing SiO_4 tetrahedra. The structure of mordenite is $(Na_2,Ca,K_2)_4(Al_8Si_{40})O_{96} \cdot 28H_2O$. Cavity dimensions are typically in the $\sim$1 nm range.

The combination of Al and Si oxides produces strong acid sites. The acidity can be controlled by the stoichiometry. By controlling the cavity side and acidity are two very strong properties to control.

Ionic intermediates are important in the isomerization and cracking of hydrocarbons. Of particular importance is the carbenium ion. If, in the presence of a catalyst, a H– is abstracted from $R - CH_2 - CH_2 - R' \rightarrow R - CH_2 - HC^+ - R'$, a carbenium ion is formed. Carbenium ions have relative stabilities $R_1R_2R_3C^+ > R_1R_2HC^+ > R_1H_2C^+$.

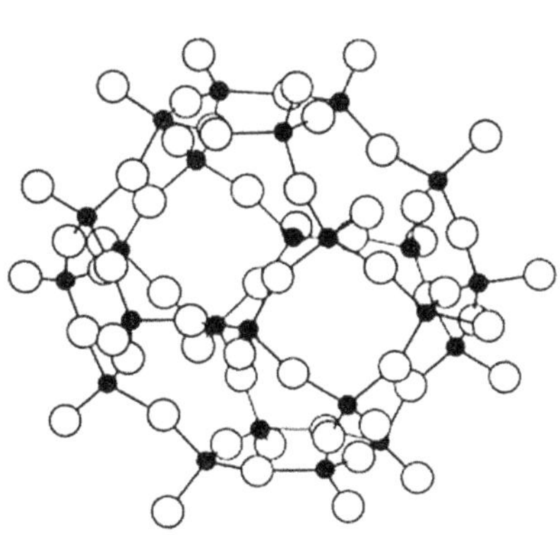

Figure 8.6. The atomic structure of a zeolite. The Si and Al atoms are depicted as solid circles, and the O atoms are depicted as open circles.

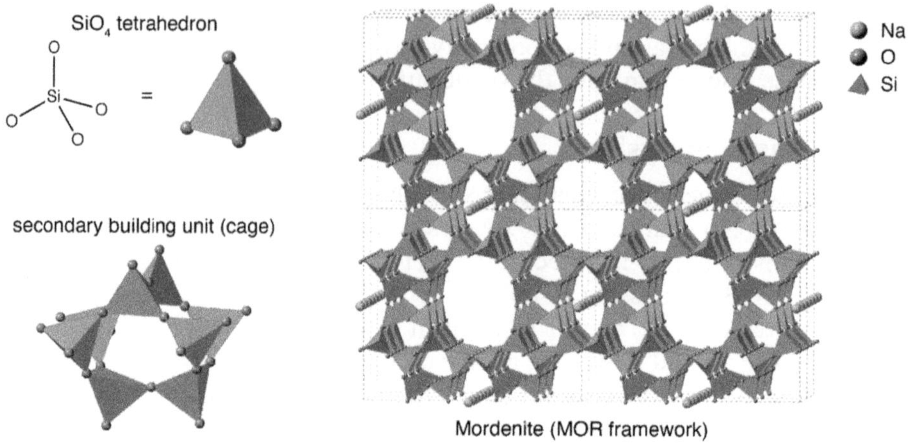

Figure 8.7. Atomic structure of mordenite, assembled from corner-sharing SiO_4 tetrahedra. Sodium is present as an extra-framework cation. Si atoms can be partially replaced by Al or other tetravalent metal. (This Zeolite structure as an assembly of tetrahedra image has been obtained by the author(s) from the Wikimedia website where it was made available by Qx8314 under a CC BY 4.0 licence. It is included within this book on that basis. It is attributed to François-Xavier Coudert [3].)

Given this relative stability, rearrangement (isomerization) is also possible, $R_1R_2R_3C{-}H_2C^+ \rightarrow R_2R_3C^+ - CH_2R_1$.

Important reactions include cracking/cyclization, cracking/addition. The final neutral product from carbenium ion reactions is formed by a loss of H^+ to produce an olefin. In some cases, if traces of H_2O are present, it may form an alcohol.

8.8 Case example of model studies for determining the mechanism of a commercial scale process: nitrogen fixation for ammonia synthesis

Ammonia is a very important chemical species for use in, for example, agriculture fertilization. The basic idea is to react nitrogen with hydrogen to form ammonia, $N_2 + 3H_2 \rightarrow 2NH_3$. For the reaction to proceed spontaneously, $\Delta G^0 < 0$. As shown in figure 8.8, which plots ΔG^0 in kcal mol^{-1} as a function of absolute temperature [4]. Clearly the free energy is less than zero over a wide temperature range.

Problem: Why do you think N_2 and H_2 do not react on simple mixing of the gases? Hint, think in terms of the basic chemical structures of the reactants and the structure of an activated state in the reaction.

A schematic of one version of the Born–Haber process to synthesize ammonia is shown in figure 8.9. In the first part of the reactor methane and water and methane and oxygen are mixed to form hydrogen. The Haber reaction feed gas chemistry is

A. Water gas shift reaction: $CH_4 + H_2O \leftrightarrow CO + 3H_2$.

B. CO and H_2 production: $2CH_4 + O_2 \leftrightarrow 2CO + 4H_2$.

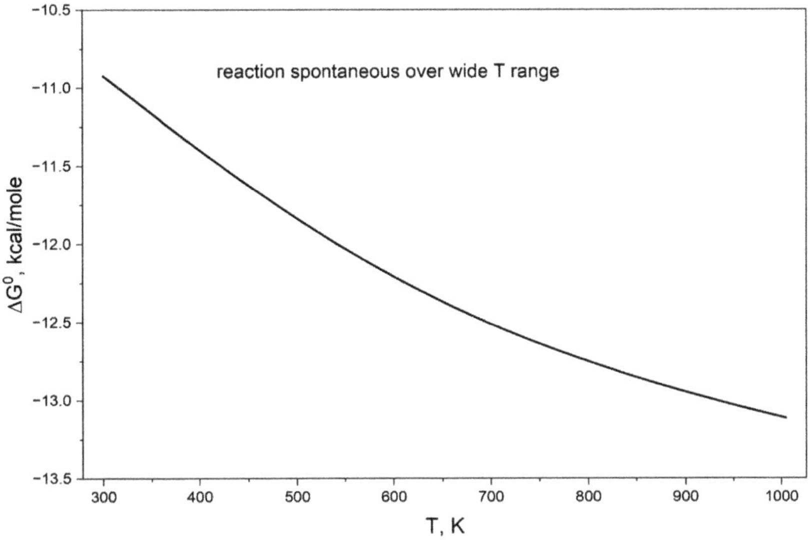

Figure 8.8. The free energy of NH_3 formation (kcal mol^{-1}) as a function of temperature (K). For more details, see reference [4].

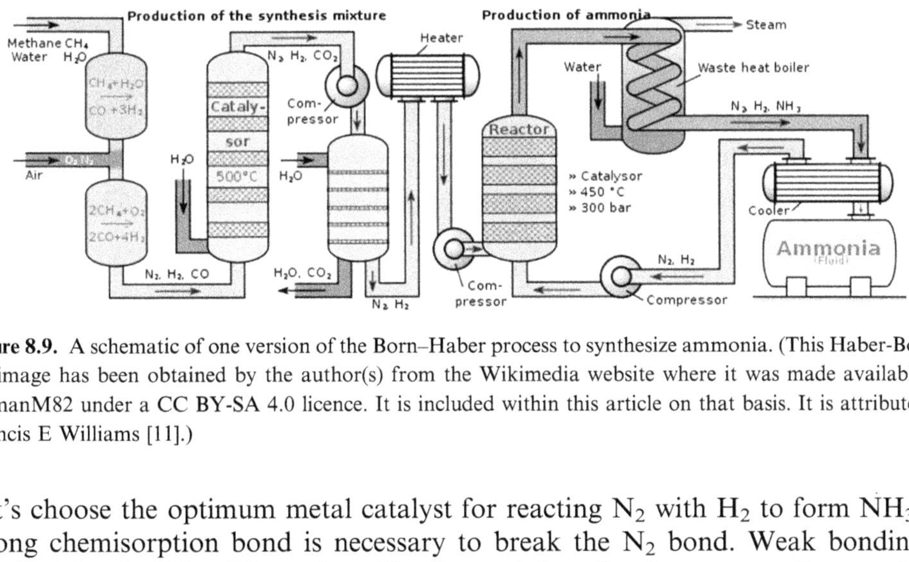

Figure 8.9. A schematic of one version of the Born–Haber process to synthesize ammonia. (This Haber-Bosch-En image has been obtained by the author(s) from the Wikimedia website where it was made available by RomanM82 under a CC BY-SA 4.0 licence. It is included within this article on that basis. It is attributed to Francis E Williams [11].)

Let's choose the optimum metal catalyst for reacting N_2 with H_2 to form NH_3. A strong chemisorption bond is necessary to break the N_2 bond. Weak bonding is required for fast N + H reaction. Finally weak bonding is necessary for easy NH_3 desorption. Theoretical calculations of a volcano plot are shown in figure 8.10 where the NH_3 concentration for a fixed set of reaction conditions as a function of the number of d electrons.

In commercial systems, see figure 8.9, an Fe catalyst is used, supported on Al_2O_3. The commercial catalyst is prepared by taking porous Fe_3O_4 powder ($\sim$10–15 $m^2\,g^{-1}$) + some combination of Al_2O_3, K_2O, CaO, MgO, and SiO_2, and fusing it under H_2.

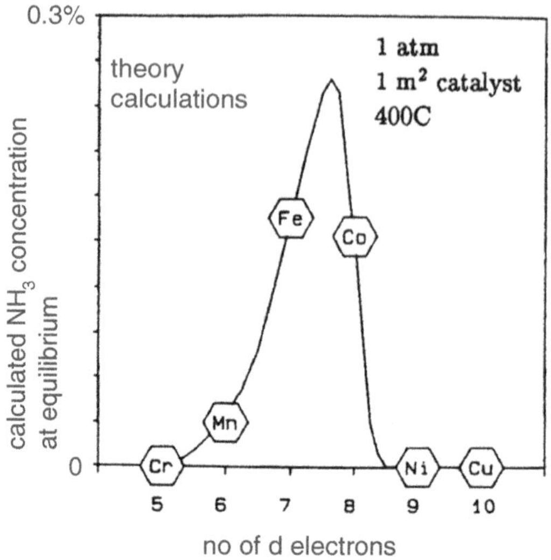

Figure 8.10. The calculated NH_3 concentration for a fixed set of reaction conditions as a function of the number of d electrons. Reprinted from [5], Copyright (1987), with permission from Elsevier.

The empirically determined best catalytic system is Fe plus K, on a Al_xO_y support with a K promoter. The best operating temperature is approximately 400 °C, and P_{N_2}, P_{H_2} are around 150–300 atm. The observed rate law for these conditions is $R \equiv \frac{dP_{NH_3}}{dt} = k \cdot f(P_{N_2}, P_{H_2}, P_{NH_3})$. P_{NH_3} is in the expression because of the back reaction to nitrogen and hydrogen. At low conversions $(P_{NH_3} \sim 0)$: $R_0 = k \cdot P_{N_2} K_A$, where K_A is a constant. The rate determining step is $N_2 \overset{Fe}{\rightarrow} 2N - Fe(s)$. The activation energy, ΔE^* is 18 kcal mol^{-1} (76 KJ mol^{-1}).

The fundamental chemical mechanism can be determined for model studies. The system schematized in figure 8.11 is combination of a high-pressure reactor and a ultra-high vacuum system to perform such a study. The system includes a gas dose (N_2, H_2, ...), a K doser (promoter), a TPD system (the mass spectrometer), HREELS, LEED, Auger, XPS, and gas chromatography. Contemporary chambers might substitute the LEED with STM or AFM and might also include infra-red spectrometry.

An example protocol for these model studies would include preparation of the Fe crystal. It would be mounted, sputtered clean with Ar^+ and annealed, then characterized with LEED (STM) and Auger electron spectroscopy. Then the Fe crystal would be moved into the reactor where it would see $P_{N_2} \sim 5$ atm, $P_{H_2} \sim 15$ atm at ~600–700 K. After a given run time, the crystal would be moved back into the ultra-high vacuum system for analysis with TPD, IR or HREELS, and/or XPS.

Typical reaction rates, R in moles NH_3 as a function of crystal orientation of the Fe substrate are shown in figure 8.12. It appears that Fe(111) shows the fastest NH_3 production.

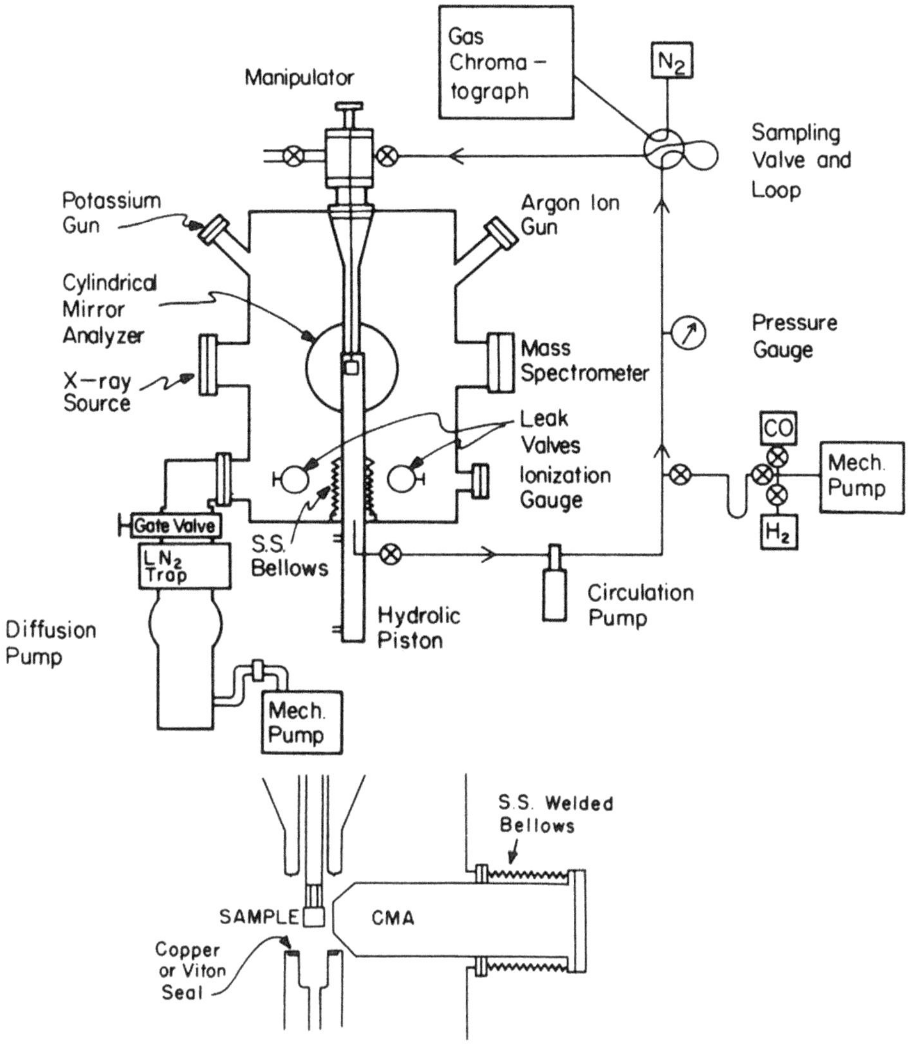

Figure 8.11. Schematic of a laboratory system for model system. The system is a combination of a high-pressure system for reaction and a low-pressure system for analysis. Reprinted from [6], Copyright (1997), with permission from Elsevier.

What is the reason for the structure sensitivity? Fe is body-centered cubic. The (111) plane is very open, compared to the (110) and (100) planes. In fact, the (110) plane is the close packed plane for body-centered cubic structures.

A follow-up experiment in ultra-high vacuum conditions studied the breaking of the strong $N \equiv N$ bond. Sticking of N_2 on Fe should only occur when dissociation occurs. So sticking was measured for (110), (100), and (111) orientations. The activation energy for adsorption was determined. The lowest activation energy for N_2 sticking should correspond to the highest rate of reaction, R_{NH_3} in the reactor at high pressure. The activation energy was found to be 0 for the open Fe(111) surface,

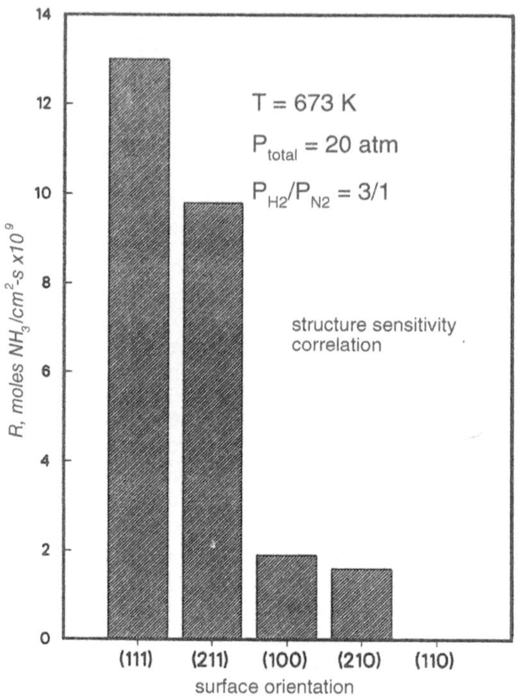

Figure 8.12. Rates of NH_3 synthesis over five Fe different crystal faces. Reprinted from [7], Copyright (1987), with permission from Elsevier.

27 kJ mol^{-1} for the close packed Fe(110) surface, and 21 kJ mole^{-1} for the intermediate Fe(100) surface.

The sticking coefficient was determined by monitoring the coverage of N_2 on the surface as a function of dosing time as shown in figure 8.13. It is clear that at 693 K that N_2 sticks most rapidly on the Fe(111) surface and least on the Fe(110) surface.

Experiments to study the effect of the Al_2O_3 substrate on the reaction. Some of the experiments suggest that Al_2O_3 can react with relatively unreactive Fe(110) and Fe(100) to form more active Fe(111) regions.

Another important aspect in catalysis of ammonia formation is the mechanism of K promotion of N_2 sticking. There are ultra-high vacuum single crystal studies illuminating the mechanism of promotion of N_2 dissociative chemisorption on Fe (100). When 1.5 × 1014 K atoms cm^{-2} are adsorbed on Fe(100) the sticking coefficient of N_2 goes from 2×10^{-7} to 4×10^{-5}, an increase by 200 times (shown below in figure 8.14). The relative N_2 dissociation rate also increases by a factor of more than 200.

The operative mechanism is that when K is adsorbed, it converts to K^+ and the electron is donated into the valence band of Fe. From there, the excess electron density is donated to the $2\pi^*$ antibonding level of N_2. This lowers the amount of energy required to break the N≡N bond.

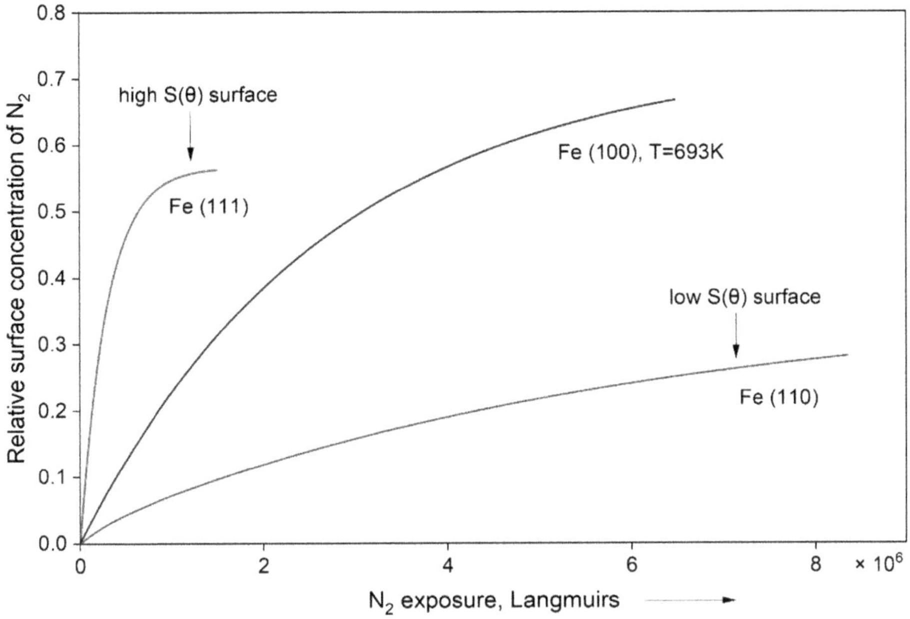

Figure 8.13. Variation of the relative surface concentration of atomic N_2 as a function of exposure. Reprinted from [8], Copyright (1977), with permission from Elsevier.

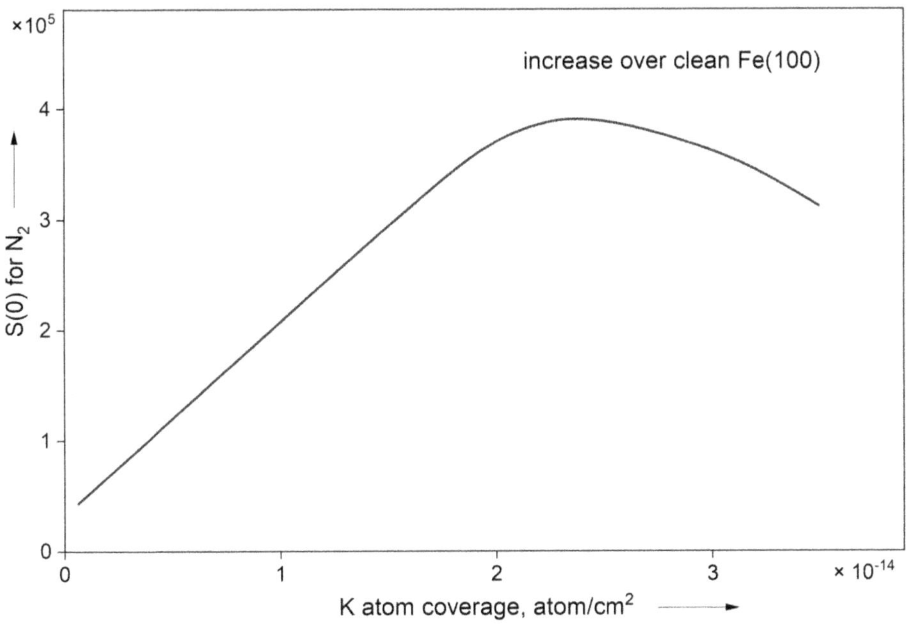

Figure 8.14. Variation in the initial sticking coefficient of with the addition of K to Fe(100) at 430 K [1].

Another effect of K adsorption is that it increases the rate of NH_3 desorption from Fe(111) surfaces, thereby leaving more space on the catalyst for adsorption of nitrogen and hydrogen. See the figure 8.15 where the temperature programmed desorption rate of NH_3 is plotted as a function of temperature, in K, for K coverages from zero (clean Fe(111), bottom) to 1.0 monolayers. Note the sharp increase of low temperature NH_3 desorption with 1 monolayer of K. It is estimated that the activation energy for NH_3 desorption has lowered by ~2.5 kcal mol^{-1}.

Thus far, ultra-high vacuum experiments have confirmed that K adsorption of Fe increases the rate of dissociative adsorption of N_2 and increases the rate of NH_3 desorption.

The enhancement in the rate of NH_3 production from N_2 and H_2 has been confirmed in reactor studies. Figure 8.16 plots the natural logarithm of the reaction rate, ln R, in moles NH_3 $cm_2 \cdot s^{-1}$ for clean Fe(100) and 0.15 monolayer of K on Fe (100) versus $1/T \times 1000$, in K^{-1}.

While the rate of reaction increases substantially, the activation of barrier height is unchanged by the adsorption of K, suggesting that general mechanism of reaction is unchanged.

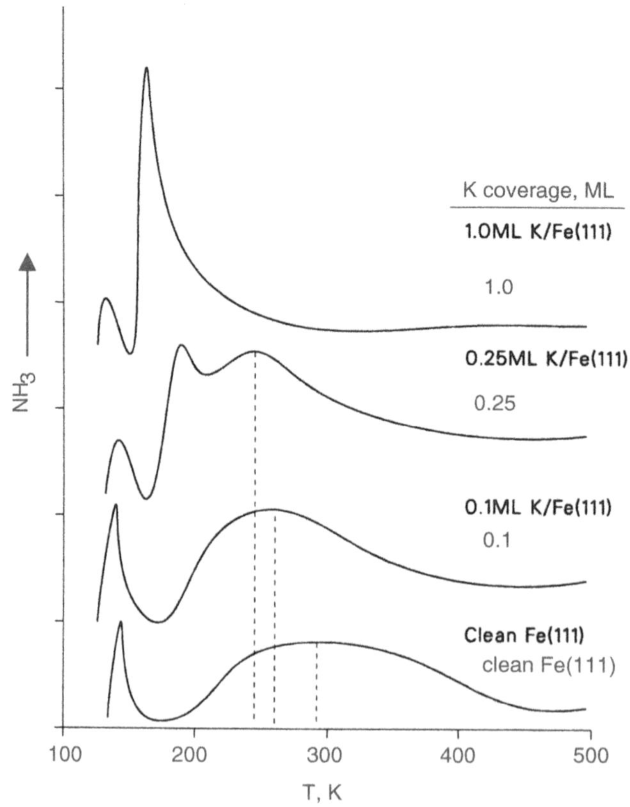

Figure 8.15. Temperature programmed desorption of NH_3 from Fe(111) and increasingly K-coated Fe(111) surfaces. Reprinted from [10], Copyright (1988), with permission from Elsevier.

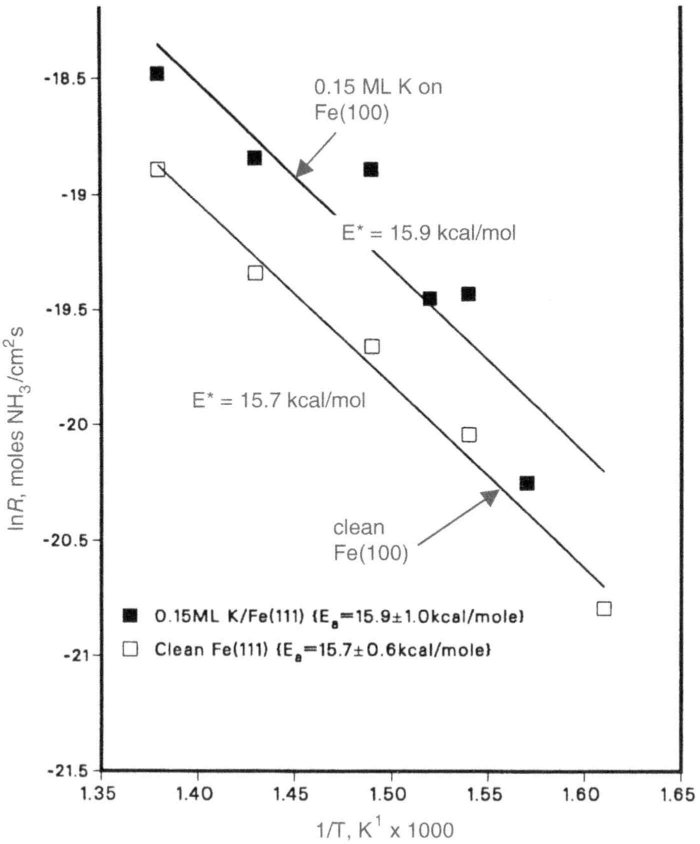

Figure 8.16. The ln R versus $1/T$. The lack of change of slope between the reaction on clean Fe(111) and K/Fe (111) suggests that the mechanism has not changed. Reprinted from [10], Copyright (1988), with permission from Elsevier.

Thus, if the overall reaction mechanism is

$$H_2(g) \xrightarrow{Fe} 2Fe - H$$

$$N_2(g) \xrightarrow{Fe} 2Fe - N$$

$$Fe - H + Fe - N \rightarrow Fe - NH$$

$$Fe - H + Fe - NH \rightarrow Fe - NH_2$$

$$Fe - H + Fe - NH_2 \rightarrow Fe - NH_3$$

$$Fe - NH_3 \rightarrow Fe + NH_3(g).$$

Then laboratory ultra-high vacuum experiments have demonstrated that K adsorption accelerates the second and sixth reaction steps, outlined in red.

Further reading

1. Somorjai G A and Li Y 2010 *Introduction to Surface Chemistry and Catalysis* 2nd edn (John Wiley). A general advanced undergraduate or beginning graduate level text focused primarily on catalysis and related UHV surface science areas. Chapter 14 covers many of the points given in the current chapter.
2. Thomas J M and Thomas W J 1997 *Principles and Practice of Heterogeneous Catalysis* (VCH Publishers).

Chapter 8 Problems

1. Calculate the reaction probability of a catalytic reaction that has a turnover rate of 10^{-3} molecules/surface site/s at 1 atm.
2. The determination of the equilibrium constant of NH_3 formation $N_2 + 3H_2 \rightarrow 2NH_3$ has been performed by Haber and Nernst. They obtained different results with the use of different catalysts. Review the literature [2] on these studies and describe the outcome of this important debate in the history of catalysis.
3. The composition of a gas emerging from the secondary reformer of a modern ammonia plant has the following composition: 39.5% H_2, 16.3% N_2, 28.3% H_2O, 4.5% CO_2, 10.7% CO, 0.2% NH_3, and 0.5% CH_4. Assuming that equilibrium is established, what would the CO content of this gas be if it were fed directly to the first stage of a shift converter employing an iron oxide/chromium oxide catalyst and operating at 450 °C? The equilibrium constant for the reaction $CO + H_2O \leftrightarrow CO_2 + H_2$ is 7.337 at 450 °C. Why is normal practice to have a low temperature (about 250 °C) shift converter as a second stage? Would the same catalyst as used for the high-temperature (first) stage be and appropriate choice for the second stage?
4. Nitric oxide leads to the depletion of the ozone layer ($NO + O_3 \rightarrow NO_2 + O_2$). Since NO is liberated in vast quantities from power stations currently in use, efforts are now underway to reduce its concentration by catalytic ammoxidation using V_2O_5–TiO_2. One such reaction is: $4NH_3 + 3O_2 \rightarrow 2N_2 + 6H_2O$. Write down other feasible reactions leading to N_2 or N_2O and H_2O as sole products. From thermodynamic data, estimate the equilibrium constants of these reactions.

References

[1] Somorjai G and Li Y 2010 *Introduction to Surface Chemistry and Catalysis* 2nd edn (John Wiley) ch 9 p 565
[2] Sinfelt J H 1991 Catalytic hydrogenolysis on metals *Catal. Lett.* **9** 159–71
[3] Coudert F X 2020 File:Zeolite structure as an assembly of tetrahedra.png *Wikimedia Commons* https://commons.wikimedia.org/wiki/File:Zeolite_structure_as_an_assembly_of_tetrahedra.png

[4] Ertl G 1991 Elementary steps in ammonia synthesis: the surface science approach *Catalytic Ammonia Synthesis: Fundamentals and Practice, Fundamental and Applied Catalysis* ed J R Jennings (Plenum)
[5] Nørskov J K and Stoltze P 1987 Theoretical aspects of surface reactions *Surf. Sci.* **189–90** 91–105
[6] Somorjai G A and Yang M X 1997 New directions of catalysis research *J. Mol. Catal.* A***115*** *389–403*
[7] Strongin D R, Carrazza J, Bare S R and Somorjai G A 1987 The importance of C7 sites and surface roughness in the ammonia synthesis reaction over iron *J. Catal.* **103** 213–5
[8] Boszo F, Ertl G, Grunze M and Weiss M 1977 Interaction of nitrogen with iron surfaces, I. Fe(100) and Fe(111) *J. Catal.* **49** 18–41
[9] Bozso F, Ertl G and Weiss M 1977 Interaction of nitrogen with iron surfaces: II. Fe(110) *J. Catal.* **50** 519–29
[10] Strongin D R and Somorjai G A 1988 The effects of potassium on ammonia synthesis over iron single-crystal surfaces *J. Catal.* **109** 51–60
[11] Williams F E 2010 File:Haber-Bosch-En.svg *Wikimedia Commons* https://commons.wikimedia.org/wiki/File:Haber-Bosch-En.svg

IOP Publishing

Surface and Interface Science

David L Allara and Robert L Opila

Chapter 9

Forces at the nanoscale—controlling interactions between nano-objects

Intermolecular forces embrace all forms of matter and determine the properties of gases, liquids and solids. These include dissolved solute molecules, small molecular aggregates, or macroscopic particles interacting in liquid or vapor. These interactions ultimately determine the behavior and properties of everyday things, soils, milk and cheese, paints and ink, adhesives and lubricants.

9.1 Overview

All atomic and molecular matter contains positive and negative charges. These charges attract and repel following Coulomb's law. These interactions roughly divide into short- and long-range, as shown in figure 9.1.

The red line shows short-range interactions such as those seen in chemical bonding. The blue line shows longer range interactions such as those seen in van der Waals interactions. For atoms and molecules these last are taken to attract like r^{-6} at longer distances, and to repel like r^{-12} at shorter distances.

The ability to stabilize and manipulate nanoparticles depends critically upon the interparticle and surface–particle attractive forces. These forces, in turn, depend upon the size, shape, materials, and surface adsorbates of the nanoparticle.

All collections of material objects have electrical forces operating between them. Ions exist in solid lattices, in solution, and in polymer ionomers. Static dipoles are present in molecules and polar solids such as mica. Molecules and polar solids can higher order multipoles (quadrupoles, octupoles). Finally charge clouds that polarize in response to nearby charges are present in ions in solution, ions in solid lattices and polymers, and polarizable electron clouds in solids, liquids, molecules, and atoms.

doi:10.1088/978-0-7503-4916-1ch9

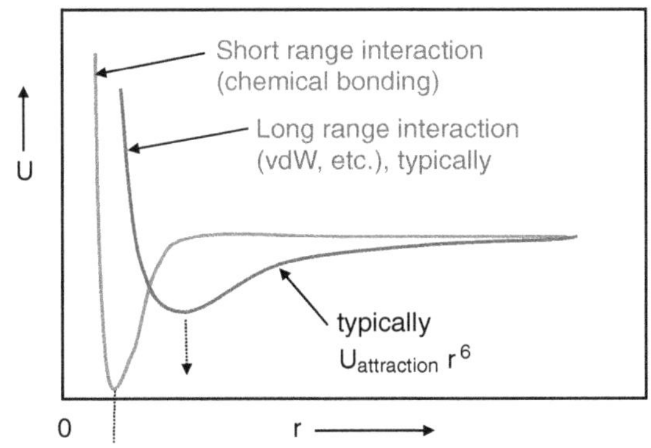

Figure 9.1. A plot of electrostatic potential energy as a function of distance. The blue curve shows long-range interactions compared to the red curve, which shows how the chemical potential might vary for interactions such as chemical bonding.

There are many kinds of electrical forces for nano-objects, including:

- Charge–charge, which attract or repel.
- Charge–static dipole, which attract and may cause the static dipole to rotate.
- Static dipole–static dipole, which can cause rotation of one or both dipoles, and may attract or repel, depending on the freedom of the dipoles to rotate.
- Static dipole–static dipole, including average over thermal energy and rotation. This force is always attractive and is called the *Keesom dipole* attraction.
- Static dipole–induced dipole, which are attractive and interact through the Debye induction force.
- Transient dipole–induced transient dipole, which are attractive and attract through the London dispersion attraction.

Multipole interactions are not included in the table. These do not generally make significant contributions to total interactions.

Dipole–dipole interactions are most important in particle-particle interactions so they will be discussed in some detail below.

Consider a population of identical molecular dipoles with thermally averaged orientation. A random alignment of a set of dipoles will yield no net electric field. But, an equilibrium alignment of molecules will be driven by stabilizing interactions, so a net attraction will result. If we take a Boltzmann distribution of all orientational states, the lowest energy states will prevail. Without derivation the average energy of this between two of these dipoles is

$$\left\langle U(\mu_1 - \mu_2)\right\rangle_\theta = U_{\text{Keesom}} = -\kappa_1^2 \frac{\mu_1^2 \mu_2^2}{3kT \cdot r^6}, \tag{9.1}$$

where μ_i is the dipole moment of i, and r is the distance between them. The dielectric constant is κ_1. This approach yields the Keesom dipole force and decreases as $1/r^6$. It is always attractive.

If a dipole is near a polarizable molecule, it will induce a dipole. The interaction of the dipole and the induced dipole are always attractive. The strength of the attraction depends on the local polarizability, α, and the distance, r, between them. Now the induced dipole may not be absolutely parallel to the permanent dipole, because the direction of polarization may not be parallel to the permanent dipole. The energy of interaction is taken as

$$\left\langle U(\mu - \mu_{\text{ind}}) \right\rangle_\theta = U_{\text{Debye}} = -\kappa_1^2 \left(\frac{\alpha \mu^2}{r^6} \right). \tag{9.2}$$

The Debye dipole induction is not temperature dependent since μ_{ind} cannot turn independently by thermal forces to average its orientations. So if μ shifts, μ_{ind} reforms instantaneously (on femtosecond or shorter time scales). Note the U_{Debye} is always attractive and decreases at $1/r^6$.

A typical theoretical prediction for the attraction of London dispersion (induced dipole–induced dipole) for atoms and molecules is

$$f_{\text{LD}}(\alpha, \nu) = \frac{3}{2} \kappa_1^2 \frac{\alpha_1 \cdot \alpha_2}{(1/h\nu_1 + 1/h\nu_2)}, \tag{9.3}$$

where α_i is the polarizability of i and ν_i is the photo-ionization frequency of i (the energy of the highest lying electron). This model reflects that the London dispersion interactions come from the outer valence electrons.

Transient dipole interactions arise from fluctuation of the outer shell electrons in atoms. Electrons in atomic/molecular electrons fluctuate, they are in hybridized atomic states. As a result, transient dipoles arise. It is excitation of the valence electrons that give rise to the largest dipoles. Valence transitions are in the optical frequency region ($\lambda \sim$100–300 nm, $\nu \sim 10^{15}$ Hz). Valence state transitions are excited with the lowest energy so they have the highest probability of fluctuate. A transient dipole $[\mu(\nu)]$ has a lifetime of $\sim 1/\nu$ ($\sim 10^{-15}$ s $=$ 1 fs). The transient dipole, $\mu(\nu)$, sustains an electric field that reaches a nearby polarizable object. On arrival, it induces a dipole, $\mu(\nu)_{\text{ind}}$ which is in opposite polarization to the original dipole. The induced dipole then radiates an electric field back to the original dipole; the dipoles in opposite polarization attract.

Now consider the distance dependence of the $\mu(\nu) - \mu(\nu)_{\text{ind}}$ attraction energy. If $\mu(\nu)$ and $\mu(\nu)_{\text{ind}}$ are close to each other, the attraction energy, U, follows the London dispersion potential,

$$\langle U[\mu(\nu) - \mu(\nu)_{\text{ind}}] \rangle = U_{\text{London}}, \tag{9.4}$$

which decreases as $1/r^6$.

As the dipole–induced dipole move further apart, the distance between them contributes a *retardation effect*. If the dipoles are separated, we need to consider the time it takes for the transiently formed electric field to travel to the polarization location and back. The signal travels at the speed of light and arrives at the site of the induced dipole in a time τ, over a distance r, to produce μ_{ind}. The return signal arrives at a time 2τ to interact with the source dipole. In the time 2τ the source

Table 9.1. Distance dependence of Coulombic interactions.

Interaction	U distance dependence	Comment
q_1–q_2	$1/r$	Sign depends on charge signs
μ–q	$1/r^3$	Sign depends on dipole direction
μ–μ	$1/r_3$	Sign depends on orientations
μ–μ	$1/r^6$	kT orientation avg.; Keesom dipole force
μ–μ_{ind}	$1/r^6$	Orientation avg.; no T-dependence *Debye induction force*
$\mu(\nu)$–$\mu(\nu)_{ind}$	$1/r^6$	Fluctuating dipole–induced dipole; *London dispersion force*
Multipole interactions	$1/r^n$	Fluctuating quadrapole, octapole, etc, interactions; $n > 6$

oscillating dipole has a new phase $\mu(\nu, \phi)$. The delay from the round trip can put $\mu(\nu, \phi)$ and returning $\mu_{ind}(\nu)$ signal our of phase and diminish the interaction force; this is the retardation effect. Typical retardation effects can occur at $r > 10$ nm.

The distance dependence of attraction/repulsion Coulomb energies are summarized in table 9.1.

The general form of the London dispersion energy is

$$\langle U[\mu(\nu) - \mu(\nu)_{ind}]\rangle_\theta = \langle U\rangle_\theta = f_{LD} \cdot \frac{1}{r^6}. \tag{9.4}$$

Because the direction of $\vec{\mu}_{ind}$ may be different than the incident $\vec{E}$, depending on the intrinsic directions along which the polarizable electrons move. The angle θ is the angle between the incident electric field and the polarization.

Various approximations to the London potential function have been made based on the behavior of the polarizable, outer valence electrons in the interacting atoms. General forms are

$$f_{LD}(\alpha_1, \alpha_2, \nu_1, \nu_2) \tag{9.5a}$$

$$f_{LD}(\alpha_1, \alpha_2, n_1, n_2) \tag{9.5b}$$

$$f_{LD}(\alpha_1, \alpha_2, X_1, X_2), \tag{9.5c}$$

where the α_i, are the polarizabilities of atom i, ν_i are the frequencies of the lowest electronic transition of atom i, n_i are the refractive indices of atom i in the optical region, and the X_i, are the diamagnetic susceptibilities of atom i.

van der Waals attraction energies are the sum of all r^{-6} interactions, including London dispersion, Debye, and Keesom interactions. van der Waals forces are present in non-ideal gases and provide an experimental method to determine the van der Waals interactions. Instead of writing the ideal gas equation as $PV = RT$ for the energy of one mole of ideal gas, one writes

$$\left(P + \frac{a}{\overline{V}^2}\right)(\overline{V} - b) = RT, \tag{9.6}$$

where $\overline{V}$ is gas molar volume, b is the correction for the volume of the constituent molecules, and a corrects for the loss of internal energy, U, compared to an ideal gas, from intermolecular attractions via van der Waals forces. Note that a has units of $1/\text{m}^6$. As a result, the van der Waals gas law provides a way to determine f_6 for inert atoms and molecules.

The attractions between atoms and molecules involves a variety of interactions with different distance dependencies. We can write a general expression

$$U_{\text{attraction}}(r) = -\sum_j f_j \cdot r^{-j} = -[\cdots\cdots + f_6 \cdot r^{-6} + f_8 \cdot r^{-8} + f_{10} \cdot r^{-10} + \cdots]. \tag{9.7}$$

Terms of order less than 6 represent interactions that include positive or negative charges. Terms of order great than 6 reflect interactions that include fluctuating multipoles. These are much smaller the terms of order 6. So for neutral particles, van der Waals contributions, of order 6, dominate the attractive interactions.

The total interaction potential includes repulsive forces as well; $U_{\text{total}}(r) = U_{\text{attraction}} + U_{\text{repulsive}}$. Typically, the attractive and repulsive energies serve as the basis of the standard 6–12 potential,

$$U_{\text{attraction}}(r) = -f_{\text{vdW}} \cdot r^{-6} \tag{9.8a}$$

$$U_{\text{repulsion}}(r) = +f_{\text{Pauli}} \cdot r^{-12}. \tag{9.8b}$$

Overall, neutral atoms have only London dispersion forces. Nonpolar molecules interact primarily through London dispersion forces. If there is an assembly of polar molecules with no hydrogen bonding London dispersion will still dominate, but Keesom and Debye forces can account for up to ~15%. For molecules that interact through hydrogen bonding London dispersion becomes secondary; hydrogen bonding dominates through special state dipole interactions that involve H atoms shifting positions to enhance charge interactions with neighboring molecules.

9.2 Hamaker constants—forces between macroscopic objects

Atoms, molecules, clusters, and colloids can be maintained in solution or gas by thermal energy. Larger particles, including larger clusters, colloids, and micrometer scale objects tend to precipitate or aggregate. Dominant aggregating forces include electrostatic (opposite charges) and van der Waals forces. Dispersing forces include electrostatic (such as charges), Pauli repulsion, and solvation.

Attractive forces between neutral (non-magnetic) objects can be calculated. The microscopic approach is to integrate over all of the atom–atom van der Waals interactions. The integration strengths are given by the van der Waals potential f_6^{1-2} between atoms 1 and 2 and the distance between them. The total interaction is given by the interaction strengths and the total number of atomic interactions, which depend on the shapes of the objects and their densities, r_1 and r_2. The final value is obtained by integration over all of the atoms. The general atom–atom potential form is

$$U(\vec{r}_{1-2}) = -f_6^{1-2} \cdot r_{1-2}^{-6} \tag{9.9}$$

and the force is

$$-\frac{\mathrm{d}U(r)}{\mathrm{d}r}. \tag{9.10}$$

Consider an atom interacting with an infinite slab in the x- and y-directions as shown in figure 9.2. If we integrate $U(\vec{r})$ from $-\infty$ to $+\infty$ in the x- and y-directions (across volume elements in a slab at fixed y) and integrate z from 0 to $-\infty$, we can calculate the van der Waals interaction between atoms in slab (1) and atom (2). The result is the physisorption energy for a single atom near a slab,

$$U(D) = -\frac{\pi}{6}\rho_n f_6^{1-2} \cdot D^{-3}, \tag{9.11}$$

where D is the distance between the atom and the surface of the slab and r_n is the atomic density in the slab. Note that

$$U(D) \propto \frac{1}{D^3}. \tag{9.12}$$

Now consider the attraction between two parallel slabs, slab 1 and slab 2, separated by a distance, D, shown in figure 9.3.

To calculate the attraction between the two slabs, select one atom inside the second slab and integrate the attraction of all atoms in the first slab. Repeat this integration for each atom in the second slab. The attraction energy per unit area is

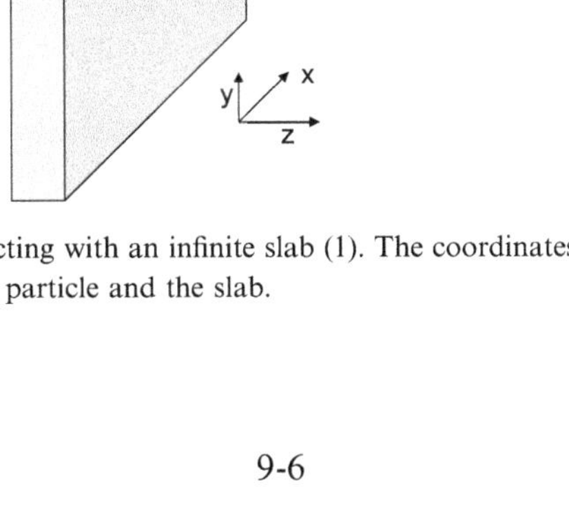

Figure 9.2. A particle (2) interacting with an infinite slab (1). The coordinates, x, y, and z are given as is the normal distance, D between the particle and the slab.

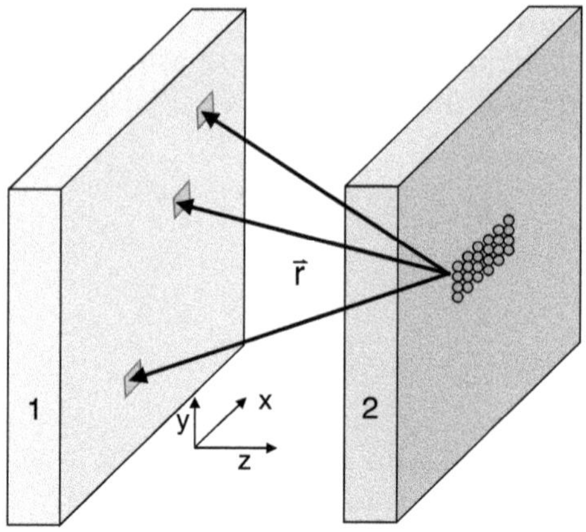

Figure 9.3. Two slabs (1) and (2) interacting. The coordinates, x, y, and z are given.

$$U(D) = -\frac{\pi}{12}\frac{\rho_1\rho_2 f_6^{1-2}}{D^2}\ \mathrm{J\,m^{-2}}. \tag{9.13}$$

For convenience we will collect all of the material related constants into a single constant, the Hamaker constant, A_{12}. Now $A_{12} = \pi\rho_1\rho_2 f_6^{1-2}$, and

$$U(D) = -\frac{1}{12}\frac{A_{12}}{D^2}. \tag{9.14}$$

Note that the distance dependence here is D^{-2} versus D^{-3} for an atom over a slab.

For a sphere, with a radius r, interacting with a slab (e.g. a colloid particle and a wall), if $D \ll r$

$$U(D) \sim -\frac{r}{6}\frac{A_{12}}{D}. \tag{9.15}$$

If $D \gg r$ the colloid-slab potential looks very much like an atom interacting with a slab,

$$U(D) \sim -\frac{\pi}{6}\frac{f_6^{1-2}\rho_1}{D^3}. \tag{9.16}$$

Similarly for two finite particles of radii r_1 and r_2, interacting, when $D \gg r_1, r_2$

$$U(D) \sim -\frac{1}{6}\frac{r_1 \cdot r_2}{(r_1 + r_2)}\frac{A_{12}}{D}. \tag{9.17}$$

At large distances, $D \gg r_1, r_2$, the interaction becomes the same as atom–atom interactions,

$$U(D) \sim -f_6^{1-2}\frac{1}{D^6}. \tag{9.18}$$

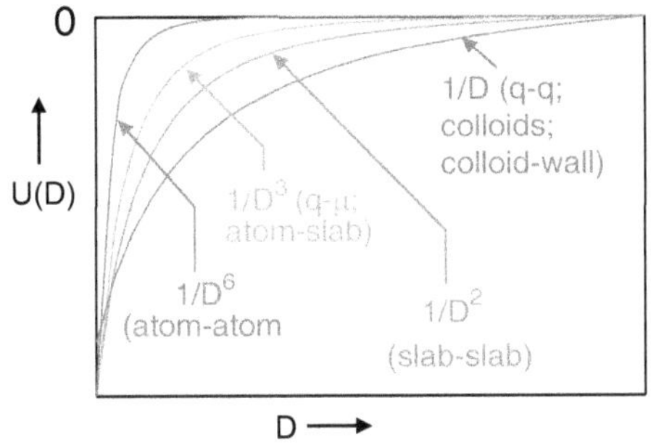

Figure 9.4. A summary of the distance dependence of selected van der Waals and Coulombic attraction energies for common objects

A summary of the distance dependence of selected van der Waals and Coulombic attraction energies for common objects is shown in figure 9.4. The $1/D$ interaction pertains to charge–charge interaction and at short distances colloid–colloid and colloid–wall interactions. The $1/D^2$ interaction is relevant for slab–slab interactions. The $1/D^3$ function pertain to charge–dipole and atom–slab interactions. Finally, $1/D^6$ is relevant for atom–atom van der Waals interactions.

There is a macroscopic approach to determine Hamaker constants. This method follows Lifshitz theory and involves integration over continuum, macroscopic interactions. This technique is a more accurate way to calculate Hamaker constants than atom by atom integration.

Condensed matter objects usually have no net dipole. Thus, static dipole-based forces between objects are negligible. London dispersion (LD) forces dominate the van der Waals interactions. The London dispersion interactions are calculated on the continuum properties of the objects. This accounts for the conduction electrons in metals, for example, that are unique properties of the collections of atoms in the object. Instead of integrating over all of the discrete atoms in the objects, continuous polarizabilities are assigned to the object on a per volume basis. These continuum properties are used to generate the f_{LD} functions for the two interacting objects. Infinitesimal volumes are integrated to the obtain the London dispersion attraction energy.

Various approximations have been made to the London potential function, f_{LD}, based on the behavior of the polarizable, outer valence electrons in the interacting atoms. General forms include

$$f_{\mathrm{LD}}\,[\varepsilon_1(\nu),\ \ \varepsilon_2(\nu)] \tag{9.19a}$$

$$f_{\mathrm{LD}}\,(\alpha_1,\, \alpha_2,\, X_1,\, X_2), \tag{9.19b}$$

where e_i, a_i, and X_i are the real and imaginary dielectric constants, polarizability, and diamagnetic susceptibilities of i, respectively. All of these properties are manifestations of the outer, valence electrons.

In the Lifshitz approach, the dielectric constant manifest in the optical spectra is used. Each object is treated as a continuum with macroscopic properties. The objects

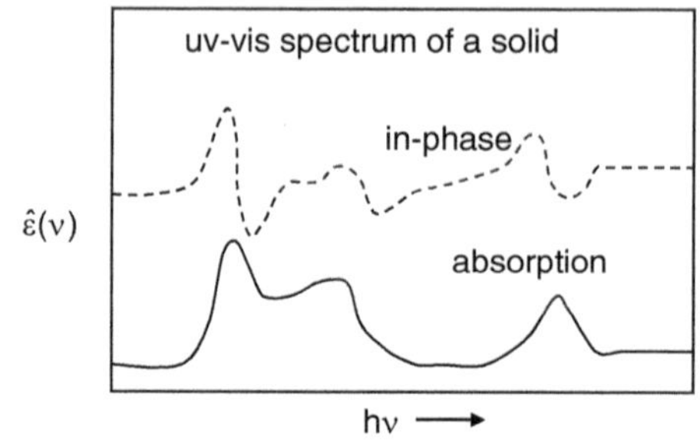

Figure 9.5. The real and imaginary parts of the dielectric constant.

Table 9.2. Hamaker constants of selected common materials [1].

Material	$A \times 10^{20}$ (J)
Acetone	4.2
Alumina	15.4
Gold	45.3
Magnesia	10.5
Metals	16–45
Rubber (natural)	8.58
Polystyrene	7.8–9.8
Silver	39.8
Toluene	5.4
H_2O	4.35

interact via internal oscillating electric fields and responsive induced fields. Oscillating fields are based on the complex dielectric constant spectrum over all frequencies $[\hat{\varepsilon}(\nu) = \varepsilon_1(\nu) + i \cdot \varepsilon_2(\nu)]$, where $\varepsilon_2(\nu)$ contains energy absorption or loss features and $\varepsilon_1(\nu)$ contains dipole and electron polarizations characteristics of the material for all electrons. See figure 9.5. Once f_{LD} is known, it can be integrated over all objects to obtain $U(z)$. A can then be obtained from $U(z)$. Since the valence electrons at the Fermi edge make the dominant contribution to e_1, A is estimated at the photo-ionization frequency, n_0. ($h\nu_0 \sim$ 4–10 eV or 125–300 nm). Lifshitz theory overcomes problems with microscopic atom integration approached, and gives better values of A.

Lifshitz theory gives a good approximation for non-metals in vacuum (less the 5%), with $A_{11} \sim 0.13\, h\nu_0 \frac{[\varepsilon_1(\nu_0) - 1]^2}{[\varepsilon_1(\nu_2) + 1]^{3/2}}$. For mica ($\lambda_0 \sim 100$ nm, $e = 7.0$ eV) $A_{11} \sim 0.94 \times 10^{-19}$ J.

Hamaker constants have been used widely to predict adhesion between material surfaces, for example, between particles and sheets. Thus, A_{11} has been determined for materials such as natural rubber, alumina, magnesia, and polystyrene (see table 9.2).

We have a few examples using Hamaker constants for materials separated by vacuum or air gaps. The first example is the adhesive van der Waals pressure between two ideally smooth, planar sheets of polyethylene in vacuum. The energy per contact area is

$$U_{\text{Area}}(D) = -\frac{1}{12\pi}\frac{A_{12}}{D^2}. \tag{9.20}$$

The pressure is just the force per unit area, or just the negative of the first derivative of the energy per contact area with respect to distance:

$$P = \frac{F(D)}{\text{Area}} = -\frac{\mathrm{d}U_{\text{Area}}}{\mathrm{d}D} = -\frac{1}{6\pi}\frac{A_{12}}{D^3}. \tag{9.21}$$

Taking the Hamaker constant for hydrocarbons, A_{11} to be approximately 5×10^{-20} J, and the contact distance, D to be about 0.2 nm gives a pressure of about -3.3×109 N m^{-2} or -3.3 kBar. The pressure is negative because the force is attractive and it about equal to 3300 atmospheres!

Another example would be the adhesive energy and force between two ideal spherical polyethylene particles (colloids) of radius 1 μm in a vacuum. The result will be compared to kT at 300 K (4.14×10^{-21} J). Again,

$$U(D) = -\frac{r^2}{12}\frac{A_{12}}{D}. \tag{9.22}$$

The force will be

$$F(D) = -\frac{\mathrm{d}U}{\mathrm{d}D} = -\frac{r^2}{12}\frac{A_{12}}{D^2}. \tag{9.23}$$

When the spheres are in contact ($D = 0.2$ nm), the energy of attraction is -2.1×10^{-17} J or more than 500 times kT at 300 k. Thermal energy is insufficient to separate the spheres. The interactive force is -1×10^{-7} N or approximately 100 nN. This force is much stronger than that in chemical bonds.

When the spheres are separated by 10 nm, the force is approximately -4×10^{-21} J, approximately the same as kT at 300 K. The attractive force is 4×10^{-15} N, or about 10^{-6} nN, far weaker than chemical bonds.

We can compare the polyethylene results to Au. Au is highly polarizable compared to polyethylene (PE). As a result, $A_{11}(Au)/A_{11}(PE) \sim 10$. The pressure between 2 Au plates in vacuum, at contact ($D = 0.2$ nm) is approximately 33 kBar, or 100 times the pressure for polyethylene plates. The adhesion energy of Au nanoparticles in vacuum for particles of 1 μm radius, with a separation of 0.2 nm is about 5000 kT. Particles of 10 nm radius, with a separation of 0.2 nm have an adhesion energy of 500 kT.

Thought question:

At what distance will Au nanoparticles with 10 nm radii have an adhesion energy of about 1 kT?

Figure 9.6. Gecko walking on a glass surface. Image by Bjørn Christian Torisson. (This Gecko foot on glass image has been obtained by the author(s) from the Wikimedia website where it was made available by Uspn under a CC BY-SA 3.0 licence. It is included within this book on that basis. It is attributed to Bjørn Christian Tørrissen. [4].)

Once particles come in close contact, van der Waals forces make them hold together tenaciously. This has implications for removing particles from clean surfaces in a semiconductor processing clean room. Part of the solution is to keep particles out of the clean room. Another part of the solution is carefully designed cleaning solutions to remove the particles from the surfaces.

Gecko's feet adhere to any reasonable smooth surface, such as glass, ceilings, trees, or GaAs (figure 9.6). Van der Waals forces support the full body weight. The gecko foot is composed of a series of pads, each of which makes intimate contact with the surface. How do they walk? By lifting the foot by removing one contact pad at a time.

9.3 Surface tension

Consider a nonpolar condensed phase held together by purely LD cohesive forces. The reversible work of de-cohesion to break molecular contact and create two separated surfaces is $W = 2\gamma$, where γ is the surface tension (see figure 9.7).

For a pure LD medium, the work to create two surfaces is just the energy needed to overcome slab–slab attraction:

$$W = -U(D_{\text{slab-slab}}) = +\frac{1}{12\pi}\frac{A}{D_0^2}, \tag{9.24}$$

where we assume that the slabs are in molecular contact with separation $D_{\text{slab-slab}} = D_0 \sim \sigma$. So, for liquids for which LD forces dominate cohesion: $A \sim 24\pi D_0^2 \gamma_\text{d}$ or $\gamma_\text{d} \sim \frac{A}{24\pi D^2}$. Here, the subscript d is used to denote dispersion forces. If other cohesive forces (Keesom, hydrogen bonding, etc) are acting then

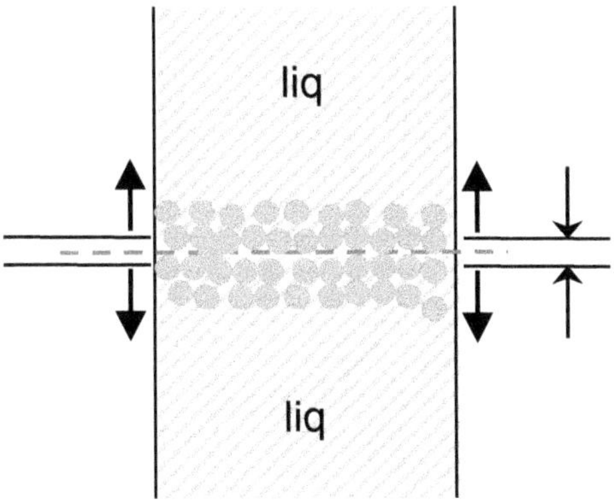

Figure 9.7. A schematic of the creation of two surfaces on a liquid.

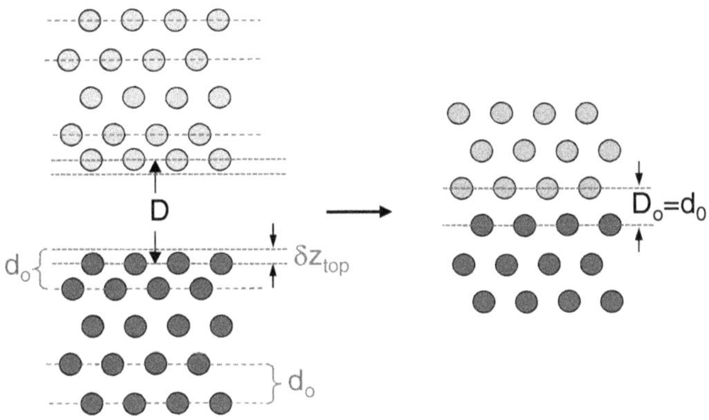

Figure 9.8. Creation of one interface from two separate surfaces.

$$\gamma = \gamma_{\text{non-LD}} + \gamma_{\text{d}} > \frac{A}{24\pi\sigma^2}. \tag{9.25}$$

How might the process of contact approach work? We must estimate D_0. Away from contact slab–slab forces are weak. The atoms in the top surface layer of the slabs are compressed to underlying layers because of the forces missing above. The top layer of atoms also has negative lateral pressure due to missing bonds.

Close to contact, slab–slab forces are nearly maximum. Atoms in the top layers of slabs are pulled back toward normal layer–layer spacings, D_0, by attractive forces from the opposite slab. The top layers of opposing slabs approach normal bulk potential energy. The negative pressure vanishes. The slab–slab spacing approaches D_0, which is about one atomic diameter, σ. Creation of an interface from two separate surfaces is shown if figure 9.8. The relationship between the potential energy and the surface tension is shown in the top of figure 9.9 In the bottom of figure 9.9, the separation between layers of atoms in the first and second layer is shown as a function of separation of the two surfaces.

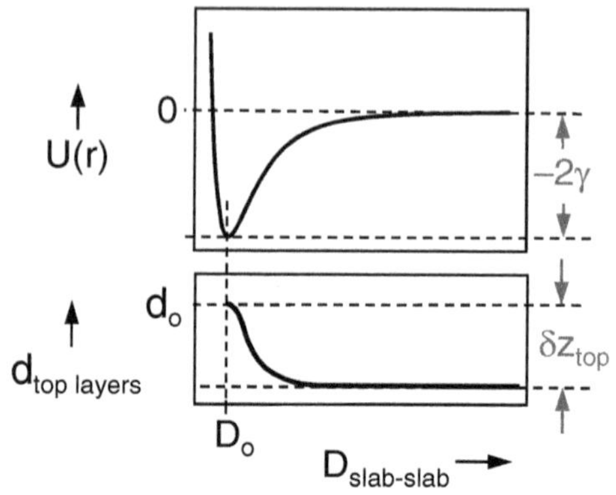

Figure 9.9. Top, the relationship between the potential energy and the surface tension. Bottom, the separation between layers of atoms in the first and second layer is shown as a function of separation of the two surfaces.

To test the relationship between A_{11} and γ_d we will:

1. Look up the A_{11} values calculated from fundamental constants (α, $h\nu_0$, etc).
2. For small symmetric molecules or atoms, we will estimate σ from the molar volumes of pure materials.
3. For larger asymmetric molecules we will estimate σ from the van der Waals radius of a constituent group (e.g. $-CH_3$ for hydrocarbons) that could represent the dividing line between two contacting slabs of material.
4. Calculate a value of γ and compare it to experimental results. For materials with cohesive forces operating in addition to London dispersion forces the calculated γ will be compared to the measured value of γ_d.

These results are tested in table 9.3. The agreement between calculation and measurement is quite good for London dispersion materials. Here the value of σ is calculated from molar volumes of pure material. For long hydrocarbons, σ is taken to be a $-CH_3$ unit. Of course, where metallic bonding or hydrogen bonding becomes important, the calculated surface tensions are not in good agreement with the measured values. Interestingly, for quartz, SiO_2, the agreement between the surface tension calculated from A_{11} and the measured value are in good agreement.

These results are tested in the table 9.3. The agreement between calculation and measurement is quite good for London dispersion materials. Here the value of σ is calculated from molar volumes of pure material. For long hydrocarbons, σ is taken to be a $-CH_3$ unit. Of course, where metallic bonding or hydrogen bonding becomes important, the calculated surface tensions are not in good agreement with the measured values. Interestingly, for quartz, SiO_2, the agreement between the surface tension calculated from A_{11} and the measured value are in good agreement.

9.4 Hamaker constants for stacked objects with intervening media

Common case has two flat plates of materials 1 and 2, separated by a liquid, 3, as shown in figure 9.10. Since forces are additive and proportional to the A_{ij}, then the

Table 9.3. Surface tension for several materials interacting through London dispersion forces and other forces.

Molecule	A_{11}[(a)] × 10^{20} (J)	σ[(b)] (nm)	Surface tension, γ (mJ m^{-2}): Calculated from A_{11}	Measured	Dispersion component, measured (γ_d)	Non-LD contribution
LD						
n-C_7H_{16}	1.05	0.40[(c)]	9	20	20	~0
n-$C_{20}H_{42}$	2.07	0.40[(c)]	17	29	29	~0
Polystyrene	2.20	0.30[(d)]	32	~40	41	~0
CCl_4	5.5	0.54	25	26.3	26.3	~0
LD+						
Hg	33	0.29	520	425	200	Metallic bonding
Glycerol	6.7	0.50	36	64.0	34	H-bonding
H_2O	3.7	0.31	51	72.8	21.8	H-bonding
SiO_2 (quartz)	6.5	0.64	75		78.0	

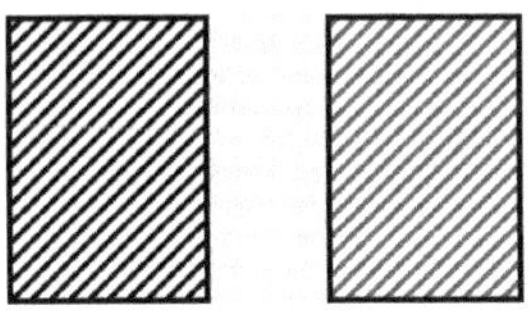

Figure 9.10. Material 1 (black) is separated from material 2 (blue) by a liquid 3. Materials 1 and 2 can approach each other, squeezing out the liquid. Similarly, materials 1 and 2 could separate further, with more liquid flowing between them.

A_{ij} are additive (for constant geometry). In the figure below we have flat materials 1 and 2 flat and separated by liquid 3. Here $F_{13} \propto A_{13}$, $F_{23} \propto A_{23}$, and $F_{12} \propto A_{12}$. Recall that forces are additive vectorially, so the A are also additive.

As materials 1 and 2 approach each other and the liquid flows out the attraction between materials 1 and 2 increases. The attraction between material 1 and the liquid and material 2 and the liquid decreases just because there is less liquid. Similarly, as materials 1 and 2 are drawn apart, the attraction between materials 1 and 2 decrease and the attraction between the solid materials and the liquid increases, because of its greater volume. Overall,

$$A_{123} = A_{12} + A_{33} - A_{13} - A_{23}. \tag{9.26}$$

Repulsive forces are possible. For examples slabs of materials 1 and 2 adhere by van der Waals forces. They are immersed in liquid 3 which can penetrate between materials 1 and 2. Suppose that liquid 3 sticks extremely well such that $A_{13} + A_{23} > A_{12} + A_{33}$. In that case $A_{123} < 0$ and objects 1 and 2 will be pushed apart when medium 3 is added between them. Essentially, liquid 3 'solvates' materials 1 and 2 by van der Waals forces.

An example of this is SiO_2 or Al_2O_3 spheres near Teflon. Forces that are initially attractive become repulsive in the presence of cyclohexane. Exactly similar approaches are used in designing cleaning steps to remove particles from semiconductor wafer surfaces during the fabrication of semiconductor devices.

9.5 Direct measurement of forces between solid objects

Direct measurement of forces between solid objects is difficult because it requires atomically flat surfaces in the absence of dust or other particulate and capillary wetting forces. Measurements have been made with quart fibers, precision balls on precision flat plates, crossed mica surface, and using atomic force microscopy (AFM). Many of these examples limit direct contact to a geometrical point.

Lee and Sigmund published an AFM study of the repulsive force (F) as a function of distance (z) for a flat Teflon surface and spheres of alumina or silica of various radii [2]. This was done in the presence of liquid medium of cyclohexane. The spheres were a 19 μm radius α-Al_2O_3 and a 21 μm radius sphere of amorphous SiO_2. The spheres were held at the end of AFM tips and the repulsive forces were noted as a function of separation between the sphere and the Teflon surface. Observations include:

1. When there is no liquid there is only attraction between the tip and Teflon.
2. In the presence of cyclohexane, the spheres are repelled from the Teflon surface.
3. The data fit the Lifshitz theory described above.

9.6 Colloidal systems

Colloidal systems are a dispersion of small (<0.5 μm) particles in a medium. The individual particles are not visible. They can pass through most filters. They are formed by grinding, chemical precipitation, or the addition of dispersion agents to precipitates or suspensions (e.g. clays).

Colloidal systems can be subclassified as:

- **Sol:**
  - Solid particles in a liquid medium.
  - Solid particles in a solid medium (e.g. colored glass).
- **Aerosol:**
  - Liquid particles in gas (fog).
  - Solid particles in gas (smoke).
- **Emulsion:**
  - Liquid particles in liquid (milk).
- **Lyophilic:** solvent attracting.
- **Hydrophilic:** water attracting.
- **Lyophobic**: solvent repelling.
- **Hydrophobic:** water repelling.

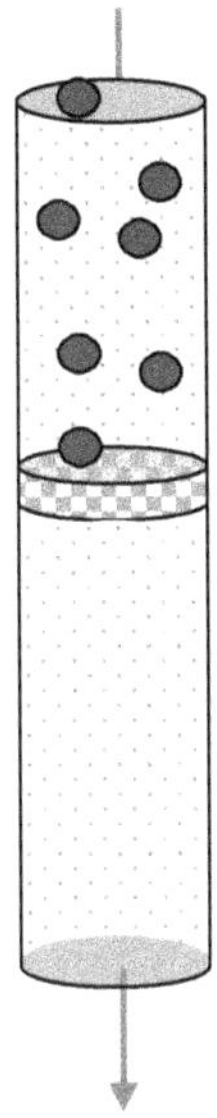

Figure 9.11. In this dialysis example there is a membrane with holes with radii less the radii of the particles. This concentrates the blue particles in the top of the tube.

Two methods of purifying theses systems are dialysis and centrifuging. In dialysis the mixture is flowed into a tube with a membrane with holes that are smaller than the particles (see figure 9.11). The solvent and ions then flow out of the tube.

Akbulut *et al.* were able demonstrate the ability to separate a distribution of sizes and shapes of gold nanoparticles on a desk-top centrifuge using aqueous multiphase systems [3]. Aqueous multiphase systems are phase-separated mixtures of water-soluble polymers and/or surfactants. Upon separation in a centrifuge the phases of the system order according to their densities. These phases often have different viscosities. The different densities and viscosities are used to separate the particles in the centrifuge.

What makes a colloid stable? Generally, small particles are unstable relative to large particles. There are two cases:

a. Molecules or atoms form a pure material.
b. Nano-objects, including particles, rods, molecular clusters agglomerate to form a heterogeneous collection of contacting individual object.

In the first case the bulk material is more stable than the component atoms or molecules. The driving force is to reduce the total area of free constituent surfaces by condensing or crystallizing and creating more favorable bonding or van der Waals interactions in the bulk. Recall that for any cluster growth that reaches a critical size, continued growth (decrease of the ratio surface/volume) is spontaneous. If surface free energy, γ, is the change in free energy with change in area, and

$$\gamma = \frac{\mathrm{d}G}{\mathrm{d}A} > 0 \text{ and } \gamma \mathrm{d}A = \mathrm{d}G.$$

If the area decreases, the surface free energy will decrease. Recall that once the surface is beyond a small but critical radius, the crystal will grow. When small objects are in equilibrium with larger objects, the larger objects will grow at the expense of the small objects. This phenomenon is known as Ostwald ripening.

Like the first case, van der Waals attractions drive aggregation of small nanoparticles by reducing free surface and replacing with favorable interparticle interactions. In the figure to the left, the particles are agglomerating. The particles may also stick on a cell wall. When agglomeration occurs in solution it is often referred to as flocculation.

Thought questions:

Because of universal dispersion forces, neutral particle by themselves will always tend to agglomerate or adhere to surfaces. So, how can small, neutrally charged particles be dispersed and stabilized?

Provide a method to separate particles to reduce the van der Waals attraction so thermal forces (kT) can maintain separation.

To prevent agglomeration particles must be coated in some way to counter attractive forces. The coating should be designed so the potential fall-off with distance is appropriate. Three possibilities are electrostatic repulsion (figure 9.12), solvation (Hamaker stabilization), or steric repulsion. Each of these effects acts to keep particle–particle distances outside of the flocculation range. The theory combining all these effects is called the DLVO theory, based on independent work by Derjaguin and Landau, and Verwey and Overbeek (see the further reading).

The stability of two charged sphere in monovalent salt solution has been studied as a function of separation of the spheres as function of ionic strength. The potential energy between two spheres with the same charge in solutions of different ionic strengths is plotted in figure 9.13. Here, the potential energy, in ergs and in values of kT, is plotted as a function of the normalized distance between the spheres. Here $s = R/a$, where R is the distance between the spheres with radii, a. K reflects the ionic strength of the solution:

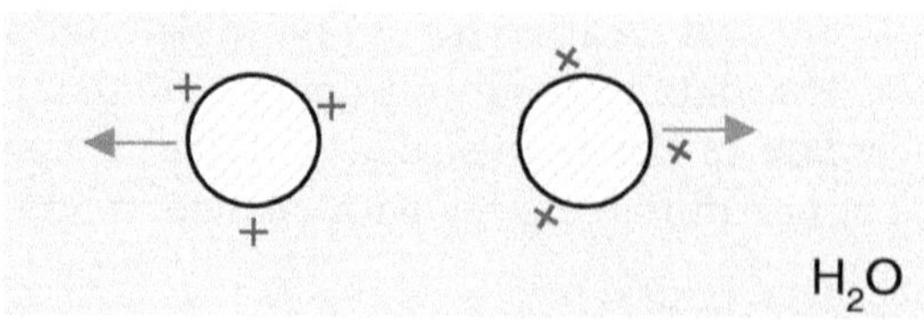

Figure 9.12. Colloidal stability through coating particles with positive charges.

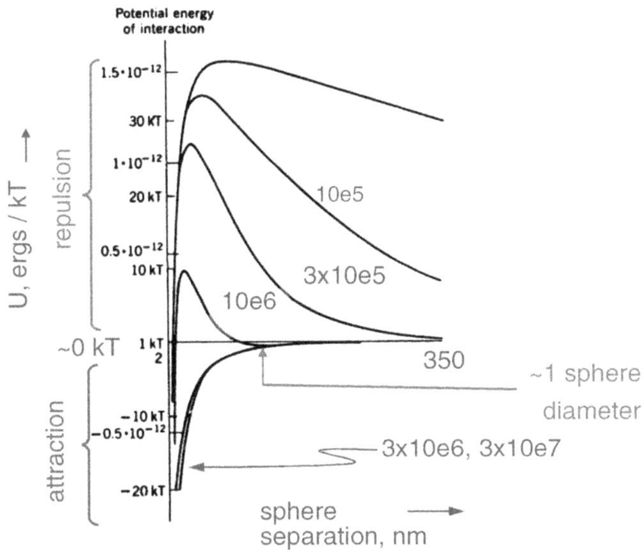

Figure 9.13. Showing the influence of the concentration of electrolyte on the total potential energy of the interaction.

$$K^2 \equiv \frac{2q_e^2}{\varepsilon k_B T} \cdot 1/2\sum_{j=1}^{J} z_j^2 c_j^0,$$

where K is the reciprocal of the Debye screening length, q_e is the elementary charge, ε is the permittivity of the medium, k_B is Boltzmann constant, z_J is the charge of ion J and C_J^0 is the concentration of ion J. The ionic strength is

$$1/2\sum_{j=1}^{J} z_j^2 c_j^0.$$

As ionic concentration increases, the ionic double layer thickness decreases, the van der Waals attraction increases, and the dispersion stability decreases. The ion double layer thickness decreases as K increases and as the Debye screening length decreases. The ion concentration threshold for flocculation occurs at approximately 1 kT of repulsion energy. So at 1 kT of repulsion energy, van der Waals forces take over and the spheres flocculate. Thus, the salt concentration can be used to tune dispersion energy.

One option for interparticle ionic (Coulomb) repulsion is to utilize intrinsic surface charge to provide particle repulsive forces. Noble metals with no surface oxide such as Au and Pt are typically negative. Metal oxides in aqueous media are usually charged; SiO_2 is negative and Al_2O_3 is positive. The surface charge of these oxides is often controlled by pH (acid-base chemistry). Chemisorbed ions on metals create surface charge (e.g. Cl^- or CN^- on Ag or Au). Au nanoparticles are often stabilized by small concentrations of strongly adsorbed citrate ions. Surface charge

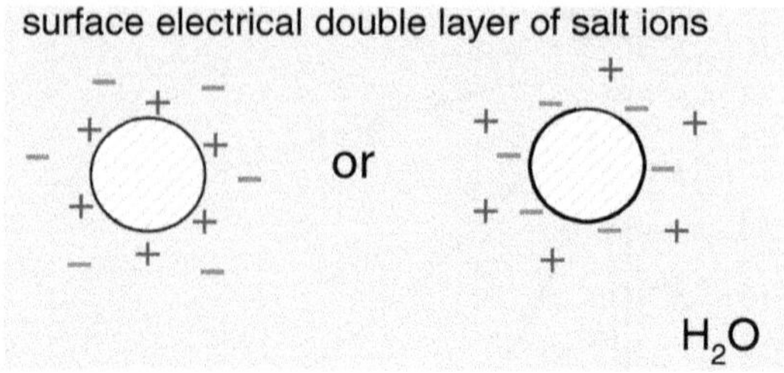

Figure 9.14. Interactions between particles with electrical double layers.

on bulk metal surfaces can be controlled by external applied voltage that prevents or encourages particle adsorption. In this case the metal is acting as an electrode.

Dissolved salts can form electrical double layers around intrinsically charged particle surface and mediate repulsion. A+ or B− goes to the particle surface depending upon the intrinsic particle surface charge. Counterions form a diffuse outer charge layer. With an inner and outer charged layer, an electrical double forms. Thus, the intrinsically charged particles still repel each other, but the presence of ion clouds reduce (or screen) the repulsion (see figure 9.14). The repulsion as a function of distance depends on the ionic strength of the salt. An increasing ionic strength increases screening of the charge and then reduces the Coulombic repulsion forces at a given distance. The best colloid stabilization occurs with low ionic strengths. Flocculation then is encouraged by high ionic strengths. Thus, salts can be used to control the stability of the colloid dispersion.

Colloid dispersions can be stabilized by the Hamaker effect. If the solvent has an affinity for the surface (lyophilic), particles may separate. Solvent affinity can involve short-range interactions, such as hydrogen bonding. It is very important to overcome long-range van der Waals attraction through the solvent polarizability (this shows up in the Hamaker contact). Recall in figure 9.10, repulsion of two flat plates of materials 1 and 2 occurs when the solvent 3 Hamaker constant yields $A_{13} + A_{23} > A_{12} + A_{33}$.

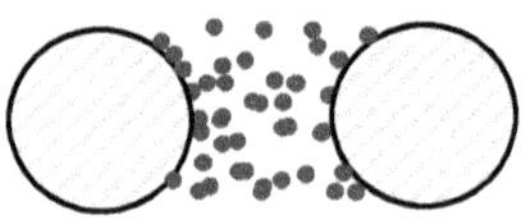

Steric repulsion is another strategy for stabilizing colloids. The basic idea is that the van der Waals attraction is much larger for a dense particle than a low-density coating. The low-density coating acts as a stop to keep the particle–particle distance outside of the flocculation threshold. Two possible methods to apply the coating are as a surfactant, where the surfactant is something like CH_3–$(CH_2)_{12}$–CO_2H, with the carboxylic acid binding to the nanoparticle. Alternatively, a polymer surfactant may act as a coating. In the example shown poly(ethylene oxide) (PEO) is acting as the coating and the oxygen atoms are closest to the surface of the nanopartice. With hydrophobic nonpolar coating, particle will repel each as the outer layers of approaching particles start to interpenetrate. This effect is often called osmotic repulsion.

Further reading

1. Verwey E J W 1947 Theory of the stability of lyophobic colloids *J. Phys. Chem.* **51** 631–6 10.1021/j150453a001. Foundational work in the stability of colloids.
2. Jacob N 2011 *Israelachvili, Intermolecular and Surface Forces* (Elsevier). A basic book on the forces between particles, in part 1, and the forces between particles and surfaces, in part 2.
3. Derjaguin B V and Landau L 1841 Theory of the stability of strongly charged lyophobic sols of the adhesion of strongly charged particles in solution of electrolytes *Prog. Surf. Sci.* **43** 30–59 10.1016/0079-6816(93) 90013-L. For completeness.

Chapter 9 Problems

1. Glass spheres of radius $R = 10$ mm are placed in a 1 mM NaCl solution in a glass beaker at 25 °C. If the glass surfaces acquire a surface charge density of 1e per 10 nm^2, show that the short-range ($D \ll R$) repulsive double layer interaction potential is $W(D) = +5.8 \times 10^{-16} e^{-D/9.61}$ J, where D is in nm.
2. Glass surfaces usually become negatively charged in water, acquiring a surface charge density of one electron charge per 10 nm^2. Small glass spheres of radius $R = 10$ μm are placed in a 1 mM NaCl aqueous solution in a glass beaker. In problem 1, it is found that this gives rise to a repulsive energy $W(D) = +5.8 \times 10^{-16} e^{-D/9.61}$ J, where D is in nm. Estimate the mean separation at which the spheres will settle (execute Brownian motion) above the flat surface of the beaker. Assume a density of 3 gm cm^{-3} for the spheres. Plot the density distribution function $\rho(D)$ of the spheres about their mean positions.
3. For molecules constrained to interact on a surface (as occurs on adsorption and on surface monolayers) there is a 'two-dimensional' van der Waals equation of state, analogous to the three-dimensional one. This may be written as $(\Pi + \frac{a}{A^2})(A - b) = kT$ where Π is the externally applied surface pressure (in units of Nm^{-1}), A is the mean area occupied per molecule, and a and b are constants. Derive this expression for molecules of diameter, σ,

interacting with a van der Waals-type interaction pair potential given by $w(r) = -C/r^6$. Find the expressions for a and b in terms of C and σ, and for C in terms of a and b. Under what conditions would a be negative, and what are the implications of this? Can this approach, which predicts the existence of a gas–liquid transition be extended to one dimension?

4. Compounds 1 and 2 are both nonpolar liquids and interact only via van der Waals dispersion forces. The Hamaker constant of 1 is larger than that of 2 ($A_1 > A_2$). A small amount of 1 (the solute) is completely dissolved in 2 (the solvent).
   a. Will the concentration of 1 below the surface of the solution be the same, greater than, or less than the bulk concentration? If there is a difference, estimate the distance range over which the concentration will be affected.
   b. Will the surface tension of the solution be the same, greater than, or less that the value for the pure solvent 2?
5. What if $A_1 < A_2$?
6. When gas dissolves in a liquid, does the surface tension increase or decrease?

References

[1] Hiemenz P C and Rajagopalan R 1997 *Principles of Colloid and Surface Chemistry* (Taylor and Francis)

[2] Lee S W and Sigmund W M 2002 AFM study of repulsive van der Waals forces between Teflon AF™ thin film and silica or alumina *Colloids Surf.* A **204** 43–50

[3] Akbulut O, Mace C, Martinez R, Kumar A, Nie Z, Patton M and Whitesides G M 2012 Separation of nanoparticles in aqueous multiphase systems through centrifugation *Nano Lett.* **12** 4060–4

[4] Tørrissen B C 2009 File:Gecko foot on glass.JPG *Wikimedia Commons* https://commons.wikimedia.org/wiki/File:Gecko_foot_on_glass.JPG

www.ingramcontent.com/pod-product-compliance
Ingram Content Group UK Ltd.
Pitfield, Milton Keynes, MK11 3LW, UK
UKHW051923100926
8556IPUK00021B/1348